SCIENCE IN THE MARKETPLACE

Nineteenth-Century Sites and Experiences

Edited by
AILEEN FYFE &
BERNARD LIGHTMAN

THE UNIVERSITY OF CHICAGO PRESS

CHICAGO & LONDON

AILEEN FYFE is lecturer in the history of science at the National University of Ireland, Galway, and author of *Science and Salvation: Evangelicals and Popular Science Publishing in Victorian Britain*. BERNARD LIGHTMAN is professor of humanities at York University, Toronto, and author of *The Origins of Agnosticism*.

The University of Chicago Press, Chicago 60637
The University of Chicago Press, Ltd., London
© 2007 by The University of Chicago
All rights reserved. Published 2007
Printed in the United States of America

16 15 14 13 12 11 10 09 08 07   1 2 3 4 5
ISBN-13: 978-0-226-27650-2 (cloth)
ISBN-10: 0-226-27650-3 (cloth)

Library of Congress Cataloging-in-Publication Data
Science in the marketplace : nineteenth-century sites and experiences / edited by Aileen Fyfe and Bernard Lightman.
    p. cm.
Includes index.
ISBN-13: 978-0-226-27650-2 (cloth : alk. paper)
ISBN-10: 0-226-27650-3 (cloth : alk. paper)
    1. Science—Social aspects—Great Britain—History—19th century—Congresses. 2. Great Britain—Intellectual life—19th century—Congresses. 3. Science musuems—Great Britain—History—19th century—Congresses. I. Fyfe, Aileen. II. Lightman, Bernard V., 1950–
Q175.52.G7S355 2007
509.41'09034—dc22

                                                            2007000102

CONTENTS

ACKNOWLEDGMENTS

The editors would like to thank all the participants at the "Nineteenth-Century Popular Science: Sites and Experiences" conference held at York University, Toronto, in August 2004. We would particularly like to thank Jonathan Rose, Sophie Forgan, and Sofia Åkerberg. We are grateful to Jim and Anne Secord for playing a key role in the conceptualization of the conference and volume at the very early stages. We would also like to express our appreciation to all the York University units which provided funding for that event: Bethune College, the Division of Humanities, the Office of the Dean of the Faculty of Arts, the Office of the Vice-President Academic, the Science and Technology Studies Program of the Atkinson Faculty of Liberal and Professional Studies, and the Ad Hoc Research Fund. Aileen Fyfe would also like to acknowledge the support of the Irish Research Council for the Humanities and Social Sciences for providing teaching relief during the period when most of the editorial work on this volume was carried out.

꿈

CONTRIBUTORS

Samuel J. M. M. Alberti is lecturer at the University of Manchester Centre for Museology and research fellow at the Manchester Museum. Trained as a historian of science, he has published on scientific cultures in Victorian Britain and on museums of anatomy and natural history. He currently is writing a history of the Manchester Museum and conducting a study of nineteenth-century pathology collections.

Richard Bellon is visiting assistant professor in the Lyman Briggs School of Science at Michigan State University, where he teaches the history of science. His current research explores the history of nineteenth-century natural history, particularly the relationship between botany and Darwinian theory in the mid-Victorian period.

Victoria Carroll is curator of history of science at the Science Museum, London. She has recently completed a PhD on "Science and Eccentricity in Early Nineteenth-Century Britain." Her research interests include the history of natural history and geology, the history of collecting, and science and communication.

Aileen Fyfe is lecturer in the history of science at the National University of Ireland, Galway. She is currently working on the role of the book trade in the provision and organization of knowledge in the mid-nineteenth century and is the author of *Science and Salvation: Evangelicals and Popular Science Publishing in Victorian Britain* (2004).

Graeme Gooday is senior lecturer in history and philosophy of science at the University of Leeds. He is the author of *The Morals of Measurement: Accuracy, Irony, and Trust in Late Victorian Electrical Practice* (2004) and editor (with Robert Fox) of *Physics in Oxford 1839–1939: Laboratories, Learning, and College Life* (2005). At present, he is completing a book on electricity, gender, and expertise in the late nineteenth century.

Bernard Lightman is professor of humanities at York University, Toronto. He is author of *The Origins of Agnosticism* (1987), editor of *Victorian Science in Context* (1997), and editor of the journal *Isis*. Currently he is completing a book on popularizers of Victorian science.

Iwan Rhys Morus is senior lecturer in the Department of History and Welsh History at the University of Wales, Aberystwyth. He has written extensively on electricity and popular culture in nineteenth-century Britain. His most recent book is *When Physics Became King* (2005). His current research focuses on magic lanterns, optics, and scientific performances.

James A. Secord is professor of history and philosophy of science at the University of Cambridge and director of the Correspondence of Charles Darwin. The author of *Victorian Sensation: The Extraordinary Publication, Reception, and Secret Authorship of "Vestiges of the Natural History of Creation"* (2000), he is currently completing a book on science in newspapers in nineteenth-century London, Paris, and New York.

Ann B. Shteir is professor of women's studies and humanities at York University, Toronto. Her publications include *Cultivating Women, Cultivating Science: Flora's Daughters and Botany in England 1760 to 1860* (1996) and *Figuring It Out: Science, Gender, and Visual Culture* (2006), edited with Bernard Lightman.

Jonathan R. Topham is lecturer in history of science at the University of Leeds. He is coauthor of *Science in the Nineteenth-Century Periodical: Reading the Magazine of Nature* (2004) and *Science in the Nineteenth-Century Periodical: An Electronic Index* (2005) and coeditor of *Culture and Science in the Nineteenth-Century Media* (2004). He is currently working on a book-length project on scientific publishing and the readership for science in early nineteenth-century Britain.

John van Wyhe is director of *The Complete Work of Charles Darwin Online* at the University of Cambridge. He is the author of *Phrenology and the Origins of Victorian Scientific Naturalism* (2004) and editor of the three-volume *Combe's "Constitution of Man," and Nineteenth-Century Responses* (2004). His current research is on the work and historiography of Charles Darwin.

CHAPTER ONE

Science in the Marketplace:
An Introduction

AILEEN FYFE AND BERNARD LIGHTMAN

Visitors to London in the nineteenth century were regularly impressed by its historic buildings, the famous cultural artifacts in its museums and galleries, and its lively theater and entertainment district. For two visitors from India, however, the most striking sites in the late 1830s were the galleries of practical science: the Royal Polytechnic Institution and the Adelaide Gallery. "We saw nothing in London,—nothing in England, half so good," they wrote, remembering the enthralling steam-powered machines and the enormous diving bell at the Royal Polytechnic Institution (Nowrojee and Merwanjee 1841, 138). Although it might be tempting to assume that there is something unique about the consumption of science that we ourselves are surrounded by in our bookstores, on our television channels, and in our toy shops, the experiences of these two visitors remind us that the sciences have long been a part of consumer culture.

Nineteenth-century scientific attractions may not have had gift shops to rival our modern science centers, but their directors were nonetheless highly skilled in the business of attracting visitors with their entertaining and instructive spectacles. The organizers of the galleries of practical science, for instance, appreciated that they needed to woo potential visitors not just from other scientific sites, such as the British Museum or the Zoological Gardens, but from the National Gallery, the Tower of London, and the West End theaters. They recognized the power of consumer choice and sought to influence it. Even directors who hoped to remain above commercial concerns recognized that they were operating in a cultural marketplace, competing for visitors with a range of other desirable, or less desirable, attractions. The notion of the marketplace is potentially useful for historians, since it directs our attention equally to those who produce and market their wares and to the consumers who choose among them. It may also help us to find a way out of the many problems which currently beset studies of "popular science."

In 1992, a conference in Manchester, UK, was devoted to the topic of "popular science." Reflecting on its proceedings, Roger Cooter and Stephen Pumfrey bemoaned how little had yet been done to study the place of science in popular culture, particularly in comparison with medicine (Cooter and Pumfrey 1994). Since then, established scholars and doctoral students alike have taken up the challenge. The history of "popular science" is now one of the major growth areas in the history of science. The nineteenth and twentieth centuries have been a particularly rich field, both because more people then had the time, money, and education to pursue an interest in the sciences and also for the practical reason that far more sources relating to nonelite groups have survived. As we realized when we began to imagine this volume, there is no longer a shortage of scholars in this area.

All the enthusiastic study, however, has generated increasing skepticism about the very workability of the concept of *popular science.* Some would now maintain that the term *popular science* should be banished from the vocabulary of all historians of science. The main characteristic that the myriad things studied under the rubric have in common—from popularization to phrenology to artisan botanists—is that they are typically regarded as lying outside the boundaries of true "science." Yet a shared exclusion does not imply the existence of any common characteristics that could be used to construct a self-coherent and independent history. Moreover, this negative definition is deeply problematic for the long period of history before the construction of a dominant scientific establishment, for a category can be defined only in opposition to something which is itself clearly defined. Thus, while historians of the twentieth century may be able to work with a definition of *popular science* as the realm of the nonexpert, early modernists have trouble doing so, and those working on the nineteenth century must grapple with the complexities implied by the gradual emergence of a robust definition of scientific expertise. If *popular* were to be defined against *expertise,* the transformation of *expertise* over the nineteenth century would make it impossible to attach any consistent meaning to *popular.* We also have to be careful not to take at face value the arguments of those nineteenth-century actors who were engaged in establishing their own authority and who had a vested interest in denigrating certain activities as "merely popular."

In the second half of the nineteenth century, Thomas Huxley and his allies fought to transform the man of science from an untrained amateur steeped in natural theology into a highly disciplined expert who pursued a methodological naturalism. Huxley wished to ensure that it would no longer be the case that "every fool who can make bad species and worse genera is a 'naturalist'!" (Huxley 1902, 1:177). His emphasis on the significance of the changed role of the man of science has been reflected by many historians of science, who cast the nineteenth century in terms of a process they label *professionalization*. The emergence of the professional scientist is seen as the link between many critical institutional and theoretical developments, while incipient professionalization is invoked to explain many events in the early part of the century. Professionalization can be seen as the underlying motivation behind the support of scientific naturalists for Darwinism and for the decline of natural theology. It would also account for the increasing emphases on laboratories as privileged spaces for producing knowledge, on the training of cadres of male experts, and on the increase in government funding. A coherent and seemingly comprehensive story of nineteenth-century British science can be built upon the professionalization thesis.

Recently, scholars have begun to find problems with this thesis. For a start, it seems that the growth of professionalization in the nineteenth century is more complex than we previously thought. It has been argued that Huxley and his allies did not understand *professional* in the twentieth-century sense of antagonism to all amateurs or to gentlemanly culture. They also disagreed among themselves on the need for government funding, while the establishment of a naturalistic science may have superseded the demand for "professional" qualifications (Desmond 2001; Barton 1998a; White 2003). Yet, an even larger problem lurks behind the emphasis on professionalization. If the rise of professional and expert science becomes the framework for analyzing the changing nature of the sciences in nineteenth-century Britain, then everything else is pushed to the margins and rendered mere detail. Huxley would have liked that notion. But historians have learned to be wary of uncritically accepting the agenda of historical actors as the foundation for understanding the past. In this volume, we consider what would happen if we emphasized the rise of audiences for science and not just the transformation of the elite. Though we acknowledge the importance of "professional" science, we would submit that a framework built upon it seriously distorts our understanding of the period.

In this volume, we offer a route toward a more balanced perspective, by reinstating the wide range of popular engagements with science that took place in nineteenth-century Britain. We have brought together some of the latest scholarship on the problem of defining *popular science* and have sought to provide some coherence by limiting the time frame and by not attempting to cover everything. We are not, for instance, dealing with what has variously been called "ethno-science" or "low science" (Cooter and Pumfrey 1994; Sheets-Pyenson 1985), terms which imply an expectation of being involved in the creation of new knowledge. Most of the audiences in this volume saw their role as consumers, whether of books, lectures, zoological gardens, or galleries of practical science. Admitting this does not mean retreating to an older view of such audiences as passive sponges. Recent scholarship on consumers (e.g., Radway 1992; Agnew 2003) emphasizes that they are fickle and selective and they appropriate and adapt what they choose to their own purposes. Individuals select from the sources available to them to create their own amalgams of knowledge about the natural world.

The chapters in this volume also present a more nuanced image of those who have so often been labeled *popularizers*. In their efforts to make certain sorts of knowledge more widely known, they acted under a variety of motivations, from the lure of financial reward to religious zeal, philanthropic educational ambitions, or the desire to supply genteel recreation. To encourage our contributors (and readers) to think about these activities in a different light, we have emphasized the roles of *sites* and *experiences*. This emphasis helps move the focus away from the scientific community and the self-confessed popularizers and encourages us to think about audiences: where might people encounter and interact with the sciences, and what sort of experiences might they have there? Emphasizing the variety of sites where the sciences might appear also helps make the point that printed books and periodicals were far from the only (or even main) sources of information for the wider public: this volume has essays dealing with country houses, museums, galleries of practical science, panoramic shows, exhibitions, lecture halls, and domestic conversations, located both in London and throughout the rest of Britain.

Sites, Experiences, and the Sciences in the Nineteenth Century

By taking "the site" as one of our points of focus, we are reminded of the need to situate activities connected with the sciences (Livingstone 2003). This need to situate activities applies even to books and periodicals, forms which have often seemed disembodied and universally available—much as science itself has, until recently, been regarded—but which book historians have now shown to be grounded in specific places, both of production and of consumption. At the start of the nineteenth century, the sites we might consider include soirées, the lectures at the Royal Institution, the expensive tomes written by gentlemen-naturalists and the private collections of those gentlemen. These sites tended to be limited to the members of polite society and were most likely to be in London. But as the century progressed, more and more sites came into play, from workingmen's lectures, penny magazines, and galleries of practical science to public gardens, arboretums, and civic museums. In general, these sites were accessible to more social groups than those of the early century, although some sites did continue to cultivate an exclusiveness in the midst of the general move toward inclusivity; social class might be one reason for this exclusiveness, but the attempt to maintain a scholarly reputation could be another.

Although London was easily the biggest city in the country, with the largest potential audience and the widest variety of attractions, a growing number of these sites could be found in other cities. At the start of the century, Edinburgh and Dublin had been the most prominent alternative centers for polite and learned society, and they followed London in gaining an increasing range of sites catering to wider audiences. But from mid-century onward, the growth of such newly industrial cities as Manchester, Birmingham, Leeds, Glasgow, and Belfast stimulated intense civic pride that encouraged the establishment of public museums, gardens, and lecture halls. The expanding railway network enabled smaller towns, which lacked formal sites of knowledge, to be visited by traveling lecturers who would transform the town hall or church hall for a night. Their example reminds us that sites are not static. Not only may a site be experienced differently by different groups of visitors at any one point in time, but visitors at different points in time are almost certain to have different experiences.

Historians of science have become well aware that both the transformation of the scientific elite and the emergence of new audiences for science took place during a period when the book trade was being transformed. Books on natural history and natural philosophy had already become a part of polite culture in the late eighteenth century, but only the wealthy could afford them and only the well educated could understand them. The technological advances and wider socioeconomic developments of the first half of the nineteenth century enabled publishers to increase the number of books produced, while also reducing average prices. As James Secord has argued, "the steam-powered printing machine, machine-made paper, public libraries, cheap woodcuts, stereotyping, religious tracts, secular education, the postal system, telegraphy, and railway distribution played key parts in opening the floodgates to an increased reading public" (Secord 2000, 30).

It was generally publishers, not writers, who were the driving forces behind the expansion of popular publishing. Both commercial publishers, such as W. and R. Chambers and Charles Knight, and charitable publishers, such as the Religious Tract Society, believed in providing instruction and information for those who had previously had little access to it, though they disagreed over the role of Christian instruction (Fyfe 2004). By mid-century, these publishers had helped create a robust market for cheap books on the sciences, particularly among a new polity of consumers consisting of a family readership among the middle and working classes. In the second half of the nineteenth century, that market expanded, helped by the repeal of the remaining taxes on knowledge, the introduction of the rotary press, and a gigantic explosion in newspaper and periodical publishing. Science continued to be featured in the increasing number of miscellaneous periodicals, but there were also ever-more publications devoted just to the sciences, for general readers as well as experts (Barton 1998b; Cantor et al. 2004).

The nineteenth century was also a period of museum growth. Existing museums were expanded and new museums were founded, especially after the Museums Act of 1845 enabled civic museums to be established throughout the provinces. As well as such great national museums as the National Gallery (founded in 1824) and the South Kensington Museum (1857, the precursor of both the Victoria and Albert Museum and the Science Museum), two hundred metropolitan, provincial, and university museums were founded in Britain (Rupke 1994, 13–15). Among them were important science museums, such as the Museum of Practical Geology (1851) and the British Museum (Natural History) in South Kensington (1881). In addition to these museums of public education, there was a wide range of more commercial enterprises, ranging from the relatively long-term—such as the Adelaide Gallery (1832–45) and the Polytechnic (1838–81)—to such decidedly ephemeral shows as the exhibition of the so-called Aztec children who took London and Dublin by storm in the summer of 1853.

At the start of the century, access to most museums had been governed by gentlemanly conventions of politeness, and some collections remained in gentlemen's private cabinets or country houses, where they were accessible only to polite or respectable classes of society. But by the end of the Victorian era, most museums had been opened up, from the country houses, which welcomed hordes of day-trippers, to the national and municipal museums, which were proudly open to all at no cost and even began to open on Saturdays and evenings. With greater accessibility came the increasing efforts of managers to control the experiences of audiences and to enforce rigorous behavior codes for working-class visitors.

Although noise and conversation were increasingly banned from museums, there was an abundance of other opportunities for listening to and talking about the sciences. Conversation remained central to the practice of science, though the conventions of late Georgian and Regency science were replaced by the display of expertise after midcentury. Throughout the Victorian period, middle-class urban audiences attended glittering social gatherings, or *conversaziones*. Science figured prominently among the displays of the fine, industrial, and decorative arts, antiquities, archaeology, literature, ethnography, technology, and music that stimulated discussion (Alberti 2003). Such events could be staged as fund-raisers, and they might also be used in efforts to improve and educate the working classes.

Public lectures were another popular social activity throughout the century, from the fashionable lectures for Regency polite society, through evangelical missionary meetings, to the philanthropic attempts to educate workingmen. Such was the demand for lectures that it became possible, in the second half of the century, to earn substantial amounts of money from lecturing. Like their competitors, science lecturers became increasingly skilled in the use of colorful visual aids, from displays of exotic specimens to spectacular optical illusions. The sciences could thus be encountered in an incredibly wide variety of venues and in a multiplicity of forms, often incorporating elements from print, display, and the spoken word at the same time. An emphasis on scientific elites and professionalization ignores all of these crucial dimensions of nineteenth-century science.

Although a focus on sites is illuminating, it is equally important to pay attention to the experiences of the audiences who came to be entertained and instructed. How did audiences respond to spectacles or to particularly dynamic lecturers? What was it like for a museum visitor to take a tour through the exhibits? Answering these questions and recovering the experiences of audiences is a daunting task. Is it even possible to get at audience reactions, especially once audiences became ever larger and more diverse? The contributors to this volume approach this difficult project from two directions. Some examine the creation and control of audience experiences, shedding new light on the relationships that writers, publishers, lecturers, and museum curators believed they had with their audiences. Although acknowledging the powerful influence of publishers, curators, and lecturers, some contributors aim to shift our attention toward the audience's own feelings and thoughts. To accomplish this task requires imaginative new ways of reconstructing audience experience out of scattered fragments from a wide array of sources, such as contemporary accounts in newspapers and periodicals, pamphlets, travel narratives, memoirs, diaries, autobiographies, letters, guidebooks, and other written traces. The examination of the creation of experiences reveals the extent to which audiences, both intended and actual, set the agenda for the public provision of scientific knowledge, while the recovery of individual visitor accounts makes us very aware of their agency in choosing how to experience a particular site or event.

Virtually all of the essays make the point that sites are not as clearly distinct as we had originally assumed. Sites often play multiple roles and support multiple forms of experiences. This assertion is most obvious in the case of those places that quite literally combine multiple sites: the Royal Polytechnic Institution, for instance, contained a lecture theater as well as display galleries. Even within a lecture theater, various sorts of activity took place. "Lecturing" involved printed materials, visual aids, and informal conversations in addition to the formal lecture. Similarly, in museums, visitors often combined the experience of looking at exhibits with that of reading labels and guidebooks. In some museums, at some times, visitors also had the opportunity to talk, to listen to a human guide, perhaps also to smell and touch the exhibits. What we originally chose as representatives of simple sites have usually turned out to be highly complex sites involving far more media than we had expected. At one level, we have the combination of words and images found in books; at another, we have the combination of music, magic lantern slides, oral deliveries, and printed descriptions at the Royal Polytechnic Institution. It may make sense to think of what went on in these sites as an early form of multimedia experience.

There is a further sense in which we can explore the multiplicity of media involved in what might seem like a simple experience when we remember that for the audiences, individual sites were not experienced in a vacuum. Obviously, they would have been chosen out of a range of possible activities for that time and place, but critically, they would never be experienced with a completely empty mind. This realization has broader applicability, for we must appreciate that visitors would view and evaluate a new lecture (or book or exhibition) in the light of previous books (or exhibitions or lectures). Reading an account of a previous visitor might influence behavior patterns or draw attention to specific items and not to others. A specific exhibit might take on a particular interest because it had been mentioned in a recent lecture or sermon. Previous exposure to one philosophical interpretation of nature might make it difficult to understand or accept a subsequent lecture based upon an opposing system. Thus, not only do multiple media come into play within particular sites, but visitors' memories may well enable them to add—mentally—an even wider, richer, and more complex array of stimuli to those which are physically present at the site (Mandler 1999).

The eventual implication of thinking about the role of past experiences on new experiences would appear to be the realization that all experiences must be utterly individual and thus infinitely different. This would be a depressing conclusion for efforts to write a history of experiences of the sciences. We can perhaps learn from historians of reading, who have already moved from assuming that a text will be understood in the same way by everyone to accepting that readers have great scope for individual creativity. The efforts to place some form of order onto histories of those interpretations have led to the ideas of *horizons of expectation* (the idea that readers' education, experiences, etc. will help determine how they react to particular texts) and of *interpretative communities* (groups of readers who see things in broadly the same way). If we extrapolate from texts to include images, exhibits, and lectures and expand the notion of readers to include viewers and listeners, then we may be able to use these notions to think about the sorts of experiences discussed in this volume.

Science in the Marketplace

The nineteenth century saw an immense growth in consumer culture, particularly in the mass market, at a time when Britain led the world. Britain may have become a consumer society in the eighteenth century (McKendrick, Brewer, and Plumb 1982; Brewer and Porter 1994), but it was not until the nineteenth century that most of the population had the opportunity to participate in this new world of goods, as products proliferated and the gap between prices and available income lessened. By the second half of the nineteenth century, we can start to talk about a mass market, in which people from all classes of society had (to varying degrees) some access to the wide range of commodities and experiences available. Among these options were many which had a connection to the natural sciences, from museums and exhibitions to lectures and books to menageries and pleasure gardens. A few of these goods or experiences were available at no cost (generally the more self-consciously educational and philanthropic enterprises), yet they were nevertheless perceived as part of a raft of options available to the consumer. It is precisely this growing range of commodities and experiences—and the growing range of people who had access to them—which makes the nineteenth century such a remarkable period to study.

Many of the essays in this volume draw attention to the enormous range of choices available to audiences, as competing publishers sought readers for their introductory science books and scientific entrepreneurs advertised their offerings amid theaters and freak shows. We believe that it might be useful to think about all the sites, products, and experiences discussed in this volume as located within the cultural marketplace. On a trip to London, for instance, visitors had an incredible choice of sites, from Wyld's Globe to the Crystal Palace and the British Museum, not to mention the competing attractions of the theaters, the historic buildings, or the bustle of the docks. By midcentury, commercial sites such as the Egyptian Hall and the Royal Polytechnic Institution were already geared up to the business of attracting visitors, whereas the directors of the British Museum were only starting to appreciate the extent of both their potential audience and their competition. By the end of the century, competing for visitors had become a familiar part of life at the British Museum.

Museums and exhibitions offer a familiar example of how sites might compete, but we can see a similar process in operation with other sites and products. In some places, visiting lecturers might find themselves the only show in town, but in the larger towns, science lecturers competed with one another as well as with evangelical preachers and literary readings. For instance, George Combe found it easier to attract an audience in Edinburgh—where he and phrenology were well known—than in some of the other towns he visited: his lectures had not changed, but the local competition was different. With books, too, publishers in the early nineteenth century were well aware that they were competing with one another to sell their wares to the very limited number of literate, affluent potential book buyers. The increase in literacy and the lowering of book prices did nothing to reduce that competition: rather, there were more books and more periodicals than ever before, only a few of which would ever become highly successful. The competition in this marketplace was not always entirely financial, for the British Museum was free, and books could be borrowed as well as bought. Rather, we might think of the products as competing for attention and to be experienced, a process that would often, but not always, involve a financial transaction.

We should also remember that lecturers competed not only with other lecturers but with books and museums: for a person with limited time and money and a particular set of constraints, the book might be preferable to the lecture because the timing of the lecture did not suit; or the lecture might be preferable because it was cheaper and would be a faster way to get the gist of the argument. We have, therefore, an image of a marketplace in which all sorts of scientific activities and experiences are on offer alongside the full range of other popular cultural forms. It is a powerful image because it gives agency not only to those who create and market their products but also and especially to those who choose between them: it recognizes the power of consumer choice as well as the shaping influence of the producers. It therefore offers a useful corrective check to studies of "popular science" that focus almost entirely upon those who produce books and lectures and devote little attention to the question of exactly how those works are then disseminated and received.

The language of the marketplace is also helpful for thinking about the changes over the course of the century. From one perspective, new social groups were gaining access to the marketplace as the genteel and polite public of the early nineteenth century became the mass audience of the late century. New audiences had different interests, needs, and resources and may thus be expected to make different choices among the available products. Meanwhile, from the other perspective, just as the director of the Royal Polytechnic Institution was continually dreaming up spectacular new exhibits and demonstrations, so publishers sought out new formulas for success, lecturers tried out new routines, and entrepreneurs set up new attractions. There were different choices available to consumers throughout the century. Novelty and innovation are important to this story, for there were waves of fashion in scientific experiences just as in personal attire. Going to the Royal Institution was the latest thing in 1810, but in the late 1840s, it was dioramas and panoramas; in 1851, it was the Great Exhibition; in 1863, it was ghost shows; and in the early 1870s, it was public aquariums.

The organizers of these sites were always playing with the balance between instruction and entertainment in their efforts to produce a form of rational recreation which would be appetizing to the consumer. Visual appeal was a key trope, as with other cultural commodities. The popularity of panoramas in the first half of the century and the rise of the illustrated weekly magazines from the 1830s onward ensured that popular culture was increasingly distinguished by its pictorial character (Altick 1978; Anderson 1991). Victorian audiences were bombarded by a stream of spectacular new visual images that changed their relationship to the world and influenced how they were to observe it. To some, the attractiveness of the sciences could be enhanced by catering to this craving for the visual and the spectacular, whether through lavish use of illustrations in books, life-size reconstructions of dinosaurs, or startling demonstrations of electricity.

Though historians have concentrated on the importance of visual culture—usually taken to mean nontextual visual stimuli—we should not forget that audiences used all five senses. Several of the contributors to this collection have taken up the challenge of the multimedia experience, reminding us of the possibilities for hearing, smelling, and touching—and very occasionally tasting—scientific specimens and displays. Thinking about the wide range of stimuli which formed part of the spectacles of entertaining instruction—or instructive entertainment—offers us an innovative angle on the variety of ways in which the sciences could be consumed in the cultural marketplace.

A few of the sites discussed in this volume (e.g., the university medical museums) tried to remain aloof from the marketplace and to encourage only scholarly and studious visitors. But many more of the sites sought to attract as many visitors as possible, whether for commercial gain or for the cause of public education or because it suited the agenda of the men of science. There were frequently more commodities on offer than just the advertised experience, since both commercial and educational enterprises increasingly offered what might be thought of as tie-in products, such as guidebooks, souvenirs, or the option of personal consultations with the lecturer.

As well as being a description of what was actually happening at a commercial level, the marketplace plays a useful role as metaphor. Consumer choice does not end at the moment the decision is made to purchase this book or to attend that lecture. Although nineteenth-century commentators typically assumed that reading a text entailed accepting its argument, the process of reading (and viewing and hearing) was far more complex. Readers were frequently aware that the particular book, lecture, or site was just one of many that they might have chosen and that the information and arguments in each might differ, however slightly. As well as choosing which products to experience, consumers had to choose which ones to trust—and how far. The argument of a particular book might be rejected as flawed, or a lecture might be dismissed as all show and no content. Intellectually, whether consciously or subconsciously, audiences had to decide what to make of the experience, to reconcile it with their existing knowledge and worldview, and to relate it to other experiences. Lecturers and authors, therefore, had to ensure not only that their product was sufficiently attractive to draw in the paying customers but also that it had the power to convince.

Being convincing meant demonstrating authority. But the style of authority needed was not necessarily that which we now associate with scientific expertise. Consider, for instance, the skills necessary to prepare and mount a realistic taxidermic specimen; or to produce a stream of best-selling textbooks; or to deliver a successful oxyhydrogen microscope lecture; or to create a tourist attraction as successful as Wyld's Globe. A wide range of expertises was needed to create a popular, successful and convincing scientific experience. Rather than seeing the tightening definitions of scientific expertise over the nineteenth century as excluding ever more of the population from the sciences, we can see that the participation of ever-more people in the full range of scientific activities actually required a wider array of skills and talents than is usually considered under the heading "scientific expertise."

From our modern perspective, or even from that of the scientific naturalists of the late nineteenth century, the skill of the showman, the writer, the lecturer, or the curator might be acknowledged but distinguished from scientific expertise by being "merely" the expertise of a performer. Yet, there was no such clear dichotomy in the nineteenth century, and to argue that there is one even now would be to ignore the importance of such performative skills to scientists who need to convince their peers—though print, through demonstration, or through oral presentation—that they have discovered something new and that they can legitimately speak with authority. In the modern scientific world, scientists are to be judged by their peers, and emphasis is explicitly placed on only some forms of expertise—for example, on experimental design or on fieldwork practice. For the activities discussed in this volume, the audience was far more extensive than the contemporary community of scientific practitioners. In this wider context, it was possible for a variety of expertises to be valued. Practitioners of science in the Regency period impressed audiences by their measured, polite conversation, not by their latest journal article or their PhD dissertation. Royal Polytechnic Institution audiences were impressed by the immensity of the electric machines demonstrated there, while Charles Waterton's visitors admired his taxidermic skills. Lecturers, from Combe to Frank Buckland and John Henry Pepper, were admired for their visual aids as well as their rhetorical performance. Presumably content mattered as well—at least to some audiences—but in a consumer culture, audiences were far more likely to judge by the performance than by paper credentials. Expertise could clearly be made, or unmade, in the public arena of the commercial marketplace.

Visiting the Marketplace

Comprehensiveness is virtually impossible when the field is so immense and when large tracts of territory are almost untouched. In this volume, we have purposely avoided sites where some work has been done, such as mechanics institutes, meetings of natural history societies, and the natural history crazes that sent Victorians into nature looking for ferns, seaweeds, and mosses. We also avoided chapters on "professional" scientists, such as Huxley and Tyndall, who lectured and wrote books for diverse Victorian audiences, and the sites in which they operated, such as the British Association for the Advancement of Science or the Royal Institution. Though fresh studies of their popularizing activities are no doubt needed, our first priority has been to deal with significant but lesser-known figures and sites. Among the sites explored in this volume are the salon, the club, the soirée, the laboratory, the lecture hall, the periodical, the book, the country house, and the international exhibition. There are many sites that cry out for further study, such as science biographies, children's toys, botanical gardens, zoos, aquariums, and schools. Among the sciences taken into account in this volume are botany (by Shteir), zoology (by Carroll, Fyfe, and Lightman), anatomy (by Alberti), phrenology (by van Wyhe), electricity (by Morus and Gooday), and optics (by Morus and Lightman). Other authors range across the scientific disciplines (Secord, Bellon, and Topham).

We initially intended to use a chronological structure for the volume to enable us to highlight developments over time and to emphasize that "popular science" was not a static entity over the course of the nineteenth century. We imagined a volume in three sections. The first, covering about 1800 to 1830, would deal with the period before expert science began to be constructed in earnest. The second, roughly from 1830 to 1860, would examine the emergence of new audiences for "popular" science, while a final section would demonstrate the consolidation of "popular science" between 1860 and 1900. The problem was not just that our contributors tended to stray across these chronological boundaries but that their essays so often problematized our intended divisions. New audiences, for instance, were emerging both before 1830 and after 1860. And to talk of "popular science" being constructed or consolidated in anything other than in very localized instances would imply a guiding hand and a coherence that did not exist in the free-flowing cultural marketplace of nineteenth-century Britain. We have, therefore, opted for a broadly thematic arrangement.

Our sections focus on three ways of engaging with the sciences: (1) conversations and lectures; (2) reading; and (3) visiting and viewing. These divisions encourage us not to focus entirely on texts—or even on texts with images. They also reflect the emphasis on the various roles of the senses, which several of our contributors have brought up. Furthermore, they emphasize the activities that were involved in the consumption of various forms of science. However, as historians of reading have shown, even the engagement with texts—so often taken to be the primary form of engagement with the sciences—is not necessarily a passive, simple, or solitary activity: it may happen in public, in groups, or while walking around a museum. Similarly, we should not think of audiences at lectures simply as listening—they are likely also to be looking and perhaps talking. Our section divisions should be regarded as useful organizing structures, but we hope they will not obscure the emphasis on active consumer engagement and multimedia experiences, which the chapters reveal.

The first section focuses on the role of orality and features three chapters on conversation and lecturing by Secord, van Wyhe, and Lightman. The importance of conversation in science has almost entirely escaped the notice of scholars while the bulk of the work on lecturing in the nineteenth century is now dated. Secord makes a persuasive case for paying close attention to talking as an integral part of the making of knowledge. But though conversation remained central to science, Secord argues that its nature was dramatically transformed over the course of the nineteenth century. In early nineteenth-century genteel society, a reputation for being knowledgeable could be based on good conversational skills displayed at parties or scientific society meetings. But by the end of the century, men of science saw substantive scientific talk as taking place in the laboratory or in the field (or in the new ritual of laboratory afternoon tea) and avoided having such conversations in general social settings, where they might be accused of talking shop. In his study of public lectures in the first half of the century, van Wyhe discusses how Johann Gaspar Spurzheim and George Combe convinced British audiences of the truth of phrenology and turned some of them into devoted practitioners. Rebutting the current suspicion of the term *diffusion* among historians of science, van Wyhe maintains that the spreading of phrenology by these two lecturers is a typical example of the diffusion of an innovation. Drawing on scholarship in the social sciences, he explains how phrenology lectures initiated a process of adoption that reflects that of many other cases of the diffusion of innovations.

Lightman tackles scientific lecturing in the latter half of the century, focusing on Frank Buckland and John Henry Pepper. He maintains that successful lecturers developed innovative styles that would attract audiences
accustomed to being entertained. Whereas Buckland relied heavily on his
extemporaneous speaking style, on his ability to find humor in natural
history, and on numerous exotic specimens to create a museum-like environment, Pepper offered his audiences spectacular optical illusions and
theatrical performances drawn from the London entertainment world.

In the middle section, Topham, Shteir, Fyfe, and Gooday have attempted
to lay out new perspectives on print culture. Topham urges us to attend to
the use of the term *popular science* by historical actors, especially as understood by those publishers involved in experiments in cheap publishing
in the early nineteenth century. While previous scholarship has centered
on the Society for the Diffusion of Useful Knowledge (1826–46), Topham
points to the existence of important commercial ventures in cheap educational publishing and cheap journalism from an earlier period. Shteir also
examines early nineteenth-century print culture and draws our attention
to ways in which female authors used scientific, literary, and popularizing
discourses in their accounts of the *Mimosa pudica*, or Sensitive Plant. She
demonstrates that in their quest to maintain their authority while communicating knowledge to their audience, these writers turned increasingly
to technical accounts and away from literary material and family-based
narratives. Fyfe considers the relation of print to museum visiting in the
middle of the century and focuses on the provision made for the large numbers of new visitors from the lower middle and working classes. Her examination of well-illustrated "virtual museums" and of guidebooks to the
British Museum show that publishers were quicker than curators to realize
that these visitors would require new forms of reading material. Gooday
brings us to the late nineteenth century and the debates about electricity
in the home. Arguing that popular and technical discourses on electricity
were symbiotic, he shows that popular concerns about risks could set the
agenda for technical writing. Gooday asserts that expertise was as much
a construction of the audience who demanded and judged it as it was the
creation of the expert.

The final section is devoted to the culture of display, from country
houses and provincial museums to urban museums of practical science
and the Great Exhibition of 1851. Carroll's chapter on Charles Waterton's
natural history collection explores the experiences of visitors. Through an
examination of firsthand accounts, she shows how the conventions governing visiting in the early nineteenth century shaped their experiences at

Walton Hall. In his chapter, Bellon emphasizes the need to understand how deeply the Crystal Palace was enmeshed in its urban context. Like the financial, entertainment, communications, and scientific institutions within walking distance, the exhibition acted as a facsimile of the globe that provided the visitor with an experience of the totality of human industry and civilization. Morus considers the significance of the performative aspect of scientific exhibitors in mid-nineteenth-century London. Stressing the spectacular scale of Victorian technologies of display, he illuminates the nature of authority in these sites as well as the reconfiguration of the audiences' sense of the boundaries between the possible and the impossible. Alberti focuses on the experiences of visitors to public museums of anatomy and natural history over the course of the nineteenth century. He tracks the changing experience of these visitors as curators increasingly sought to control what they saw, heard, and smelled, and he examines their reactions of awe and disgust. Taken together, the four chapters in this section provide striking new insights into the complex range of experiences of visitors to specific museum and exhibition sites.

If we are really to understand the role of the sciences in nineteenth-century Britain, we must thoroughly reexamine the many activities lumped together under the now pejorative label "popular science." They were not just frills, nor curiosities, nor even necessarily a by-product of "professional science." Exhibitions, lectures, books, periodicals, and museums opened crucial windows on the meaning of the sciences for the British public. During the nineteenth century, an increasing number of such opportunities became available, and a greater proportion of the population had access to at least some of them. In this volume we have recovered but a sample of the richness and variety of the experiences of audiences who took advantage of these opportunities. Certainly, these activities need to be placed in the context of other, more academic and specialist scientific activities. But equally, and crucially, they must be placed alongside the myriad other popular activities and commodities available in the expanding cultural marketplace. Only by doing so can we begin to understand both the strategies of competing entrepreneurs and popularizers and the power of consumer choice.

NOTES

We are indebted to Sophie Forgan and Jim Secord for their suggestions on earlier drafts of this introduction.

REFERENCES

Agnew, Jean-Christophe. 2003. "The Give-and-Take of Consumer Culture." In *Commodifying Everything: Relationships of the Market*, ed. Susan Strasser, 11–39. London: Routledge.

Alberti, Samuel J. M. M. 2003. "Conversaziones and the Experiences of Science in Victorian England." *Journal of Victorian Culture* 8:208–30.

Altick, Richard. 1978. *The Shows of London*. Cambridge, MA: Harvard University Press, Belknap Press.

Anderson, Patricia. 1991. *The Printed Image and the Transformation of Popular Culture, 1790–1860*. Oxford: Clarendon Press.

Barton, Ruth. 1998a. "'Huxley, Lubbock, and Half a Dozen Others': Professionals and Gentlemen in the Formation of the X Club, 1851–1864." *Isis* 89:410–44.

———. 1998b. "Just Before 'Nature': The Purposes of Science and the Purposes of Popularization in Some English Popular Science Journals of the 1860s." *Annals of Science* 55:1–33.

Brewer, John, and Roy Porter. 1994. *Consumption and the World of Goods*. London: Routledge.

Cantor, Geoffrey, Gowan Dawson, Graeme Gooday, Richard Noakes, Sally Shuttleworth, and Jonathan R. Topham. 2004. *Science in the Nineteenth-Century Periodical*. Cambridge: Cambridge University Press.

Cooter, Roger, and Stephen Pumfrey. 1994. "Separate Spheres and Public Places: Reflections on the History of Science Popularisation and Science in Popular Culture." *History of Science* 32:237–67.

Desmond, Adrian. 2001. "Redefining the X Axis: 'Professionals,' 'Amateurs' and the Making of Mid-Victorian Biology—A Progress Report." *Journal of the History of Biology* 34:3–50.

Fyfe, Aileen. 2004. *Science and Salvation: Evangelical Popular Science Publishing in Victorian Britain*. Chicago: University of Chicago Press.

Huxley, Leonard. 1902. *Life and Letters of Thomas Henry Huxley*. 2 vols. New York: D. Appleton.

Livingstone, David N. 2003. *Putting Science in Its Place: Geographies of Scientific Knowledge*. Chicago: University of Chicago Press.

Mandler, Peter. 1999. "'The Wand of Fancy': The Historical Imagination of the Victorian Tourist." In *Material Memories*, ed. Marius Kwint, Christopher Breward, and Jeremy Aynsley, 125–42. Oxford: Berg.

McKendrick, N., J. Brewer, and J. H. Plumb, eds. 1982. *The Birth of a Consumer Society: The Commercialization of Eighteenth-Century England*. Bloomington: Indiana University Press.

Nowrojee, Jehangeer, and Hirjeebhoy Merwanjee. 1841. *Journal of a Residence of Two Years and a Half in Great Britain*. London: Wm. H. Allen.

Radway, Janice A. 1992. *Reading the Romance: Women, Patriarchy and Popular Literature*. 2nd ed. Chapel Hill: University of North Carolina Press.

Rupke, Nicolaas A. 1994. *Richard Owen: Victorian Naturalist.* New Haven, CT: Yale University Press.

Secord, James A. 2000. *Victorian Sensation: The Extraordinary Publication, Reception, and Secret Authorship of "Vestiges of the Natural History of Creation."* Chicago: University of Chicago Press.

Sheets-Pyenson, Susan. 1985. "Popular Science Periodicals in Paris and London: The Emergence of a Low Scientific Culture, 1820–75." *Annals of Science* 42:549–72.

White, Paul. 2003. *Thomas Huxley: Making the "Man of Science."* Cambridge: Cambridge University Press.

SECTION I

Orality

CHAPTER TWO

How Scientific Conversation Became Shop Talk

JAMES SECORD

With the English, conversation is a languid silence broken by occasional monosyllables, and by the water flowing every quarter of an hour from the tea urn. —E. Jouy, *Encyclopédie Moderne* (Jouy 1823–32)

Writing about talking is a paradoxical activity. The apotheosis of print in the nineteenth century has led other forms of communication to seem feeble and ephemeral. This is especially true in science, where print has dominated ideas of what it means to make a contribution to knowledge. Historians have assumed that scientific talk is ornamental and supplementary, mattering only on the way to producing a published result or as nothing more than a remnant of a fading oral culture. "Words are but wind," as one etiquette manual noted in 1861, "and when spoken leave no mark behind" (Anon. [1861], 44).

Oral performance, however, has been and remains at the heart of the making of knowledge. Although little has been written on the nineteenth century, scientific talk has been extensively examined in other settings. One body of literature is focused on the salon culture of Paris, London, and other European cities of the eighteenth century, when leading figures came together in salons and coffeehouses for discussions of natural philosophy along with literature, politics, and the arts (Fara 2004; Sutton 1995; Terrall 1996; Walters 1997). The other is the scientific laboratory of the twentieth century. Beginning in the 1970s, anthropologists and sociologists recorded the discourse of technicians, experimenters, and others as part of a process of understanding science in the making (Collins 2004; Latour and Woolgar 1979; Traweek 1988). An important tradition has grown out of work in ethnomethodology, exemplified by Michael Lynch's *Art and Artifact in*

Laboratory Science: A Study of Shop Work and Shop Talk in a Research Laboratory (1985).

By comparison, the nineteenth century has been an uneasy borderland. This failure to understand scientific talk in this period is in large part because so many studies have depicted it in terms of decline from the supposedly golden age of conversation in the eighteenth century. On the positive side, this stress on polite civility, coffeehouses, and salons has enabled historians of Enlightenment natural philosophy to recover the intellectual life of an era that had been dismissed as an awkward pause between the scientific and industrial revolutions. Yet an idealized image of the eighteenth century as a conversational golden age is far from unproblematic (Cowan 2005). Most significantly, the political implications of enlightened conversation have typically been understood in terms of social theorist Jürgen Habermas's concept of a bourgeois public sphere, arising in the late seventeenth and eighteenth centuries and based on rational discussion between private individuals in a social realm free of state control. On this conception, the nineteenth century appears as a period of fragmentation and commodification, when the public sphere was transformed under the pressures of the emerging industrialized mass media. Habermas aimed to use history to sketch an ideal type of rational public discourse, and his account is thus primarily a social critique rather than a history (Habermas [1962] 1989; Calhoun 1992). Although Habermas's views have been decisively criticized, eighteenth-century conversation continues to hold a revered place in our own culture, in large part as a nostalgic reaction against the forces of modernity and technological progress (S. Miller 2006).

My aim, in contrast, is historical: to put conversation back at the center of our understanding of science in the nineteenth century. Rather than charting the transformation of a utopian Enlightenment ideal, I will look at the ways in which scientific talk was managed and how this changed. Science is of particular significance in this context, as it has been hailed on one hand as a model for unconstrained rational discussion and lamented on the other as a force for specialization leading to the fragmentation of public intellectual life. In tackling these issues, I will deal with a variety of settings for talk, from papers delivered at scientific meetings to the rituals of courtship. My focus will be on informal conversation in the clubs, salons, and soirees of the English social elite. This was a world characterized by the few hundred elite families that constituted "Society" and dominated the annual "Season." Among these circles, conversation—defined as the unplanned, easy talk of the well bred—was central to sociability (Klein 1994, 3–14). Conversation, as I will suggest, remained central to science, although what could be said to whom, and under what circumstances, was transformed. The conventions of late Georgian and early Victorian science gave way in the final decades of the nineteenth century to a situation in which expertise played very new and different roles.

The story sketched in this chapter is built upon a reassessment of what has usually been labeled "gentlemanly science," a phrase that can better be rendered as "polite science." The term *polite science* makes sense for reasons of gender, as many practitioners were women (Shteir 1996), but more fundamentally it brings out the significance of broader ideals of civil society and social deference that underpinned the pursuit of knowledge in England throughout much of the nineteenth century. The ethos, etiquette, and attitudes of those engaged in scientific pursuits have been closely studied for the late seventeenth and eighteenth centuries but far less thoroughly for the nineteenth. Historians of Victorian science have devoted a great deal of attention to gentlemen since the pathbreaking work of Jack Morrell and Arnold Thackray in *Gentlemen of Science* (1981), but only more recently have they begun to analyze the practices of gentility (Barton 2003; Winter 1998; White 2003; Green Musselman 2006). In consequence, the continuing significance of polite sociability for the aims and practice of science has often been minimized, with gentlemanly status simply reinforcing the dominance of a relatively modern-looking core of experts (Rudwick 1985, 16). This too easily leads to a division between "specialist" and "popular" science that is inappropriate to understanding a period when people differing in gender, rank, and depth of expertise not only talked about science but in doing so contributed directly to its making.

Talking and Publishing

To understand polite science and the significance of conversation within it, we need to begin by examining the broader relationship between oral performance and publishing (Livingstone 2005, 96–97). It is still too often assumed that the publication patterns of science took on a quintessentially modern form with the foundation of the Royal Society's *Philosophical Transactions* in the 1660s as a means for exchanging information among those engaged in the pursuit of natural knowledge. On this basis, Charles Bazerman's pioneering study of the origins of the scientific article could leap from the seventeenth to the late twentieth century.[1] Yet throughout the eighteenth and the first half of the nineteenth century, the routes to making public a novel finding were as many and diverse as the range of practitioners engaged in natural philosophy and natural history. Take a list of what became identified as major discoveries and see where they were first announced. Some were made public in lectures or demonstrations of new instruments, others in letters to friends or for publication in the *Gentleman's Magazine* and in newspapers, some in books of travel or philosophy (Golinski 1992; Schaffer 1986). Still others were shared among a small circle of friends and patrons, as with the botanist Robert Brown's work on particulate motion or the canal surveyor William Smith's findings about the role of fossils in determining the order of strata.

These publication patterns were part of a shared culture involving a wide variety of practitioners, from learned theologians and genteel women to quarry workers and skilled artisans (D. Miller 1986; A. Secord 1994; Shteir 1996). The networks which brought these groups together were founded on prevailing ideals of a hierarchical order of gender, rank, and status. Novelty was not identified with discovery events announced in scientific papers, a genre which became defined only during the later nineteenth century, but depended instead on a broad spectrum of communication. Contributions ranged from newly observed nebulae and rare mosses to philosophical and theological frameworks for understanding nature. The work of science passed between various domestic settings and public meeting rooms, taverns, and theaters. Collecting, traveling, experimenting, and observing were important activities, but they were always valued in a system grounded in heredity and hierarchy.

Presentation in print was only one moment, and not necessarily the most significant, in the broader process of the making of a matter of fact within polite science. If we look at the groups central to the ideals of natural philosophy and natural history in this period, notably the aristocracy, high gentry, and traditional professions, we find that knowledge was communicated primarily though the spoken word. Certain knowledge was intimate knowledge, grounded in private conversation, discussions at meetings, and lectures by celebrated speakers. At Oxford and Cambridge the professors of science of the eighteenth century published little: enlightened university life was about local reputation and high table conversation—not about rushing into print or joining the world of commerce. Those who did publish extensively, such as Newton's successor William Whiston, were often on the margins of university life; after a series of theological scandals, Whiston moved to London and made a living through his books and lectures. Publishing was only one part of the broad spectrum of ways to establish a position as a possessor of knowledge.

In medicine, too, physicians gained reputation not so much by publishing technical works on spa waters or the classification of diseases but

primarily through performance at the bedside, with skill in medical consultations demonstrating true mastery. Physicians needed Latin learning
and traditional medical knowledge as signs of gentility: to focus too single-
mindedly on publication could harm a practice, by highlighting those aspects of medicine that made it a less desirable career than other professions
such as the law, the clergy, and the military (Jewson 1974). To stress new
science-based research of the kind carried out in France, as Thomas Wakley
did in founding the *Lancet* in 1823, was a new and radical idea (Desmond
1989). The surgeon Gideon Mantell certainly felt his prospects had been
damaged by his reputation as a practitioner of science.

In an intellectual culture in which oral performance held such high
status, one could reach the pinnacle of scientific success with virtually no
publications at all. The most notable case is that of the naturalist Sir Joseph
Banks. President of the Royal Society and leader of British science for several decades, Banks published almost nothing. The flora from his Pacific
travels was never issued (it was too expensive even for the tiny audience
who traditionally would have underwritten such works), nor did he publish a book on his Icelandic travels or his researches into plants. Instead,
his reputation was founded on his collections and library, his experience
as a traveler, and his status as a gentleman. He maintained his position
not through publishing but through personal contacts and correspondence.
Banks and his sister Sarah Sophia welcomed numerous visitors to his great
house on Soho Square (see figure 2.1), and his vast library, which held his
books and collections, became a place to meet, eat, and talk for those with
natural history interests (Gascoigne 1994). Conversation was central to
this way of doing science.

From this perspective, the mathematician Charles Babbage's notorious
polemic *Reflections on the Decline of Science in Britain* (1830) missed the
point in decrying the large number of fellows of the Royal Society who
had not published a paper in its *Transactions*. Publishing in this way was
a sign of merit, but it was by no means the only one. Status could derive
from the possession of a fine collection of mineral specimens or dried plant
specimens; a position as a patron of science; or through extensive reading
and travels.

The foundation of specialist societies in London, starting with the
Geological Society of London in 1807, is often taken as marking a fundamental shift in the pursuit of science and a challenge to what David
Miller (1983) has called "the Banksian learned empire." Although there
was real conflict, it is important to recognize just how much those advocating a new order shared with their opponents. Although publication

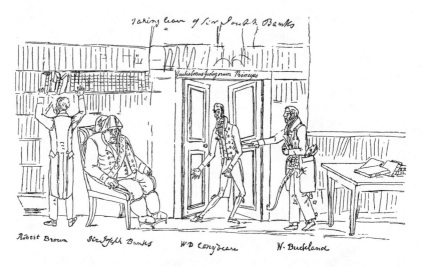

Figure 2.1. "Taking leave of Sir Joseph Banks," 1820, from a drawing by the Austrian Count Joseph Breunner. The geologists William D. Conybeare and William Buckland (labeled the "princes of geology" in Latin over the doorway) depart deferentially from Banks, who is holding court at Soho Square while the botanist Robert Brown organizes the books in the library. Banks died shortly afterward. Reprinted from Gordon 1894, 39.

was given a higher priority, it remained embedded within a gentlemanly culture that placed a high value on face-to-face contact. Most of the new specialist societies grew out of informal meetings of enthusiasts in taverns and the premises of publishers. Their rooms were located in London's West End, near the leading metropolitan clubs. The most radically innovative of the new groups, the Geological Society of London, had been founded as a "little talking Geological Dinner Club" (Rudwick 1963, 328), and although it soon expanded its aims, the subject as a whole remained focused on verbal discussion and conversational contact. Its first president, George Bellas Greenough, published only one short book and devoted most of his life to compiling maps and organizing his impressive personal collection. Like-minded gentlemen could meet in the society's rooms at Somerset House on the Thames; membership was expensive, and women, as in most other clubs, were not allowed. Or the society's members could visit, often with wives and sisters, Greenough's mansion in Regent's Park, which was purpose-built by the architect Decimus Burton for natural history research. These were the places where visiting foreigners really learned about what was going on in geology. Scientific practitioners inhabited a familiar landscape of upper-class metropolitan intellectual life, embedded in the regular calendar of "Society" and "the Season." Annual charts, widely distributed and displayed, showed the main schedule of scientific meetings running from November through June (see figure 2.2). This was when professional men were in town and when most books were published. Once Parliament opened in the new year, the town experienced its largest influx of aristocrats and people of fashion: this was when most private parties were given, when the opera was in season, and when tailors and dressmakers were at their busiest.[2]

Of course, societies such as the Geological, Astronomical, and Linnean viewed publishing papers read at their meetings as one of their chief aims. But it is significant that until the 1840s, quarto transactions were the main forum for publication—a luxurious format with fine paper, wide margins, good spacing between the lines, and a stately publication schedule (Rudwick 1995). These were works fit for a gentleman's library. To supplement them, most of the societies also issued proceedings, which recorded the basic contents of the papers read at each meeting. Their purpose, however, was as much to tell those who had not been at the meetings what they had missed as it was to communicate findings within the wider European community of learned specialists (Rudwick 1985). Priority was not defined as being primarily a matter of publication until the middle decades of the century. That was when discoveries such the planet Neptune

MEETINGS OF THE LITERARY AND SCIENTIFIC INSTITUTIONS OF LONDON, 1838-39.

Societies.	November.	Dec.	January	February.	March.	April.	May.	June.	Time of Meeting.
ROYAL SOCIETY	15, 22	6,13,20	10,17,24,31	7,14,21,28	7, 14, 21	11, 18,-25	2, 9, 16, 30	6, 13, 20	Thurs. 8½ P.M.
SOCIETY OF ANTIQUARIES	15,22,29	6,13,20	10,17,24,31	7,14,21,28	7, 14, 21	11,18, 25	2, 9, 16, 30	6, 13, 20	Thurs. 8 P.M.
ROYAL INSTITUTION	...	...	18, 25	1,8,15,22	1,8,15,22	12,19,26	1,10,17,24,31	7, 14	Friday, 8½P. M.
ROYAL GEOGRAPHICAL SOC.	12, 26	10	14, 28	11, 25	11, 25	8,22	13, 27	10, 24	Monday, 9 P.M.
GEOLOGICAL SOCIETY	7, 21	5, 19	9, 23	6, 27	13, 27	10, 24	8, 22	5	Wednes.8½P.M.
*LINNEAN SOCIETY	6, 20	4, 18	15	5, 19	5, 19	2, 16	7	4, 18	Tuesday,3 P.M.
*HORTICULTURAL SOCIETY	6	4	15	5, 19	5, 19	2,16	7, 21	4, 18	Tuesday,3 P.M.
ZOOLOGICAL SOCIETY	1	6	3	7	7	4	2	6	Thursd. 3 P.M.
ZOOLOGICAL SOCIETY	13,27	11	8, 22	12, 26	12,26	9, 23	14, 28	11, 25	Tuesday 8½P.M.
*ENTOMOLOGICAL SOCIETY	5	3	7	4	4	1	6	3	Monday, 8 P.M.
STATISTICAL SOCIETY	19	17	21	18	18	15	20	17	Monday, 8 P.M.
ASTRONOMICAL SOCIETY	9	14	11	8	8	10	10	14	Friday, 8 P.M.
PHRENOLOGICAL SOCIETY	5,19	3,17	7,21	4,18	4,18	1, 15	6, 20	...	Monday, 8½P.M.
SOCIETY OF ARTS	7,14,21,28	5,12,19	9,16,23,80	6,13,20,27	6, 13, 20, 27	3,10,17,24	1,8,15,22.29	5, 12	Wednes 7½P.M.
SOCIETY OF ARTS	13	11	8	12	12	9	14	11	Tuesday,8 P.M.
ROYAL SOC. OF LITERATURE	22	13, 27	10, 24	14, 28	14, 28	11, 25	9, 23	13 27	Thursd. 4 P. M.
ROYAL ASIATIC SOCIETY	...	1,15	5,19	2,16	2, 16	6, 20	11	15	Saturday,2P.M.
ROYAL COL. OF PHYSICIANS	...	...	...	...	...	...	...	...	Monday, 9 P.M.
ROYAL MEDICAL and CHIRURGICAL SOCIETY	13, 27	11	8, 22	12, 26	12, 26	9, 23	14, 28	...	Tuesd. ,8½ P.M.
MEDICO-BOTANICAL SOCIETY	14, 28	12	9, 23	13, 27	13, 27	10, 24	8, 22	12	Wednes 8 P.M.
MEDICAL SOCIETY	5,12,19,26	3,10,17	14,21,28	4,11,18,25	4,11,18,25	1,8,15,22,29	6,13,20,27	...	Monday, 8 P.M.
HARVEIAN SOCIETY	3,17	1,15	5,19	2,16	2, 16	6, 20	4,18	...	Saturd. 8½ P.M.
WESTMINS. MEDICAL SOC.	3,10,17,24	1,8,15	5,12,19,26	2,9,16,23	2,9,16,23	6,13,20,27	4,11,18,25	1,8,15,22,29	Saturd. 8 P.M.
INSTIT.OF CIVIL ENGINEERS	...	...	8, 29	5,12,19,26	5,12,19,26	0,16,23,30	7,14, 28	...	Tuesday,8 P.M.
ARCHITECTURAL SOCIETY	6,20	4,18,26	1,15,29	12. 26	12, 26	9, 23	7, 21	4,18	Monday, 8 P.M.

ANNIVERSARIES.—Royal, Nov. 30, 11 A.M.—Antiquaries, April 23, 2 P.M.—Royal Institution, May 1.—Royal Geographical, May 21, 1 P.M.—Geological,Feb.15, 1 P.M.—Linnean, May 24, 1P.M.—Horticultural, May 1, 1 P.M.—Zoological, April 30th, 1 P.M.—Entomological, Jan. 22, 8 P.M.—Statistical Society, March 15, 3 P.M.—Astronomical, Feb. 8, 3 P.M.—Phrenological, March 31, 8 P.M.—Royal Society of Literature, April 26, 4 P.M.—Asiatic, May 5, 1 P.M.—College of Physicians (Harveian Oration), June 25, 4 P.M.—Royal Medical and Chirurgical Society, March 1.—Medico-Botanical, Jan. 16, 8 P.M.
The BRITISH ASSOCIATION will meet in August next, at Birmingham.
* These Societies continue their Meetings throughout the year.

Figure 2.2. "Meetings of the literary and scientific institutions of London, 1838–39." As with intellectual life more generally, the institutions of metropolitan science were integrated into the seasonal rhythms of society and politics. Reprinted from [Timbs] 1839, 282.

and the composition of water emerged as subjects of fierce partisan rivalry (D. Miller 2004). The stress on publication was accompanied by the rise of new, heroic ideas of authorship, with discovery attributed to individual men of science whose celebrity was marked by their appearances in print.

Many aspects of the relation between publication, oral performance, and scientific practice usually thought to be characteristic of the eighteenth century thus remained in place for much longer than is often assumed. Notably, the verbal presentation of a paper—even when not accompanied by immediate discussion—remained central to the presentation of new scientific work. Right through the 1840s, the great moment for any paper at the Geological Society was not when it appeared in print but when it was read and debated at one of the meetings (Thackray 2003). On his return from the *Beagle* voyage, Charles Darwin was struck by the way in which publishing in geology seemed secondary to verbal performance. "Geology is at present very oral," he told a friend in 1846, "& what I here say is to a great extent quite true" (quoted in Burkhardt 1987, 338).

Even the publication patterns of Charles Babbage, the astronomer John Herschel, the geologist Adam Sedgwick, and other scientific reformers bear this assertion out. Within the metropolitan setting for intellectual life, they were seen as "great men," characters of exceptional ability and talent. Their characteristic writings were not articles in specialist journals, although they wrote many of them; instead, these men were known for encyclopedia essays, observational records, reflective treatises, and speeches at the Royal Institution of London (founded in 1799) and the British Association for the Advancement of Science (founded in 1831), which rapidly emerged as key sites for oral performance in the new sciences. Contemporaries saw such men as possessing genius or talent not so much through publications but through their persons. Their position was signaled by their status as "lions," whose presence could give intellectual depth and sparkle to a social gathering (J. Secord 2000, 410–16). Conversability was one of the key attributes of the gentleman.

The more formal, institutional platforms for oral scientific performance tended to be reserved for men, although women engaged fully in certain kinds of fieldwork, experiment, and conversation. These activities could lead to an international reputation, especially in subjects (such as mineralogy and botany) dominated by private collections. In European natural history circles of the 1820s, Charlotte Murchison was better known as a conchologist than her husband Roderick was as a geologist (Kölbl-Ebert 2004).

Polite Science

The importance of informal associations and conversation in establishing reputation is well illustrated by the case of Mary and William Somerville after they moved to London in 1816.[3] Married in 1812, they belonged to a circle of learned couples who became central to metropolitan science. Alexander and Jane Marcet—the latter's *Conversations on Chemistry* ([1806] 2004) had recently appeared—were key figures, as were other couples such as the Katers, the Murchisons, and the Youngs. Mary Somerville was a self-taught mathematician who had been encouraged as a young woman in Edinburgh partly through the networks of military practitioners during the wars with Napoleonic France. Her second husband, William Somerville, was an experienced traveler who had served as an army physician in North America and in South Africa.

For the Somervilles, science offered a vital niche in urban society. They participated in a lively intellectual culture centered on informal gatherings that involved both men and women. They collected minerals, made observations, discussed new books, and welcomed visiting foreigners. "All kinds of scientific subjects were discussed," Mary Somerville recalled, "experiments tried and astronomical observations made in a little garden in front of the house" (M. Somerville 1873, 130). When William's army position was abolished after the end of the Napoleonic Wars, he and Mary traveled to the Continent, where they could live more cheaply until a new post turned up. Here they were welcomed to similar gatherings in the best scientific circles of Paris and Geneva. The Somervilles were not aspiring amateurs nor potential patrons (even with Mary's substantial inheritance from her first marriage, they did not have the money for that); rather, they were valued participants in an enterprise which was open to many different kinds of practitioners. The scientific activities pursued by the Somervilles after their return to London in 1818 grew out of this predominantly domestic milieu. In mineralogy, for example, they devoted their efforts to what all serious mineralogists considered as the science's core activity: they formed a cabinet, filled with superb specimens given by friends and purchased on their travels. Their house was filled with paintings by Mary and with books on literature, art, and science.

The Somervilles were thus at the very heart of Regency science in London. This standing was not simply because of their social status—they were high gentry but not in a position either to purchase the best specimens or to employ others in scientific work. And during the 1810s and 1820s, their status was certainly not based on their reputations as original discoverers or writers. Mary Somerville had published prizewinning contributions to the *New Series of the Mathematical Repository* both before and after her marriage to William, but she did so under a pseudonym ("A Lady"). The primary significance of her pseudonymous mathematical puzzle solutions did not derive from publication but rather their status as marks of participation in a network of practice that ranged from workingmen to John Herschel and, in 1819, the thirteen-year-old John Stuart Mill.[4]

Although Mary Somerville's status as a genteel woman might seem to involve special issues when publication was at stake, this is true only in part. The case of William Somerville shows that many of the same concerns applied to both sexes. When they married, he was a seasoned traveler, a friend of the poet Sir Walter Scott, and the first European to visit parts of the Cape Colony. Using his position as a medical man, he had conducted researches on the much-debated anatomy of the female sexual organs of the Khoi-khoi, a subject that soon came to be much in the news through the exhibition of Sara Baartman, the so-called Hottentot Venus (Qureshi 2004; Schiebinger 1993). He presented a paper to the Royal Society of London, which was published in the *Medico-chirurgical Transactions* in 1816, translated into Latin to limit its readership (W. Somerville 1816). The illustrations were too explicit to be published in this context and could be viewed only in the Royal Society's library, where the manuscript of his paper remains to this day (W. Somerville 1979). Although this article was all that William Somerville ever published, he became a full participant in the learned life of London and was elected a Royal Society fellow in 1817.

It was, in fact, with the aim of meeting William and gaining access to Chelsea Hospital that the opinionated American physician Charles Caldwell presented himself at the Somervilles' house in the winter of 1821–22. Like many foreign travelers, he recorded numerous customs that natives took for granted. Caldwell was especially interested in conversation, and, left in the company of Somerville's wife while his host attended to his patients, Caldwell engaged her in talk. They began by discussing "the polite literature of the day," a subject upon which she seemed completely at ease. Noticing some volumes on natural history in an opened bookcase, he found that she knew far more about mineralogy than he did. A volume of Pierre Simon Laplace's abstruse masterpiece, the *Méchanique Céleste*, suggested astronomy as a subject and in this "she conversed with such a familiarity and compass of knowledge as might have led to a belief that she had just returned from a tour among the heavenly bodies." After realizing that she was also a skilled experimentalist and an artist of some skill, he asked the crucial question: "Who are you?" "I am Mrs. Somerville, sir," she purportedly replied. "I know that, madam, but who were you before you became Mrs. Somerville?" (Warner 1855, 378–80).

Mary Somerville then is reported to have said that she had been "a little Scotch girl" and "a pupil" of the Edinburgh natural philosopher John Playfair. In Caldwell's novelized memoir, the reader has been moved from a commonplace conversational opening about literature, through a series of revelations of increasing understanding about science, and then down again by a remark that was simultaneously modest but also revealed Somerville's connection to contemporary centers of expertise. The entire narrative depends on the fact that it reports an encounter with Somerville at a time when she was well known in London but before she had become famous in the United States; it also depends on the implication that the reader has low expectations of the conversation of the women of England.

Caldwell clearly said as much to Mary Somerville, and a few days later she sent him an invitation to a "conversation party of ladies" at her home. The gathering was explicitly designed to prove the conversational powers of women in general and of English ones specifically. Here, at a single evening party, were Elizabeth Fry, Elizabeth Lamb, Maria Edgeworth, and two of Edgeworth's sisters.

> And when the introduction was over, my kind cicerone said to me playfully, but in an undertone, ["]I have caught you; prepare yourself; you are about to have female conversation enough to convince you that the ladies of England can talk as well as those of your own country—a truth which you seem to doubt." . . .
>
> And, in a moment, she disappeared, and "left me alone in my"—no, not in my "glory"—but in my half dismay, at the task of entertaining half a dozen of talk-loving ladies. "But," said I to myself, "they are all, I hope, so anxious to hear the vibrations of their own tongues, that they'll talk and entertain each other, and I shall have nothing to do but listen."
>
> Very soon, however, I discovered my mistake. (Warner 1855, 381–83)

One by one, the women conversed with Caldwell in such a way as to convince him that Mary Somerville was not exceptional: English ladies could converse with all the propriety, skill, and interest of American ones. Other than her husband, Caldwell was the only man present.

Caldwell noted that none of his interlocutors at this party, other than his hostess, had spoken to him on deep or scientific subjects; but that was not a reflection of the interests of women such as Fry or the Edgeworths, who were very conscious of moving in circles where such issues were regularly discussed. A typical evening party of the kind they frequented is described in a letter Maria Edgeworth wrote to her stepmother while staying at the country house of the agricultural improver Sir John Sebright in 1822:

> Mrs. Somerville and I were sitting on the opposite sofa all night and Dr. Wollaston sitting before us talking most agreeably and giving us a clear account of the improvement of refining sugar which Mr. Howard who established his patent for the same just before his death has left five thousand a year to his children—a good reward for one invention!— for an improvement in a common process. Observe it is on an article of universal consumption. (Edgeworth 1971, 322)

There was, in fact, a striking continuity between the kind of topics that could be discussed at such intimate gatherings and the contents of the conversational genre of science books that became such a dominant feature of the educational literature of the 1820s and 1830s. For example, William Hyde Wollaston's lively description of sugar making became the basis for a passage in Edgeworth's improving children's book, *Harry and Lucy Concluded*, in 1825 (Edgeworth 1971, 322). The most famous of these scientific dialogues, Jane Marcet's *Conversations on Chemistry*, first published in 1806, grew out of actual conversations with a "friend," probably her husband, Alexander Marcet, the Swiss chemist, mineralogist, and physician (Marcet [1806] 2004; Bahar 2001). The book also stressed that the experiments had been carried out by the author at home, thus demonstrating that science was suitable part of domestic life. Works such as those by Marcet and Edgeworth became a major feature of the publishing scene in the 1820s and 1830s. The title—*Conversations*—that came to be most closely, but not exclusively, associated with the genre (Fyfe 2004) clearly involved a claim that scientific knowledge could be a polite accomplishment.

The Etiquette of Science

Of course, even at the beginning of the century, not all science would have been deemed appropriate among mixed company or in general discussion. This was the tension surrounding the notion that, in principle, science was available to all. Some subjects were simply unfashionable. Mathematics, other than in the form of puzzles and conundrums, was generally seen as too obscure: Babbage was regularly quizzed for his use of abstruse mathematical analogies in everyday speech (Lyell 1881, 1:363–34). Phrenology appears to have been out of bounds in most London circles from the 1830s, after phrenological bores had been widely parodied and pilloried in the press. Major General John Mitchell, author of a pseudonymous semisatiric *Art of Conversation* (1842), described his surprise on hearing an English lady express interest when an Italian marquis asked her if she had heard of phrenology:

> "The Italians," continued my companion, "know so little of these things themselves that they think other people equally ignorant; it was therefore kind of him to give us the information." "Then you would have voted an Englishman, who should have lectured you on this exploded old subject, a regular bore?" "To be sure I would, and would have cut him and his lecture fast enough." "I am glad to hear it," said I, "for it shows how much more you really expect from your own countrymen than from all these foreigners, much as you praise and admire them." "Hem," said Lady C., "I wish you would employ your philosophy in getting me a good partner for the next quadrille." ([Mitchell] 1842, 21–22)

Other subjects were in bad taste. It was improper to pose certain kinds of direct questions: ask a physician what his specialty was, and he might be forced to tell the company that he was a male midwife (Anon. 1847, 48).

Certain topics were seen as opening the possibility of political or religious controversy. This was definitely the case with evolutionary debate in most circles before publication of *Vestiges of the Natural History of Creation* in 1844 and in some afterward (J. Secord 2000, 155–66). Even among men of science, open expression of heterodoxy was often seen as bad manners. Take the following exchange between the leading London physiologist William Sharpey, the mathematician Augustus De Morgan, and the ambitious government geologist Andrew Ramsay, who recorded it in his diary in 1848:

> "You geologists" said Sharpey "shock our religious feelings dreadfully, you make so light of all established chronologies as regards the age of the world." "I was not aware" quoth I, ["]you had any religious feelings, or I might have managed my expressions better." "Oh" said he "its not that, but you see we are of all sorts here, some of us have religious feelings, & others have none" waving his hand to De Morgan. "Well" said I "I can't stop to make an expression always that will please all, & must

speak the plain truth." "I don't see the necessity" growled De Morgan "if it be heterodox"!!![5]

The most common problem was pedantry: sparkling conversation could be hard to combine with the close study and intense devotion demanded by an independent scientific vocation. This could lead men of science to seem monomaniacal, self-absorbed, small-minded, or eccentric (Shapin 1991). Learning, no matter how profound, was "unwelcome and tiresome" unless accompanied by good breeding and a proper sense of occasion (Anon. 1861, 14). An army friend who had known Roderick Murchison in his foxhunting days lamented his turn toward geology: "an excellent fellow and a most agreeable comrade" now seemed able to talk only of dull graywacke and the "Silurian system" ([I. Campbell] 1906, 1:349).

Aware of the problem, those engaged in science recognized that conversation was only one kind of talk, and they carefully distinguished among various kinds of spoken discourse. What might be said in an annual presidential address differed from what was appropriate for a report of new research; the verbal sparring used in the Geological Society's parliamentary-style meeting room was not the same as private chat in the corridor beforehand. (Worries that the society's lively debates might be misconstrued led to defenses that they were "conversations.") Even the most expert practitioners sometimes refrained from talking about science at all. Gathering together until two o'clock in the morning in Lord Cole's rooms after the anniversary meeting, the gentlemanly Charles Lyell and his friends drank "fines inflicted of bumpers of cognac on all who talked any 'ology" (Thackray 2003, 81).

Early Victorian novels often feature men of science obsessed with priority and insignificant detail. A classic example is the chapter "The Man of Science" in Charles Kingsley's *Alton Locke,* first published in 1850. The episode involves the novel's eponymous hero and a clerical professor of natural history at Cambridge, but the caricature was intended to criticize the conversational faults of men of science more generally. The man of science proposes to carry out an experiment on Alton Locke, to see if a common workingman can be turned into a naturalist.

> "And what have you read on these subjects?" he asks.
>   I mentioned several books: Bingley, Bewick, "Humboldt's Travels," "The Voyage of the Beagle," various scattered articles in the Penny and Saturday Magazines, &c., &c.

"Ah!" he said, "popular—you will find, if you will allow me to give you my experience—".

That experience turns out to be a complete knowledge, based on thirteen years' study, of a single obscure reptile from the Balkans. The man of science claims that the fashionable world was "overstocked with the *artem legendi*—the knack of running over books, and fancying that it understands them, because it can talk about them." In the end, Alton Locke does not find mastering the details of science so difficult after all: his mentor's attempt to denigrate learning by calling it "popular" has failed the test of experience (Kingsley [1850] 1883, 179–84).

The case of Mary Somerville brings the increasing dangers of pedantry into clear relief. Her most widely read books, especially *On the Connexion of the Physical Sciences*, published in 1834, were neither for narrow specialists nor for novices only; they were reflective treatises targeting an inclusive readership, based on her experiences in London in the 1810s and 1820s. As time went on, and especially after she became a potential "lion" after publication of the *Mechanism of the Heavens* in 1831, she seems increasingly to have regulated her scientific talk, avoiding displays of knowledge except in conversations with those she knew to have an interest. People who met her hoping for a feast of witty repartee went away disappointed. As the celebrated talker Sydney Smith said in 1843, "as to herself I could never get anything out of her beyond what you might get from any sempstress. She avoids all depth in converse—and no pretty superficial—very amiable no doubt" (quoted in Edgeworth 1971, 601). Clearly Somerville feared being seen as a bizarre specimen, a bluestocking. As a young girl, Julia Clara Byrne remembered being terrified of a mathematical grilling from the author of the *Mechanism of the Heavens*, though recognizing that her concern was groundless, "as no one could be less pretentious in manner, or more amiable and gentle in conversation" (Byrne 1892, 1:260).

Somerville's own writings were explicitly part of the vast effort that went into facilitating talk about science. Good conversation was not chit-chat, gossip, or small talk; it had to be planned for and carefully orchestrated. This belief is evident in the extensive machinery of review journals that emerged during the nineteenth century. The weeklies, monthlies, and quarterlies served as an adjunct to the literature of conduct, suggesting to readers what might be appropriate to say about particular books. The reviews selected passages and quoted them at length, pointing out "beauties" and phrases that might be useful in discussion. Etiquette manuals were usually despised for communicating rules that everyone was supposed to know but not follow slavishly; but everyone read periodical reviews. Reading informed opinion about the latest books was part of the process of preparing for conversation.

Like any fine art, the best conversation usually required carefully planned circumstances. A hostess could play a crucial role in harmonizing talents, restraining the overexuberant or gauche, and drawing out the shy. In a widely reprinted essay, "Pic-nic," published posthumously in 1826, the improving writer Anna Barbauld stressed the skills required by the accomplished hostess. Just as each guest was supposed to furnish a dish to make up a picnic, so too could everyone attending a party contribute to the conversational feast. The narrator of Barbauld's tale commended the fictional Lady Isabella's efforts in creating opportunities for scientific conversation:

> I noticed particularly her good offices to an accomplished but very bashful lady and a reserved man of science, who wished much to be known to one another, but who would never have been so without her introduction. As soon as she had fairly engaged them in an interesting conversation she left them, regardless of her own entertainment, and seated herself by poor Mr. ——, purely because he was sitting in a corner and no one attended to him. You know that in chemical preparations two substances often require a third, to enable them to mix and unite together. Lady Isabella possesses this amalgamating power:—this is what she brings to the pic-nic. (Barbauld 1826, 194–95)

Here, as in the Goethe's 1809 *Elective Affinities*, a chemical metaphor characterizes human interaction, which involves in turn an "attraction" based on science.

From the 1830s onward, a flood of etiquette manuals aimed at nouveaux riches from the middle classes attempted to codify the rituals of conversation. These manuals tended to lay down formal rules, stressing that manners were the mask that made society function (Morgan 1994). Initially this literature recommended science as a key element in the social game. As one evangelical guide told young ladies preparing for the marriage market, botany and botanical conversation were especially favored topics:

> On the whole, of the sciences now, happily, so much in vogue for ladies, and which constitute such innocent and laudable pursuits, I am inclined . . . to place Botany on the pinnacle. It is more easily attainable

than Mineralogy or Conchology—it is a cheap pleasure—its materials are scattered invitingly by a liberal hand—its books are in the fields—it is of all seasons—even in the winter there are mosses and lichens, the very study of which is almost enough for a lifetime. (Anon. 1845, 18)

The recommendation of botany as a feminine study was a commonplace (as in Mrs. Sarah Ellis's best-selling books), but it is revealing to see an approach recommended that could lead to deep expertise. The recommendation here is not only of the pursuit of an ornamental knowledge of flowering plants, easily accessible and easily obtained, but also of the potential of botany as a lifelong interest, involving difficult subjects such as the cryptogams. Expertise about plants was particularly likely to be shared between women and men.

Etiquette manuals aimed at aspiring gentlemen recognized the role that the sciences could play in socializing with the opposite sex. The more cynical authors even encouraged male readers to exploit these topics as tactical openings and pickup lines; an interest in the heavens could be the first step toward a seduction. *Etiquette for Gentlemen: With Hints on the Art of Conversation* (1847) told young men to think through topics and tactics before attending parties. Science could have a much more useful role than was often assumed.

> It is a common practice with men to abstain from grave conversation with women. And the habit is in general judicious. If the woman is young, gay and trifling, talk to her only of the latest fashions, the gossip of the day, etc. But this, in other cases is not to be done. Most women who are a little old, particularly married women,—and even some who are young,—wish to obtain a reputation for intellect and an acquaintance with science. You therefore pay them a real compliment, and gratify their self-love, by conversing occasionally upon grave matters, which they do not understand, and do not really relish. You may interrupt a discussion on the beauty of a dahlia, by observing that as you know they take an interest in such things, you mention the discovery of a new method of analysing curves of double curvature. People who talk only trifles will rarely be popular with women past twenty-five. (Anon. 1847, 51)

From this worldly-wise point of view, women had to feign an interest in scientific knowledge, while men had to humor them by pretending that their interest was genuine.

Shop Talk

Of course, moving a conversation in a single step from the most ornamental parts of botany to the most complex realms of mathematics was not really a recommended strategy: it was a joke. The example does, however, illustrate the ambiguities conversation about science was beginning to pose in a society increasingly organized around formal rules rather than tacit agreement. The central distinction involved aspects of a topic suitable for all and those that could be brought up for discussion among those engaged upon it as a paid professional career. Not least, this discussion marked a widely regarded distinction between women and men. As one of the manuals put it, "Professional men must carefully avoid professional wit, as well as professional subjects, unless when called upon to bring forward some peculiar case or illustration; for general society tolerates no university, mess-table, or law-court stories" ([Mitchell] 1842, 110).

This issue became known as the problem of "shop talk," which emerged as a major concern from the 1850s, when periodical essays, novels, and etiquette manuals highlighted the difficulties of dealing with integrating a large influx of professionals, businessmen, and industrial entrepreneurs into the higher ranks of society. The term had originated as slang, with *shop* or *shoppy* signifying talk of commerce or business. It carried strong class overtones, for to introduce money and finances into everyday conversation was bad form and a sign of low origins. As Fanny Trollope wrote in her industrial novel, *The Life and Adventures of Michael Armstrong, the Factory Boy*, "the gay fatherly phrase, 'Don't talk of that, for God's sake, my dear!—it smells of the shop,' has turned away many innocent eyes from contemplating that, which had they looked upon it, could hardly have endured so long" (Trollope 1840, 137–38). Shop brought the politics of the machine into the salon.

During the first half of the century, scientific talk might be pedantic or inappropriate in other ways, but it was not "shoppy" because a knowledge of nature was not considered the province of a specific profession or business pursued for pay. In fact, the lack of connection to "trade" had long been one of the chief attractions of science as a topic for general conversation. Early in *Conversations on Chemistry*, Jane Marcet had one of the characters doubt the worthiness of chemistry as a subject for general conversation on the grounds that it was part of the trade of pharmacy. The authoritative Mrs. B, however, corrects this misconception. Chemistry was a significant aspect of human learning, appropriate for all, while pharmacy "properly belongs to professional men, and is therefore the last that I should advise you to study" (Marcet [1806] 2004, 1:2–3). On the same grounds, attendance at scientific lectures could be encouraged, even when they were part of a medical course—as long as women and other members of the general public skipped those parts that were concerned with therapeutics or the more intimate aspects of human anatomy (A. Secord 2002, esp. 40–45).

From the 1860s, the increasing dominance of the professions in intellectual life was one of the factors leading to a broadening of the debate about "shop talk" beyond medicine, law, and similar fields. An article in the *Cornhill Magazine* for 1865 cautioned the new class of literary professionals against assuming that talk of literature was inherently of general interest.

> But what we wish to impress upon the literary class is that such writing *is* shop, and just the same in principle as the parish talk of parsons, or the barrack talk of soldiers. All these relate equally to the mechanical routine—we had almost said drudgery—of the three professions; and contain nothing either to please or to improve persons who are not already familiar with them. Among a party of literary men seated round a club table, or enjoying a tavern dinner, such talk is natural, and perhaps profitable. . . . In a word, the outside public cares not for professional topics except when they rise above the lower level of the workshop into that broader region where they are to some extent common property. In the case of literature this region is wider, and extends lower than in the case of other professions. But literature, too, like them, has its mere mechanical sphere, its "shop," in fact; and this, we say, can be interesting only to the workmen. ([Kebbel] 1865, 494)

This broader definition was being applied to areas from philosophy and philology to geography and geology. An article in the *Saturday Review* for 1875, for example, mentioned Egyptian antiquities and the origin of language as having potential for shop talk (Anon. 1875, 42).

The blurring of boundaries is subtly brought out in Charles Dickens's *Our Mutual Friend* ([1864–65]). The indolent solicitor Mortimer Lightwood apologizes for using "the phraseology of the shop" in employing the term *client* to refer to Mr. Boffin. His excuse is that Boffin is likely to be the only client he shall ever have and hence is "a natural curiosity probably unique." Although the apology for shop talk is wholly unnecessary in the circumstances, it leads Lightwood to develop an analogy that does go into another kind of jargon, that of natural history. Not only is Boffin "the natural curiosity which forms the sole ornament of my professional museum" but Boffin's secretary also becomes "an individual of the hermit-crab or oyster species, and whose name, I think, is Chokesmith—but it doesn't in the least matter—say Artichoke" (Dickens, [1864–65], 357). Lightwood has worried so much about the rules of etiquette in talking about the law that he fails to avoid shop talk about science.

As the polite science characteristic of the period up through the 1840s disappeared, the opportunities for using science in conversation among the social elite began to change. New discoveries and inventions remained suitable topics for discussion, but they were more closely tied to current affairs as reported in the daily newspapers and weekly press. In many circles, for example, Darwin's *On the Origin of Species* began to be widely debated only after Paul du Chaillu returned from Africa with his gorillas (Rushing 1990). Such spectacles offered novelty and general interest in a form that was self-consciously aimed at a nonspecialist audience. In 1867, Hugo Reid, the author of mathematical texts and self-help guides, listed the appropriate topics for talk:

Prepare, then, for conversation by storing the mind with interesting matter on subjects calculated for general discourse. Amongst the principal of these are, History, not forgetting the history going on at the present time; Biography, particularly of recent celebrities; Anecdotes and curious traits of human nature; Remarkable Crimes and Trials; Adventures; Voyages and Travels; Manners and Customs of different Nations; Antiquities; Geography; Curious Facts in Physical Science; Natural History; Commerce and Manufactures; Inventions; Statistics; and, above all, a knowledge of the Lives, Works, Opinions, and Sayings of Great Men of all ages. (Quoted in St. George 1993, 58)

This seems an impossibly ambitious list, but in fact it was not. All that was needed was a familiarity with current affairs, a liberal education, and handy outlines of the kind that men such as Reid were writing. Controversial, difficult, or specialized subjects could to be repackaged for ready use as specific "topics," with a stress on the telling anecdote and the remarkable story. News events became increasingly central to introducing science and other forms of knowledge into public discussion. This development can be seen in a cartoon from *Punch*, in which a young woman drawn by John Leech as self-consciously "pretty" tells a young man attempting to engage her in conversation that she has already talked enough of the gorilla, the rebuilt Crystal Palace at Sydenham, and the other regular topics of the day. She is thereby labeled a "HORRID GIRL!" who has deprived an

HORRID GIRL!

Mild Youth. "Have you seen 'The Colleen Bawn'?"
Horrid Girl (with extreme velocity). "Seen 'The Colleen Bawn'! Dear, dear!
Yes, of course. Saw it last October! and I've been to the Crystal
Palace, and I've Read the Gorilla Book!"   [*Mild Youth is shut up.*

Figure 2.3. "HORRID GIRL!" In this cartoon by John Leech from *Punch,* a studious young man asks a self-consciously pretty society woman if she has seen Dion Boucicault's sensational melodrama, "The Colleen Bawn." She sharply replies that she not only has seen the play but also has been to the Crystal Palace at Sydenham and read Du Chaillu's book on gorillas. Reprinted from *Punch,* June 1, 1861, 226.

awkward admirer of his planned conversational openings (see figure 2.3). *Punch* became the great repository of bemused reflection on the problem of shared knowledge, as everyone could laugh at social gaffes that were just a bit more ridiculous than those going on all around them.

The sciences were by no means the only area affected by fears that society, as it admitted new classes of experts, would be swamped by shop talk. In fact, the elite social world became more segmented and specialized,

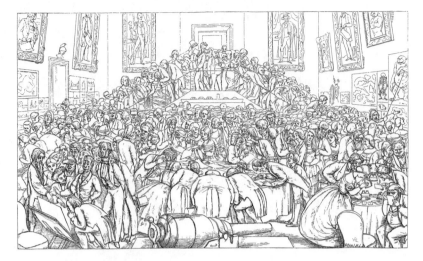

Figure 2.4. "Conversazione: Science and Art," part of a series illustrating the
increasingly distinctive spheres of mid-Victorian society. Reprinted from
Doyle 1864, 45–46.

with specific occasions organized to appeal to the various "sets"—artistic,
literary, sporting, political. These sets are brilliantly depicted in a number of entertaining surveys of the contemporary scene, notably George Augusta Sala's *Twice Round the Clock* (1859) and Richard Doyle's *Bird's Eye
Views of Society* (1864). Both men were survivors of the competitive world
of journalism and keen observers of metropolitan life. Their books gave
readers a sequence of panoramic views of the highly diversified settings
of mid-Victorian London society: "At Home" musical evenings, political
banquets, art sales, clubland dinners, and so forth. Doyle's "Conversazione:
Science and Art" (figure 2.4) shows a room packed with "all kinds of novel,
curious, interesting, and instructive objects": antiquities, microscopes,
telescopes, portraits, batteries, diving bells, and weapons of war. The scientific soirée, which appears to have originated in France and was imported
by Babbage in the 1820s, became one of the chief ways in which scientific
discussion could take place on a more sustained basis within polite society
(Alberti 2003; J. Secord 2000, 410–21).

Such occasions were attractive to the wider social world in that they
offered a carefully mediated way of encountering scientific novelties. For
those pursuing studies in the laboratory, field, and museum, they increasingly provided opportunities for explaining science in a setting that was, in

effect, halfway between a lecture and an informal conversation. As Doyle's cartoon shows, the role of the practitioner in these settings had changed: there is more overt didactic presentation and a stress on celebrity and individual discoverers. (One of the paintings on the wall, for example, recalls a well-known portrait of the anatomist Richard Owen grasping a fossil femur.) There are many women present, but they are part of the audience, with the exception of a lone "female savant" talking to no one, identifiable by "a decided tendency she has to part her hair on one side," standing alone in the middle of the right-hand side of the picture. She does not belong to any of the groups of women taking a look through the telescopes or other stereoscopes, but neither is she engaged in the more specialized talk enjoyed by the scientific men. The implication is that learned women do not fit in to this new form of diffusing popularized science. As the accompanying text notes, the men of science "chat to one another, exchanging ideas, or criticising some new invention, or drinking tea." The esoteric quality of their talk is signaled by their strange appearance: "Some with a half mild, half wild, slightly eccentric look, others eager and thoughtful, a good many with spectacles and long hair" (Doyle 1864, 45–46).

The talk of such celebrities, even those without long hair, was seen to be worth recording. Those who knew the leading men of science well recognized the skill it took to manage conversations on topics that were increasingly seen as specialized and esoteric. A widely acknowledged master in such social settings was the lower-middle-class schoolmaster's son, Thomas Henry Huxley. Darwin thought him "the best talker I have known," remarking that he was skilled in adjusting his conversation to the listener, retaining his fierce anticlerical polemics for print or for conversation with intimates (Barlow 1958, 106). Huxley's son's biography made a special point of describing his style of conversation. A friend sent the following recollection:

I was in some trepidation, because I didn't know anything about science or philosophy; but when your mother began to talk over old times with my wife, your father came across the room and sat down by me, and began to talk about the dog which we had brought with us. From that he got on to the different races of dogs and their origin and connections, all quite simply, and not as though to give information, but just to talk about something which obviously interested me. I shall never forget how extraordinarily kind it was of your father to take all this trouble in entertaining a complete stranger, and choosing a subject which put me

at my ease at once, while he told me all manner of new and interesting
things. (Huxley 1900, 2:429)

Such eulogies, often found in late Victorian and Edwardian lives and letters, were effectively recommendations about how a man of learning could
be a moral example to others in his consideration for the needs of others
(White 2003). Huxley simultaneously encouraged his visitor to relax and
showed that there was nothing to be feared about scientific conversation
after all. In preserving such exchanges, biographers felt that they were recording a style of talk that was disappearing.

Notably, it was in the waning years of the Victorian era that antiquar-
ians developed the myth of the early eighteenth-century coffeehouse as the
apotheosis of civil conversation (Cowan 2005). Certainly Huxley felt that
the intellectual tone of high society was declining. In an essay written just
before his death in 1895, he raged against the "refined depravity among the
upper classes," who were taking in spiritualism and superstition rather
than science (Desmond 1997, 227). He spoke against the "half-cretinised
products of over-civilisation," who had become soft and weak-minded
(Desmond 1997, 227; Lightman 1997). In this vividly imagined world of fin
de siècle decadence, there seemed little place for the kind of serious, morally informed discourse on science that Huxley had attempted to bring
into the highest realms of metropolitan social life.

Although Huxley's reaction was extreme, the final decades of the nineteenth century do mark a significant change in the relations between elite
society and what was rapidly becoming known as "the scientific class." As
Stefan Collini (1991, 223–24) has pointed out, this is the period when *academic* began to be used in a pejorative sense, as "not leading to a decision;
unpractical; theoretical, formal, or conventional." In many of the trendset-
ting groups, the detailed pursuit of science became relatively unfashionable, replaced by a new emphasis on pleasure and partying for its own sake.
Leonore Davidoff has shown in *The Best Circles* ([1973] 1986, 59–70) how
society became focused on the rituals of presentation to the queen, the
marriage market, and renewing the upper classes with American and industrial money—and less the matrix for discussing politics and ideas. Parties became more elaborate, with fantastic costume balls, more relaxed
sexual attitudes, and a stress on witty repartee. As Lady Dorothy Nevill
disapprovingly recalled at the end of her life,

In the old days Society was an assemblage of people who, either by
birth, intellect or aptitude, were ladies and gentlemen in the true sense

of the word. For the most part fairly, though not extravagantly, dowered with the good things of the world, it had no ulterior object beyond intelligent, cultured and dignified enjoyment, money-making being left to another class which, from time to time, supplied a select recruit for this *corps d'élite*. Now all this is changed, in fact, society (a word obsolete in its old sense) is, to use a vulgar expression, "on the make." (Nevill 1906, 355, quoted in Perkin 1989, 65)

Although her contrast is drawn too sharply, many commentators deplored the loss of moral earnestness as the century drew to a close. The social elite continued to learn (and talk) about scientific, technological and other "improving" issues, but with a few distinguished exceptions,[6] such knowledge was increasingly acquired not through contact with original investigators but through articles by science journalists and professional writers. Topics for conversation were taken not from long articles in the *Quarterly* and *Edinburgh* but from short, punchy pieces in the dailies and weeklies. In effect, the best circles increasingly had their own "shop," focused on high fashion, topical sensation, and social ritual.

A *Punch* cartoon published in 1873, two years after Darwin's *Descent of Man*, shows what could happen when the social elite confronted its own forms of "shop" with those of men of science (see figure 2.5). The dialogue begins, like dozens of others in *Punch*, with a pedantic "Dr. Fossil" commencing a didactic lecture, using Latin terminology for characteristic features of a primate skeleton. The tables, however, are immediately turned. A well-bred lady immediately interrupts by arguing that because men have a funny bone in the same place, they could never have been monkeys. Her military escort agrees, noting that "that—aw—fellah" could not hold a gun and hence could not have been a man. The strong-minded woman, while clearly aware of the Darwinian theory, is shown as being more concerned with making a point than with understanding the truth; her lowbrow companion is capable of talking of nothing other than subjects related to his own professional training in the military. We never even get to hear what the man of science thinks of Darwin's theory, or if he believes the skeleton is relevant to the question. The social elite is shown to be more obsessed, pedantic, blinkered, and dogmatic than the bewildered doctor of science, who retreats from the salon, "a sadder if not a wiser man." Not surprisingly, commentators around this time acknowledged that within certain limits, shop was "the best talk of all," as it simply meant conversation informed by an understanding of the subjects under discussion.

Figure 2.5. "Baffled Science Slow Retires," an illustration in which the artist William Newman offers a middle-class perspective on the increasing distain of the aristocracy for scientific learning. Reprinted from *Punch*, January 4, 1873, 10.

Laboratory Conversation

For their part, men of science at the end of the century increasingly sensed that the best conversations were to be had with those who possessed similar training. The typical practitioner was frequently pursuing science for pay, either as an academic on behalf of the state or as a professional writer (Collini 1991; Heyck 1982). The new breed of academically trained practitioners tended to look not to genteel society for models of behavior but to middle-class ideals of propriety, or even to alternatives in scientific bohemia. One sign of this change in sensibility, found in many late Victorian scientific lives and letters, is the way in which biographers lamented what appeared to them to be their subjects' inordinate fondness for rank. Charles Lyell, Roderick Murchison, and Mary Somerville seemed to belong to a different scientific world, one which was seen to have paid undue obeisance to social rank.

Even Huxley's mastery of social chitchat with the great, as we have seen, was considered remarkable and worth recording. Of course, such connections could be hugely important for public support and the funding of science. A quiet word in the corridors of the Athenaeum or Reform Club could be worth hours of lobbying at Westminster; ties maintained through family or college connections could smooth over differences about the religious consequences of, say, evolution or materialism. But the generation of scientific men who came to power in the 1860s, however patient they might have been in conversing with novices and potential patrons, thought of substantive scientific talk as occurring much more exclusively in their places of work. Memories of talk in general social settings are taken over by recollections of lively exchanges at the laboratory bench or in the field.

Younger scientists found such interactions intensely liberating. Talk of this kind, in their view, was just one of the many benefits that flowed from their campaign for professional careers. From this perspective, discipline boundaries were not Foucauldian prisons but ways of maximizing pleasurable occasions for the free exchange of ideas and a recognition of shared understandings (Kohler 1999). At the most general level, proponents of science argued that the Enlightenment ideal of conversation as rational discussion among equals was realized most completely at the lab bench.

One sign of the new order was the increased significance of tea in the making of science. The situation in Cambridge, where the tensions of introducing science into a genteel culture of learning were especially acute, gives a good idea of the issues. Throughout most of the nineteenth century, tea had usually been served at evening parties. Some professors invited everyone around to their house or college rooms, part of the way in which scientific issues were part of more general patterns of academic life. John Stevens Henslow and his wife, Harriet Jenyns Henslow, hosted evening events of this kind in their home which the young Charles Darwin regularly attended (Browne 1995, 123). Tea in the afternoon, an innovation introduced into London aristocratic houses during the 1840s (Pettigrew 2001), does not appear to have become a university tradition for several decades. When new laboratories began to sprout up on the New Museums site during the 1870s and 1880s, they had no common rooms or facilities for refreshment.

So it was only in 1879 that the Cavendish professor, Lord Rayleigh, began to sponsor laboratory teas as a way of combining a relaxing break with possibilities for free discussion. It was an informal occasion, using a teapot with a broken spout. His wife, Lady Rayleigh, often attended, as did his sister-in-law Eleanor Sidgwick, who worked on determining the

Figure 2.6. Parody coat of arms for the Botany School in Cambridge, with rampant
helianthuses facing a cup of tea. Reprinted from *Tea Phytologist* x, no. 1 (January 23,
1934): 1, University Archives, BG 66, by permission of the Syndics of Cambridge
University Library.

fundamental units of electricity (Gould 1997). As Rayleigh's son recalled,
others in the laboratory soon joined in, so that tea afforded "a valuable
opportunity for intercourse and exchange of ideas on scientific subjects"
(Strutt 1924, 128–29). Afternoon tea, a fashionable innovation of the Victorian aristocracy, was thus reinvented by Lord and Lady Rayleigh as a
means for ensuring that scientific discussion did not just focus on specific
experiments but retained the openness considered essential to innovation.
At first the teas were held in the professor's rooms, but early in the twentieth century, those engaged in planning new facilities began to set aside
staff rooms and other spaces for scientific talk (Munby 1921).

Many of the new research schools that developed during these years
crystallized as much around the tea room as the lab bench or field. A good
example is archaeology, where new ideas that made the department famous
from the early decades of the twentieth century were hatched over tea
(Smith 2004). In 1908 the students at Cambridge's Botany School dubbed
their magazine the *Tea Phytologist,* a spoof on Arthur Tansley's *New Phytologist,* founded six years before (Godwin 1977, 3). In a mock coat of arms
from the 1934 issue (figure 2.6), the teacup took its rightful place as the
central symbol of modern scientific practice. Parodying the university's
own celebrated motto, it proclaimed, "Hinc lucem et pocula theae" ("from
here, light and cups of tea").[7] Informal scientific talk did not decline in the
new circumstances of the twentieth century, but it moved into new places
and assumed new forms.

Conclusion

There is no easy divide apparent in any part of this story between "popular" and "expert" science. The marketplace for science has always involved exchange and never simply passive consumption. In the late Georgian and early Victorian era, the talk of the scientific elite had been an integral part of the conversational practice of polite society. This was even true of accounts of original research, as indicated by the case of the Geological Society and similar groups in London's West End clubland. By the end of the century, the situation had changed. Practices in science still drew upon those of polite society, but often in distinctive settings such as the laboratory tea room. As I have suggested, rituals associated with the genteel world of polite conversation became key elements in the academic research schools that dominated science in the late nineteenth and twentieth centuries. Such borrowings are no less significant or pervasive just because they were less obvious.

There could be a temptation to read the story of this chapter as yet another indication of the impact of specialization and professionalization within science. And clearly the creation of new roles, new academic structures, and new forms of paid careers is part of what made the "scientist" a less likely dinner companion in a Mayfair townhouse in 1900 than his counterpart three-quarters of a century earlier. Science, with its arcane terminology and finely honed disciplinary categories, has often been depicted as a leading force in breaking down the possibilities for a coherent intellectual culture (Heyck 1982). In his *Absent Minds*, Collini (2006, 451–72) shows how central the issue of specialization became to debates about the character of intellectual life in early twentieth-century Britain.

Yet as Collini points out, discussions of the subdivision of labor in the pursuit of knowledge have characteristically been accompanied by a sense of decline: the world of the generalist is always the world we have lost. Accusations of scholasticism and overspecialization are as old as knowledge itself, and they are certainly not peculiar to late Victorian and Edwardian Britain (Shapin 1991). If anything, the arrival in domestic settings of science-based technologies such as electric lighting made the topicality of science more evident than ever before (Gooday, this volume). Moreover, the emergence of science as a paid career was the outcome of a campaign, by Babbage and others, to associate the pursuit of knowledge with the gentlemanly status long enjoyed by those in the existing professions of medicine, the military, and the law. It was not intended to break the link between science and gentility but to strengthen it.

There is no reason, then, to see the drive for professionalization within science as the primary reason for the transformation in practice I have sketched here. Instead, the growing tensions in talking about science toward the end of the nineteenth century can best be understood as part of more general changes in the social organization of elites in Britain (Perkin 1989). These changes had as much—probably more—to do with the specific needs of genteel society than with any changes in the comparatively small-scale world of science. During the agricultural depression of the 1870s through the 1890s, new money from the United States and the wealthy commercial classes had to be brought into the marriage market, and this infusion of nonelite guests made learned company at the dinner table potentially awkward. Men of science were increasingly excluded from the highest social circles not because their knowledge was more specialized than it had been but because the purpose of conversation among the elite had been transformed.

Paradoxically, then, men of science began to gain the professional status some of them had craved, at the very moment that status was losing its cachet as an entry into genteel society. The new plutocrats, increasingly focused on creating a potent fusion between power and wealth, welcomed the money flowing into late Victorian and Edwardian London from the United States; but they did not have much of a taste for what appeared to be another American import, the "scientist." The word *scientist* had been invented at a meeting of the British Association for the Advancement of Science in the 1830s but became regularly used in Britain only from the 1890s onward (White 2003, 4–5). With a few exceptions, the fashionable world did not embrace the growing class of experts associated with industry, state education, and government bureaucracy. The unwritten rules of talk had changed. To discuss the electron theory or fossil botany at ordinary dinners and soirees, other than when these subjects were in the news, was (with rare exceptions) seen to be talking shop.

The taboo reached its apotheosis around the time of the debate over the novelist C. P. Snow's 1959 essay "The Two Cultures," which in many ways was a meditation on scientific conversation. Snow puzzled about the kinds of subjects deemed appropriate for discussion at London parties and Cambridge high tables. "Oh, those are mathematicians!" he reported one college head as saying, "We never talk to *them*" (Snow [1959] 1993, 3). Snow was appalled that so many otherwise well-educated people could complacently admit that they knew nothing of nonparity conservation or the second law of thermodynamics. The problem, in his partisan diagnosis, involved a divide between science and all the other main institutions of British public life, from the Westminster establishment to the haunts of the self-proclaimed literary avant-garde in Chelsea. The distance from those places to Imperial College and the Science Museum in South Kensington was the conversational equivalent of an ocean, "as though the scientists spoke nothing but Tibetan" (2–3). The separation, Snow suggested, had been bad at the start of the twentieth century, but in the intervening decades, it had only become worse (17). Snow's book created a controversy that has continued for half a century, although the situation it described has since been transformed out of all recognition through television, the Internet, and the impact of science on everyday life. The "two cultures" debate was itself one sign that things were beginning to change, as the problem of how to talk about science emerged as the topic of lively conversation it remains today.

NOTES

I am grateful to Patricia Fara, Aileen Fyfe, Jan Golinski, Ludmilla Jordanova, Larry Klein, Bernie Lightman, Margaret Meredith, Don Opitz, Simon Schaffer, Anne Secord, and Paul White, all of whom have commented extensively on this essay.

1. Bazerman 1988. Bazerman himself has gone some way to examine the intervening period in *The Language of Edison's Light* (Bazerman 1999), as have Gross, Harmon, and Reidy (2002) in their work on the rhetoric of the scientific article. Johns 2000 offers a more realistic view of the *Philosophical Transactions.*

2. The best discussion I have found on the timing of "the Season" for the early period is the discussion of Mayfair in Greater London Council 1977; for the later nineteenth century, see Davidoff [1973] 1986.

3. The first half of following section draws on my general introduction to the *Collected Works of Mary Somerville* (J. Secord 2004). For more on Mary Somerville, see Neeley 2001; Patterson 1983; and especially Brock 2006.

4. These problem solutions, which predate her first experimental paper by fifteen years, are included in the first volume of Mary Somerville's *Collected Works* (M. Somerville 2004).

5. Andrew Ramsay, diary entry, 1 February 1848, KGA Ramsay 1/10, fol. 20r, Imperial College Archives, London.

6. There were, of course, notable practitioners of science among particular families within the social elite, such as the Balfours and the Rayleighs (Opitz 2004; Schaffer 1997). These important instances of "country-house science" need to be balanced with an understanding of the role of science within upper-class households more generally, and especially in metropolitan settings—the possibilities for what could be called "town-house science."

7. *Tea Phytologist* x, no. 1 (January 23, 1934): 1; see also *Tea Phytologist* x + 1, no. 1 (November 27, 1939): 1; *Tea Phytologist* T42, no. 24T (March 1984): 1. The motto was a variant on the university's famous maxim, "Hic lucem et pocula sacra," literally translated as "from here, light and sacred draughts."

REFERENCES

Alberti, Samuel. 2003. "Conversaziones and the Experience of Science in Victorian England." *Journal of Victorian Culture* 8:208–30.

Anon. 1845. *The English Gentlewoman: Or, Hints to Young Ladies on Their Entrance into Society.* London: Henry Colburn.

Anon. 1847. *Etiquette for Gentlemen: With Hints on the Art of Conversation.* 30th ed. London: David Bogue.

Anon. 1861. *Etiquette for All, or Rules of Conduct for Every Circumstance in Life: With the Laws, Rules, Precepts, and Practices of Good Society.* Glasgow: George Wilson.

Anon. [1861]. *Etiquette for Ladies.* London: B. Blake, Family Herald Office.

Anon. 1875. "The Art of Conversation." *Saturday Review* (9 Jan. 1875), 42–43.

Babbage, Charles. 1830. *Reflections on the Decline of Science in England and on Some of its Causes.* London: B. Fellowes.

Bahar, Saba. 2001. "Jane Marcet and the Limits to Public Science." *British Journal for the History of Science* 34:29–49.

Barbauld, Anna. 1826. *A Legacy for Young Ladies, Consisting of Miscellaneous Pieces, in Prose and Verse.* London: Longman.

Barlow, Nora, ed. 1958. *The Autobiography of Charles Darwin, 1809–1882.* London: Collins.

Barton, Ruth. 2003. "'Men of Science': Language, Identity and Professionalization in the Mid-Victorian Scientific Community." *History of Science* 41:73–119.

Bazerman, Charles. 1988. *Shaping Written Knowledge: The Genre and Activity of the Experimental Article in Science.* Madison: University of Wisconsin Press.

———. 1999. *The Languages of Edison's Light.* Cambridge, MA: MIT Press.

Brock, Claire. 2006. "The Public Worth of Mary Somerville." *British Journal for the History of Science* 39:255–72.

Browne, Janet. 1995. *Charles Darwin: Voyaging.* London: Jonathan Cape.

Burkhardt, Frederick H., and Sydney Smith, eds. 1987. *The Correspondence of Charles Darwin.* Vol. 3. Cambridge: Cambridge University Press.

Byrne, Julia Clara. 1892. *Gossip of the Century.* 2 vols. London: Ward and Downey.

Calhoun, Craig, ed. 1992. *Habermas and the Public Sphere.* Cambridge, MA: MIT Press.

[Campbell, Ina], Dowager Duchess of Argyll. 1906. *George Douglas, Eighth Duke of Argyll, K. G., K. T.: Autobiography and Memoirs.* 2 vols. London: John Murray.

Collini, Stefan. 1991. *Public Moralists: Political Thought and Intellectual Life in Britain*. Oxford: Clarendon Press.

———. 2006. *Absent Minds: Intellectuals in Britain*. Oxford: Oxford University Press.

Collins, Harry. 2004. *Gravity's Shadow: The Search for Gravitational Waves*. Chicago: University of Chicago Press.

Cowan, Brian. 2005. *The Social Life of Coffee*. New Haven, CT: Yale University Press.

Davidoff, Leonore. [1973] 1986. *The Best Circles: Society, Etiquette and the Season*. London: Cresset Press.

Desmond, Adrian. 1989. *The Politics of Evolution: Morphology, Medicine and Reform in Radical London*. Chicago: University of Chicago Press.

———. 1997. *Huxley: Evolution's High Priest*. London: Michael Joseph.

Dickens, Charles. [1864–65]. *Our Mutual Friend*. London: Hazell.

Doyle, Richard. 1864. *Bird's Eye Views of Society*. London: Smith, Elder.

Edgeworth, Maria. 1971. *Letters from England*. Ed. Christina Colvin. Oxford: Clarendon Press.

Fara, Patricia. 2004. *Pandora's Breeches: Women, Science and Power in the Enlightenment*. London: Pimlico.

Fyfe, Aileen. 2004. Introduction to *Conversations on Chemistry*, by Jane Marcet, xxi–xxvii. Bristol: Thoemmes Continuum.

Gascoigne, John. 1994. *Joseph Banks and the English Enlightenment: Useful Knowledge and Polite Culture*. Cambridge: Cambridge University Press.

Godwin, Harry. 1977. "Early Development of *The New Phytologist*." *New Phytologist* 100:1–4.

Golinski, Jan. 1992. *Science as Public Culture: Chemistry and Enlightenment in Britain, 1760–1820*. Cambridge: Cambridge University Press.

Gordon, E. O. 1894. *The Life and Correspondence of William Buckland, D. D., F. R. S.* London: John Murray.

Gould, Paula. 1997. "Women and the Culture of University Physics in Late Nineteenth-Century Cambridge. *British Journal for the History of Science* 30:127–49.

Greater London Council. 1977. "The Social Character of the Estate: The London Season in 1841." In *Survey of London*, vol. 39, *The Grosvenor Estate in Mayfair, Part I (General History)*, 89–93. London: Athlone Press.

Green Musselman, Elizabeth. 2006. *Nervous Conditions: Science and the Body Politic in Early Industrial Britain*. Albany: State University of New York Press.

Gross, Alan G., Joseph E. Harmon, and Michael Reidy. 2002. *Communicating Science: The Scientific Article from the 17th Century to the Present*. Oxford: Oxford University Press.

Habermas, Jürgen. [1962] 1989. *The Structural Transformation of the Public Sphere: An Inquiry into a Category of Bourgeois Society*. Trans. Thomas Berger. Cambridge, MA: MIT Press.

Heyck, T. W. 1982. *The Transformation of Intellectual Life in Victorian England*. London: Croom Helm.

Huxley, Leonard, ed. 1900. *Life and Letters of Thomas Henry Huxley*. 2 vols. London: Macmillan.

Jewson, Nicholas. 1974. "Medical Knowledge and the Patronage System in Eighteenth-Century England." *Sociology* 10:369–85.

Johns, Adrian. 2000. "Miscellaneous Methods: Authors, Societies and Journals in Early Modern England." *British Journal for the History of Science* 33:159–86.

Jouy, E. 1823–32. "Conversation." In *Encyclopédie Moderne*, ed. M. Courtin. Paris: Bureau de l'Encyclopédie. English translation appears in [Mitchell] 1842, 67.

[Kebbel, Thomas E.] 1865. "Shop." *Cornhill Magazine* 11:489–94.

Kingsley, Charles. [1850] 1883. *Alton Locke, Tailor and Poet. An Autobiography*. New ed. London: Macmillan.

Klein, Lawrence E. 1994. *Shaftesbury and the Culture of Politeness: Moral Discourse and Cultural Politics in Early Eighteenth-Century England*. Cambridge: Cambridge University Press.

Kohler, Robert E. 1999. "The Constructivists' Tool Kit." *Isis* 90:329–31.

Kölbl-Ebert, M. 2004. "Charlotte Murchison." In *The Dictionary of Nineteenth-Century British Scientists*, 3:1440–42. Ed. Bernard Lightman. Bristol: Thoemmes Continuum.

Latour, Bruno, and Steve Woolgar. 1979. *Laboratory Life: The Social Construction of Scientific Facts*. Beverley Hills, CA: Sage.

Lightman, Bernard. 1997. "'Fighting even with Death': Balfour, Scientific Naturalism, and Huxley's Final Battle." In *Thomas Henry Huxley's Place in Science and Letters: Centenary Essays*. Ed. A. Barr. Athens: University of Georgia Press.

Livingstone, David. 2005. "Text, Talk and Testimony: Geographical Reflections on Scientific Habits: An Afterword." *British Journal for the History of Science* 30:93–100.

Lyell, K. M., ed. 1881. *Life Letters and Journals of Sir Charles Lyell, Bart.* 2 vols. London: John Murray.

Lynch, Michael. 1985. *Art and Artifact in Laboratory Science: A Study of Shop Work and Shop Talk in a Research Laboratory*. London: Routledge.

Marcet, Jane. [1806] 2004. *Conversations on Chemistry: In Which the Elements of That Science Are Familiarly Explained and Illustrated by Experiments*. 2 vols. Bristol: Thoemmes Continuum.

Miller, David Philip. 1983. "Between Hostile Camps: Sir Humphry Davy's Presidency of the Royal Society of London, 1820–1827." *British Journal for the History of Science* 16:1–47.

———. 1986. "The Revival of the Physical Sciences in Britain, 1815–1840." *Osiris*, 2d ser., 2:107–34.

———. 2004. *Discovering Water: James Watt, Henry Cavendish and the Nineteenth-Century "Water Controversy."* Aldershot, UK: Ashgate.

Miller, Stephen. 2006. *Conversation: A History of a Declining Art.* New Haven, CT: Yale University Press.

[Mitchell, John] Orlando Sabertash. 1842. *The Art of Conversation, with Remarks on Fashion and Address.* London: G. W. Nickisson.

Morgan, Marjorie. 1994. *Manners, Morals and Class in England, 1774–1858.* Houndmills, Basingstoke, UK: Macmillan.

Morrell, J. B., and A. Thackray. 1981. *Gentlemen of Science: Early Years of the British Association for the Advancement of Science.* Oxford: Clarendon Press.

Munby, Alan E. 1921. *Laboratories: Their Planning and Fittings.* London: G. Bell.

Neeley, Kathryn A. 2001. *Mary Somerville: Science, Illumination, and the Female Mind.* Cambridge: Cambridge University Press.

Nevill, Dorothy. 1906. *The Reminiscences of Lady Dorothy Nevill.* Ed. Ralph Nevill. London: E. Arnold.

Opitz, Don. 2004. "'Behind Folding Shutters in Whittinghame House': Alice Blanche Balfour (1850–1936) and Amateur Natural History." *Archives of Natural History* 31:330–48.

Patterson, Elizabeth C. 1983. *Mary Somerville and the Cultivation of Science, 1815–1840.* Boston: Martinus Nijhoff.

Perkin, Harold. 1989. *The Rise of Professional Society: England since 1880.* London: Routledge.

Pettigrew, Jane. 2001. *A Social History of Tea.* London: National Trust.

Qureshi, Sadiah. 2004. "Displaying Sara Baartman, the 'Hottentot Venus.'" *History of Science* 42:233–57.

Rudwick, Martin J. S. 1963. "The Foundation of the Geological Society of London: Its Scheme for Cooperative Research and Its Struggle for Independence." *British Journal for the History of Science* 1:325–55.

———. 1985. *The Great Devonian Controversy: The Shaping of Scientific Knowledge among Gentlemanly Specialists.* Chicago: University of Chicago Press.

———. 1995. "Historical Origins of the Geological Society's *Journal.*" In *Milestones in Geology: Reviews to Celebrate 150 Volumes of the "Journal of the Geological Society."* Ed. Michael John Le Bas, 5–8. London: Geological Society.

Rushing, Homer. 1990. "The Gorilla Comes to Darwin's England." MA thesis, University of Texas at Austin.

Sala, George Augusta. 1859. *Twice Round the Clock.* London: Houlston and Wright.

Schaffer, Simon. 1986. "Scientific Discoveries and the End of Natural Philosophy." *Social Studies of Science* 16:387–420.

———. 1997. "Physics Laboratories and the Victorian Country House." In *Making Space for Science: Territorial Themes in the Shaping of Knowledge,* ed. C. Smith and J. Agar, 149–80. Basingstoke, UK: Macmillan.

Schiebinger, Londa. 1993. *Nature's Body: Gender in the Making of Modern Science*. Boston: Beacon Press.

Secord, Anne. 1994. "Corresponding Interests: Artisans and Gentlemen in Nineteenth-Century Natural History." *British Journal for the History of Science* 27:383–408.

———. 2002. "Botany on a Plate: Pleasure and the Power of Pictures in Promoting Early Nineteenth-Century Scientific Knowledge." *Isis* 93:28–57.

Secord, James A. 2000. *Victorian Sensation: The Extraordinary Publication, Reception, and Secret Authorship of "Vestiges of the Natural History of Creation."* Chicago: University of Chicago Press.

———. 2004. General introduction to *Collected Works of Mary Somerville*, ed. James A. Secord. 9 vols. Bristol: Thoemmes Continuum.

Shapin, Steven. 1991. "'A Scholar and a Gentleman': The Problematic Identity of the Scientific Practitioner in Early Modern England." *History of Science* 29:279–327.

Shteir, Ann B. 1996. *Cultivating Women, Cultivating Science: Flora's Daughters and Botany in England, 1760–1860*. Baltimore: Johns Hopkins University Press.

Smith, Pamela J. 2004. "A Splendid Idiosyncrasy: Prehistory at Cambridge, 1915–50." PhD diss., University of Cambridge.

Snow, C. P. [1959] 1993. *The Two Cultures*. Ed. Stefan Collini. Cambridge: Cambridge University Press.

Somerville, Mary. 1873. *Personal Recollections, from Early Life to Old Age*. London: John Murray.

———. 2004. *Collected Works*. Ed. James A. Secord. 9 vols. Bristol: Thoemmes Continuum.

Somerville, William. 1816. "Observationes quaedam de Hottentotis." *Medico-chirurgical Transactions* 7:154–60.

———. 1979. "On the Structure of Hottentot Women." In *William Somerville's Narrative of His Journeys to the Eastern Cape Frontier and to Lattakoe, 1799–1802*, ed. Frank Bradlow. Cape Town: Van Riebeeck Society.

St. George, Andrew. 1993. *The Descent of Manners: Etiquette, Rules and the Victorians*. London: Chatto and Windus.

Strutt, Robert John. 1924. *John William Strutt, Third Baron Rayleigh*. London: Edward Arnold.

Sutton, Geoffrey V. 1995. *Science for a Polite Society: Gender, Culture, and the Demonstration of Enlightenment*. Boulder, CO: Westview.

Terrall, Mary. 1996. "Salon, Academy, and Boudoir: Generation and Desire in Maupertuis's Science of Life." *Isis* 87:217–29.

Thackray, John C., ed. 2003. *To See the Fellows Fight: Eye Witness Accounts of Meetings of the Geological Society of London and Its Club, 1822–1868*. Faringdon, Oxfordshire, UK: British Society for the History of Science.

[Timbs, John]. 1839. *Year-book of Facts in Science and Art.* London: Simpkin.

Traweek, Sharon. 1988. *Beamtimes and Lifetimes: The World of High Energy Physicists.* Cambridge, MA: Harvard University Press.

Trollope, Frances. 1840. *The Life and Adventures of Michael Armstrong, the Factory Boy.* London: Henry Colburn.

Walters, Alice N. 1997. "Conversation Pieces: Science and Politeness in Eighteenth-Century England." *History of Science* 35:121–54.

Warner, Harriot W., ed. 1855. *Autobiography of Charles Caldwell, M. D. with a Preface, Notes, and Appendix.* Philadelphia: Lippincott.

White, Paul. 2003. *Thomas Huxley: Making the "Man of Science."* Cambridge: Cambridge University Press.

Winter, Alison. 1998. *Mesmerized: Powers of Mind in Victorian Britain.* Chicago: University of Chicago Press.

CHAPTER THREE

The Diffusion of Phrenology
through Public Lecturing

JOHN VAN WYHE

Phrenology was one of the most widespread sciences in Victorian Britain, yet it did not come from the universities or great scientific societies. It was mostly practiced by ordinary people. By 1845 there were dozens of phrenological societies and journals throughout the country as well as itinerant bump-reading "professors" crisscrossing the land making money from head readings (see figure 3.1). There is abundant evidence that phrenology was accepted to some degree by a large percentage of the population and was familiar to practically everyone. Yet in 1814, when the German-born phrenologist Johann Gaspar Spurzheim (1776–1832) first came to Britain to lecture on the new science, there were no societies, journals, or indeed any phrenologists at all in Great Britain. How did the outré new science touted by a single man become such a widespread cultural phenomenon practiced and known by hundreds of thousands?

This chapter will demonstrate how the knowledge and practice of phrenology was diffused throughout Britain by public lecturers. In doing so, it will be shown that we must think of lecturing in an expanded sense as more than just a speaker addressing an audience. Much more activity was involved in a lecture tour than just the brief hours before an audience. For example, there were advertisements in newspapers announcing the lectures. There were often dinners and other meetings with the speaker, and there were tours of local institutions. This essay will also show that the spread of phrenology is the same kind of phenomenon as that studied by the interdisciplinary field called *diffusion of innovations*. It will be argued that historians of science have much to learn from this vast body of scholarship on the diffusion of cultural innovations.

Unlike many sciences of the early nineteenth century such as astronomy or botany, phrenology did not have a long history but began quite discretely

Figure 3.1. "A phrenological lecture for ladies." Illustration reprinted from
Anon. 1865, 108; in author's collection.

around the turn of the century. The science was created by the Viennese
physician Franz Joseph Gall (1758–1828) in the 1790s from elements available in his medical and philosophical contexts in Vienna and the literature
available to him (van Wyhe 2002). These elements were combined with
Gall's personal interests and derived vigor from his own ambitions. Gall
came to believe that the brain is the organ of the mind and that the mind
is composed of twenty-seven innate faculties. Because each faculty is distinct, each must have a separate seat or "organ" in the brain. A faculty's
power would therefore depend on the size and condition of its corresponding
physical organ. Embryological studies had shown that the skull takes its
shape from the underlying brain; therefore the surface of the head could

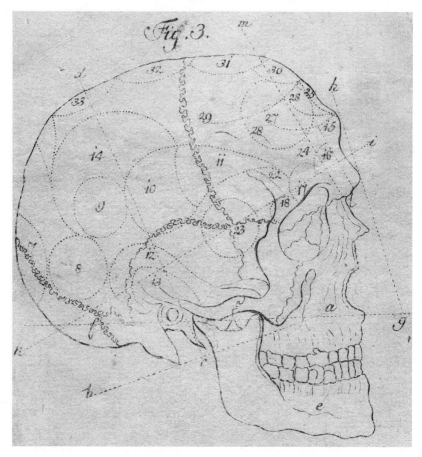

Figure 3.2. Engraving showing Gall's early markings on the skull with numbering.
Reprinted from Arnold 1805; in author's collection.

be read as an accurate indicator of the underlying brain organs and their
respective psychological aptitudes and tendencies. Gall also numbered all
of the mental faculties (see figure 3.2). Although one can find precedents
for nearly every aspect of Gall's science in the writings of predecessors and
contemporaries, his science was altogether a highly innovative creation.
The reading of human heads by sight, measuring device, or manipulation
with the hands was new. This and other features of Gall's science remained
barely altered as the core of phrenology even into the twentieth century.

Knowledge of Gall's mental science, or *Schädellehre*, as he initially
called it, began to spread when he offered public lectures on the subject at
his home in Vienna beginning in 1796. These lectures were organized as a

course, with the same people attending each subsequent lecture to learn a new aspect of the science over a week or so. In this sense Gall's lectures were no different than the many other extramural lectures providing medical instruction in Vienna (Oehler-Klein 1990). However, because Gall's system had a popular psychology side as well as a brute anatomy side, many of his auditors were not men of science but elite literati, aristocrats, ambassadors, and even ladies. Gall employed a "dissectionist" who would demonstrate anatomical points or bumps on skulls while Gall lectured to his polite audience.

In 1799 the first publication on Gall's system appeared (Froriep 1799). There were two more publications in 1800 and four the following year (van Wyhe 2004). In December 1801 Gall's lectures were banned by the ruling Holy Roman emperor Francis II. Legend has attributed this incident to political or religions reactions to Gall's doctrine. It may also have stemmed from the personal jealousy of the emperor's conservative physician. In any case, the ban greatly increased public interest in Gall's system. Twice as many pamphlets on his system were published in 1802, following the ban, and the number doubled again in subsequent years. Later, in 1805, when Gall embarked on a very successful lecture tour of Europe, the number of publications per year shot up higher than ever before. When Gall ceased his itinerant lectures in late 1807, new discussions of his doctrine essentially ceased to appear. In the following years he published his major works on his science (Gall and Spurzheim 1810–19; Gall 1822–25).

From the perspective of British phrenology, one feature of Gall's story is curious. He made no practicing converts after lecturing in more than fifty cities throughout Europe (van Wyhe 2002; 2004, Appendix A). No societies or journals for the study and propagation of his science were set up anywhere in Europe. As far as is known, no interested disciples imitated Gall and began to lecture on his system themselves. Yet Johann Gaspar Spurzheim lectured in half the number of cities in Britain and left many converts and societies in his wake.

This is a curious pivotal moment in the history of public phrenology lectures. Spurzheim's lectures were based on Gall's. Spurzheim seems to have delivered his in a very similar manner, with specimens such as skulls and busts to illustrate various organs and dissections when fresh brains could be procured. Like Gall's, Spurzheim's lectures were offered as a course over a number of days. Even Spurzheim's naturalistic rhetoric was largely borrowed from Gall (van Wyhe 2004).

There are several reasons why Spurzheim's lectures made active converts and Gall's did not. Gall was not lecturing in Britain but on the Continent, and we will never be able to know what would have happened had Gall also lectured in Britain instead of Spurzheim. Gall also lectured ten years before Spurzheim, during the Napoleonic Wars. So we are comparing not only different peoples with a different culture but also a different time. But perhaps most important of all, Gall was not trying to create practitioners. Gall wanted his audiences to believe him, but he did not teach them about something they could do as he did. The personal status of Gall's audiences was also often too high to copy actively and tour themselves. Ambassadors and aristocrats were hardly the sort of audience that would start giving lectures on the new science of the skull. Nevertheless, there is no doubt that many of Gall's auditors believed his claims (Oehler-Klein 1990).

Spurzheim, on the other hand, wanted to convert others to become practicing phrenologists. His version of the science was tailored for others to adopt and use and preach again themselves. His audiences were less elite than Gall's: largely middle-class professionals such as physicians, anatomists, lawyers, and clergy; the sort of people who certainly could lecture on the subject. The diffusion of Spurzheimian phrenology through Britain began quite distinctly when he arrived on British shores in March 1814. Before Spurzheim's arrival there were no phrenologists in Britain. There were no societies, no journals, no practical head readers. Thirty years later Spurzheimian phrenology was everywhere.

This rapid dissemination is the broadest and most palpable feature of phrenology's social history; it is a classic example of the diffusion of an innovation. Phrenology proceeded from one man in 1814 to a few auditors and personal converts and from each of these few to a few more and eventually, by the same process of word of mouth, to countless thousands. And along the way phrenology evolved and diverged into many things for many people. This is exactly the process which sociologists call *diffusion*. At no step do we find that an identical copy of the knowledge of phrenology was passively absorbed by openmouthed audiences, absorbing spongelike what the lecturer pronounced. Diffusion, as will be discussed in this chapter, is simply the description of the spread of a cultural element through a population.

The Lectures of Johann Gaspar Spurzheim

Spurzheim's venues were similar to Gall's. In London Spurzheim lectured at his lodging, the first floor of 11 Rathbone Place near Oxford Street, which had a room "large enough for fifty auditors," Spurzheim wrote to his fiancée, Honorine Pothier, on June 21, 1814 (Spurzheim Letters, folder 1). In Bristol Spurzheim took "a lodging consisting in a large drawing room and a bedroom. In the former I shall deliver my lectures," he told Pothier on January 25, 1815 (Spurzheim Letters, folder 3). He also gave private lectures and demonstrations to noble patrons such as Katharina Pawlowna, the grand duchess of Oldenburg and sister of Czar Alexander I of Russia, who was staying in London during her Europe-wide hunt for a new aristocrat husband (Jena 2003). In Dublin Spurzheim sought "permission to lecture at a public institution . . . [because] it will be more honorable than in a private house." "I like such places the best," he wrote to Pothier about October 20, 1815, "because other scientific [subjects] are taught there, as minerology, chemistry, natural history" (Spurzheim Letters, folder 4). This opinion on Spurzheim's part suggests several important considerations for the assessment of popular science lectures in the nineteenth century. What did it mean to an individual to attend lectures in Britain at this time? What other sorts of lectures had Spurzheim's audiences already experienced? And what difference did it make where the lectures were held? (Spurzheim is illustrated in silhouette lecturing in figure 3.3.)

One of the aims of this chapter is to stress that we need to have an expanded understanding in mind when using the term *lectures*. The word *lecture* is an abbreviation which stands for, but at the same time conceals, many forms of behavior and phenomena. For example, Spurzheim, like Gall before him, advertised his lectures in advance in London newspapers like this one from the *Times:*

> The PHYSIOGNOMICAL SYSTEM of Drs. GALL and SPURZHEIM.—On Wednesday, the nineteenth inst. at 8 o'clock in the evening, Dr. Spurzheim, Coadjutor and Joint-Lecturer with Dr. Gall, will COMMENCE, at his residence, No. 11 Rathbone-place, a COURSE of LECTURES on their SYSTEM of PHYSIOGNOMY explaining the Manifestations of the Propensities, Sentiments, and Intellectual Faculties of Man, as indicated by the External Configuration of the Head. The lectures will be continued every Monday, Wednesday, and Friday. Gentlemen's tickets, three guineas; Ladies' two; together with a Syllabus of the Lectures, may be had at Dr. Spurzheim's residence. (Spurzheim 1814)

Spurzheim also invited the editors of newspapers and journals to attend gratis, knowing that a mention in print would more than compensate for a few free tickets by bringing more auditors to subsequent lectures. From London, on September 1, 1814, Spurzheim wrote to Pothier, "I have already a good number of prosolites who speak for me, and engage other persons to attend my lectures" (Spurzheim Letters, folder 2). He was quite conscious

DR. SPURZHEIM LECTURING.

Figure 3.3. "DR. SPURZHEIM LECTURING." n.d. Illustration reprinted from
Wells 1896; in author's collection.

of the steps he took to ensure the propagation of his doctrine. "I . . . make
acquaintances with different persons who can be useful in inserting different articles in scientific journals," he explained to Pothier on September 20, 1814 (Spurzheim Letters, folder 2). And, of course, the lecturer spent
the majority of his time in a given location not actually lecturing, which

would occupy a few hours a day at most, but dining and visiting with local people. Spurzheim, like Gall before him, often visited local institutions such as schools, prisons, and asylums. These tours were often conducted with a local gentleman or the director of the institution and a small entourage of interested locals. Spurzheim was able to diagnose psychological problems and demonstrate his science on various individuals. So instead of thinking only of a speaker addressing an audience, we must remember that the actual lecture was the peak of related activity beginning with the anticipation and discussions in a given location before a lecture, engendered by the advertisements, culminating in the lecture itself, and continuing with further social interaction in dinners or touring local institutions followed by local reviews, discussion, and debate after the departure of the lecturer.

Spurzheim's audiences learned of his lectures through advertisements in newspapers, from review articles, from his publications, through his careful introductions to the influential, and by word of mouth. Spurzheim, like Gall before him and George Combe and others after him, required a minimum number of subscriptions to be collected by those who wished to invite him to speak in their city. As he put it in a letter to Pothier on April 17, 1816, "I am disgusted to lecture in any town of England without being invited and knowing a certain number of subscribers" (Spurzheim Letters, folder 6). Of course we should never forget that most popular science lectures were not free. George Combe paid Spurzheim two pounds and two shillings for his twelve one-hour lectures in 1817. Combe recalled years later that Spurzheim "never lectured under £1, 1s. for 12 lectures" (Gibbon 1878, 2:10).

Spurzheim described his lectures as "demonstrative lectures." Like Gall, Spurzheim offered dissections when he could procure fresh brains of humans or other animals and always made use of skulls, busts, models, and diagrams. As Spurzheim wrote to Thomas Forster on November 5, 1816, "I have made several private dissections and made converts each time" (Forster Papers). In the remains of the Henderson Trust Collection at the University of Edinburgh Medical School are large diagrams used by Spurzheim, which are thought to have been drawn by his wife (Follen 1834, 319). Spurzheim would tell his audience about a particular organ and then offer some examples of it from his large collection of skulls and casts or point to a chart to make an anatomical point clear. The use of such aids made his lectures not just an audible but also a visual experience. Auditors would be able to buy charts or books with diagrams from Spurzheim or a bookseller. In this way the experience and something like the feel of the lecture could be taken away and used further, and shared with others. Spurzheim wrote to Pothier from London on July 26, 1814, "I lecture three times every week on Mondays, Wednesdays and Fridays from 1–three o'clock. . . . I have invited several professors who attend and seem to be pleased, because they come regularly. In general our doctrine is not at all known in this country and the English are very circumspect and slow. Moreover it is not the season for lecturing, every one is gone or goes into the country" (Spurzheim Letters, folder 1). Spurzheim referred to his intended audience as "people of high rank."

Spurzheim also intended his first book (Spurzheim 1815) to be part of the lecturing experience. He wrote to Pothier on July 26, 1814, "The book will constitute at the same time a book of reference for my lectures, so that the book will invite auditors and these will buy the book" (Spurzheim Letters, folder 1). So Spurzheim envisioned that his audiences would read his words, hear his lectures, and view his specimens and charts. This combined sensory cocktail was meant to interest, intrigue, impress, and, ultimately, convert.

In the case of his lectures in Dublin in 1815, one of his auditors was impressed and converted. The lawyer Andrew Carmichael later wrote of Spurzheim's lectures:

> I listened to his first lecture, expecting it to breathe nothing but ignorance, hypocrisy, deceit, and empiricism . . . [but] in place of a hypocrite and empiric, I found a man deeply and earnestly imbued with an unshaken *belief* in the importance and value of the doctrines he communicated.
>
> I listened to his second lecture, and *I adopted his belief.* I was satisfied of the importance and value of those doctrines, and exulted in participating those treasures of knowledge. . . . I listened to his third lecture, and perceived, with all the force of thorough conviction, that there was nothing of any value in the metaphysics of ancient or modern schools, except so far as they coalesced and amalgamated with the new system. From that hour to the present, I have regarded the science with increasing confidence and unalterable devotion. More certain or more important truths the divine finger has not written in any of the pages of nature, than those which Spurzheim, on this occasion, unfolded to our examination—our study—our admiration. (Carmichael 1833)

Spurzheim believed a course of lectures was "the most effectual way to form a school of practical phrenologists," he wrote to George Combe on

January 9, 1826 (George Combe Collection, folder 93). In the case of Carmichael's experience, this tactic certainly succeeded.

Spurzheim, like Gall, was an accredited physician, and his science was by its very nature significantly medical. Naturally phrenology was first known and discussed in medical circles because these were the sorts of people personally known by, and most similar to, Spurzheim. He first came to London's medical circles, no doubt through acquaintances he had made in Paris. He carried letters of introduction to further medical practitioners when he traveled through England, Ireland, and Scotland. But Spurzheim wanted to extend his audience beyond the medical world. On January 23, 1815, he wrote to Pothier from Bath, "I am sure that I have already a greater number of zealous pupils in England than Gall in France. I foresee that I shall travel from town to town in this united kingdom and teach our doctrine. The only thing which ought to be destroyed; is the prejudice, that our doctrine belongs particularly to medical men. On account of this error, at the beginning in any town, the auditors are far the greatest number physicians or surgeons. So it was at London and here in Bath" (Spurzheim Letters, folder 2).

As Spurzheim achieved greater popularity, he expanded his range of lectures to cater to different audiences. Again he wrote to Pothier, on January 23, 1815, "The greatest interest is excited, and every one seems to be anxious to understand something of the doctrine. There is all probability that I shall have a large class in Nov. I intend to lecture every day, to distribute the audience into classes. Ladies and idle people prefer the day, for them I shall lecture at two o'clock, for professional and scientific gentlemen I lecture at 8 o'clock in the evening. In this way everyone who wishes to attend will be accommodated" (Spurzheim Letters, folder 2). Spurzheim also tailored his lectures to the interests of his audiences. Medical audiences received more anatomy and dissection. His public lectures in Edinburgh seem to have been more philosophical to suit the audiences he found there, as he wrote to Pothier on August 16, 1816, "In the united kingdoms I have not lectured to any audience who were so interested about the philosophical part of our doctrine as they generally were in the little course here [in Edinburgh]" (Spurzheim Letters, folder 7).

Lecturing on Phrenology: A Head Start

By 1817 Spurzheim had personally made several zealous converts, especially in Edinburgh and London. These were men he met during the time surrounding the occasion of his lectures. The spread of phrenology beyond the medical world also seems to have been via word of mouth. For example, Spurzheim met an advocate named Dundas on the coach to Edinburgh. Dundas recommended a hotel and invited Spurzheim to dinner at his home. It appears from a letter Spurzheim wrote to Pothier on August 20, 1816, that Dundas may have introduced Spurzheim to other Edinburgh advocates such as James Brownlee, who later introduced Spurzheim to George Combe (Spurzheim Letters, folder 7). The Edinburgh Phrenological Society was founded largely by jurists through this connection.

The occasion of lectures like Spurzheim's inspired local printed reviews and handbills and increased sales of phrenological items such as busts and prints. In most cases the lecturer was the guest of local physicians or converts, and many personal contacts were made before and after the lectures themselves. These more intimate interactions often resulted in more profound effects or influence, as in the conversions of Thomas Forster and George Combe. The increase in the number of active phrenologists was primarily through such personal contacts on the lecture circuit in addition to and with the aid of phrenological artifacts (including texts).

Spurzheim wrote to Pothier from Edinburgh on August 30, 1816, "I have delivered six Lectures. In the first I dissected another brain, which was excellent and exhibited every part quite distinctly. There were many auditors from the beginning, but this number increased every day, because one told it to the other, so that in the last lecture the room was scarcely large enough. They stood close till to the staircase. They were quite enthusiastic. I have spoken only of six organs which are certain and easily discovered, in order to make them convinced by their own observations that we speak from experience" (Spurzheim Letters, folder 7). On December 30 of the same year, he wrote to her, "It is astonishing how I have gained in the public opinion since my lecturing in Edinburgh. Everyone is now anxious to know some thing of our doctrine; it is no longer quackery, but an important science" (Spurzheim Letters, folder 7).

Active practice of phrenology diffused through Britain via personal contact. By *active practice* I distinguish between mere awareness of the fact that there was something called phrenology and adoption and practice of it. Spurzheim's lecture tour left converts and phrenological societies in its wake. The Edinburgh Phrenological Society was founded in 1820 by converts of Spurzheim's 1817 visit. The London Phrenological Society was founded in 1823 by converts of Spurzheim's 1814–17 stay there. George Combe's 1829 lectures in Dublin and Glasgow led to the foundation of phrenological societies in both cities in the same year (Gibbon 1878, 1:216). These societies often served as the base from which subsequent phrenological lecturers, such as George Combe or David Goyder, branched out to offer further lectures. These lecturers made new converts who founded further societies. However, Spurzheim did not suggest or invent phrenological societies. Society formation was a local response by interested local men. Creating a scientific society was their way of pursuing their interest in the science further. It was also a particularly British addition to phrenology.

The distinction between awareness, even acceptance of phrenology and practice of it is clear in the case of David Goyder (1796–1878). Goyder was a Bristol- and later Glasgow-based Swedenborgian minister and Pestalozzian schoolteacher. He first encountered phrenology in the form of the itinerant lectures of William Henry Crook in 1822 (Crook 1827). Goyder believed in phrenology as a result of attending Crook's lectures but did not become an active phrenologist until more than a decade later, when he encountered the Glasgow Phrenological Society. With the contact and encouragement of practicing phrenologists, Goyder became a practitioner himself (Goyder 1857). Goyder then went on to become a phrenological lecturer and writer. The majority of phrenological principles could be gleaned from books, but it took the detailed instruction of those in the know to become a phrenologist. We cannot today take up a phrenological book or bust and begin to read heads as a phrenologist would have done. Too much of the practice required tacit knowledge that was not written down and indeed could not be written down.

The meaning of public lectures for us as historians should not be limited to a speaker and an audience at a specific time in a particular place but should be seen as an occasion for increased thought and talk about a science over a period of several days before and after the actual lectures. Individuals who experienced the talk and array of artifacts displayed at lectures were more likely either to be converted or to reject phrenology because they were confronted with it. Some of those who attended lectures were inspired by the experience to form local phrenological societies or to tell others about phrenology.

Beyond Spurzheim: George Combe

Certainly Spurzheim's most influential convert was the Edinburgh barrister George Combe (1788–1858). Of almost equal importance, we have detailed evidence for his introduction to and practice with phrenology. Combe first met Spurzheim at the home of a fellow barrister, James Brownlee, in the aftermath of Spurzheim's famous confrontation with his great antagonist, the Edinburgh anatomist John Gordon (1786–1818; see Cantor 1975; Kaufman 1999). There was great excitement to see and hear Spurzheim because the *Edinburgh Review*'s damning critique of his book was the talk of the town ([Gordon] 1815). It was said that Spurzheim had faced down the anonymous reviewer, widely known to be Gordon, at a demonstrative dissection. Combe was impressed by Spurzheim's talk and dissection of a human brain for the dinner guests. Combe later ordered plaster casts of heads from London, read about the subject further, and discussed it with his friends (Gibbon 1878, 192–93). In February 1820 Combe, together with his brother Andrew, Brownlee, Sir George Steuart Mackenzie, William Waddell, Lindsey Mackersey, and David Welsh, founded the Edinburgh Phrenological Society (EPS; see *Minute Book of the Phrenological Society*). By 1826 there were 120 members (Anon. 1824; Anon. 1826; van Wyhe 1999–). This phrenological cell evolved its own strain of Spurzheimian phrenology. In addition to the British scientific society structure and trappings, the Edinburgh phrenologists' science was slightly different from the master's in rhetoric, in head reading, and in social uses of the science (Shapin 1975; Cooter 1984).

Members of the EPS were soon spreading the word through such publications as the *Transactions of the* [Edinburgh] *Phrenological Society* (1823), a direct emulation of the *Philosophical Transactions of the Royal Society of Edinburgh*. The phrenologists' *Transactions* was superceded in December 1823 by the *Phrenological Journal and Miscellany* (*PJ*), a four-shilling quarterly which continued as the main organ of the phrenologists, and especially those of Edinburgh, until October 1847 (Cooter 1989, 263). The journal's sales never exceeded about 550 copies, though it may have been read by as many as 3,000 (*PJ* 12 [1839]: 292).

More relevant to this analysis were the lectures by these Edinburgh phrenologists. George Combe was the most active lecturer and proselytizer. It is important to keep in mind what lecturing meant to Combe and his peers before he became a phrenologist. Since childhood Combe had heard sermons in church and lectures at school, at the Edinburgh high school, and at the university. He also "attended the lectures on conveyancing delivered under authority of the Society of Writers to the Signet in 1806" (Gibbon 1878, 1:71). As a young man he was a member of one of the Edinburgh Young Men's Debating Societies. Here he first spoke in public, on the topic of capital punishment. Then he attended the lectures of Spurzheim in 1817. With the exception perhaps of the debating society, all the lecturing experiences Combe and his contemporaries knew were those of a high authority speaking to an audience of less authority on a subject more familiar to the lecturer than his audience. So when Combe prepared to speak about phrenology, we must assume that some legacy from all these other forms of public speaking which he had experienced throughout his life influenced his writing and preparations before and his manner and address during lectures. (Combe is illustrated in silhouette lecturing in figure 3.4).

Before his first public phrenology lectures, Combe practiced at home by rehearsing in his dining room twice a week for about a dozen friends (Gibbon 1878, 1:117). In May 1819 he moved to Brown Square, where he kept his growing collection of phrenological casts in the attic. His biographer recounts that "there he gave his demonstrations once or twice a week, or whenever any man of eminence in science or philosophy desired to have an opportunity of estimating the value of the system from the exposition and illustrations of its chief apostle. The apartment was prepared for the purpose; it served as a lecture-room, and on high occasions the drawing-room chairs were transported aloft for the accommodation of visitors." (Gibbon 1878, 1:139). In 1821 Combe purchased a building on Clyde Street (today apparently occupied by a bus station) which contained "the hall in which Spurzheim had lectured during his visit to Edinburgh in 1816" (Gibbon 1878, 1:144). This building became the headquarters of the Edinburgh Phrenological Society. Here Combe offered his first course of public lectures on phrenology from May to July 1822. Members of the society were admitted gratis and nonmembers at two guineas for the course.

Combe estimated to his friend and fellow phrenologist the Reverend David Welsh that "seven-tenths of my lectures will be new," that is, not drawn from his published essays on phrenology (Gibbon 1878, 1:152; Combe 1819). Combe's introductory lecture drew an audience of about seventy and his second sixty-five, but by the third lecture, as the novelty wore off, attendance had dropped to about thirty-two. The following lectures had only about ten. His brother Andrew Combe returned from studying medicine in Paris and was able to dissect brains while Combe lectured. Perhaps this was a Scottish version of the original Gall and Spurzheim performance? The important question is not whether the Combes and other British phrenologists perpetuated or created a tradition of phrenological lecturing because both are true. They obviously perpetuated the phrenological lecturing taught by Spurzheim, yet at the same time they could not, even had they wished to, duplicate him utterly and completely. Just as societies and journals came from the British scientific culture, so did a thousand subtle characteristics in the way the science was described, lauded, and promulgated.

Combe's biographer remarked, "The system was much talked about in the scientific and social circles of [Edinburgh], and the lectures, besides

Figure 3.4. Illustration of George Combe lecturing. Reprinted from
Combe 1839; in author's collection.

affording great pleasure to himself, had the effect of attracting a large measure of serious attention to the subject and of adding to the ranks of its disciples. He consequently announced a winter course, to begin in November" (Gibbon 1878, 1:153). These lectures had nineteen auditors. Things started off rather slowly and locally for Combe, and he continued to offer public lectures in Edinburgh intermittently until 1840. He also eventually lectured in London (1823, 1824), Dublin (1829, 1831), Glasgow (1824, 1829, 1836), Newcastle (1835), Aberdeen (1836), Manchester (1837), Birmingham (1838), Bath (1838), the United States of America (1838–40), and Germany (1842).

Like Spurzheim, Combe also required a minimum number of subscribers before agreeing to lecture in a particular city. His audiences were, for the most part, hardly distinguishable from Spurzheim's. Combe's contacts, however, were more often lawyers or businessmen rather than medical men. Combe's lectures also sometimes resulted in the founding of a local phrenological society. Combe, although a well-known author, was never as famous as Spurzheim, who had managed to associate his name, in the English-speaking world, indelibly with Gall's as cofounder of the science. Spurzheim had also, for a time, been the subject of national review and ridicule with the *Edinburgh Review* and *Quarterly Review* reviews of his first book ([Gordon] 1815; [D'Oyly] 1815). Only later would Combe's fame eclipse that of Spurzheim's in many circles as the author of *The Constitution of Man* (1828). However, *Constitution* was not always received or admired because of its phrenological content or background (van Wyhe 2004).

Lecturing was not always a successful business of dissemination. The correspondence of George Combe is replete with examples of lectures with only a handful of auditors or the sting from the laughter of ridicule at the outré science. Combe wrote to a fellow phrenologist about his 1835 lectures in Newcastle:

> I have given six lectures on phrenology here, which are attended by all that the lecture-room will contain, said to be 300 and 20 standing. . . . I find the intellectual condition of the audience here lower than any I have ever lectured to. Out of the 300 apparently one-half are very intelligent, and the rest ignorant. This ignorant section laughs in the silliest way at facts of the gravest importance, *e.g.*, at a man stealing after being trepanned at the organ of acquisitiveness, &c. It is not a laugh of contempt, but of foolish wonder . . . (Gibbon 1878, 1:311)

In Aberdeen in April 1836 Combe complained, "[The people of Aberdeen] are extremely alive to ridicule, and it appears that no man with the rank of a gentleman would act as my assistant, through fear of ridicule. I have got a workman [instead]. They seem afraid to know Mrs Combe or me in private. We have lived in entire solitude with the exception of two families" (Gibbon 1878, 1:314).

William Henderson, later remembered for his posthumous patronage of phrenology and Combe's *Constitution of Man*, lectured on phrenology in Leith circa 1826, but a speech impediment apparently amused his audience more than phrenology, and he abandoned lecturing. Combe also worried about his speech. Writing to the reformer Richard Cobden in 1837, before his Manchester lectures, Combe explained that "I have a very broad Scotch accent, with a total absence of grace and eloquence" (Gibbon 1878, 1:317). Similarly, Combe wrote to James Adam, editor of the *Aberdeen Herald*, in preparation for lecturing in Aberdeen:

> Phrenology is treated with ridicule or contempt. . . . My only motive, therefore, is to knock that ridicule and contempt on the head. The best way of doing so, is to make the science appear interesting and respectable. The grand element in producing this effect is to secure a *large and respectable* audience to hear my exposition of its merits. . . . To go and lecture to a small audience would increase the prejudice, damage the interests of the science, waste my time and labour, prove injurious to my reputation, and disagreeable to my feelings. (Gibbon 1878, 2:9)

Attendance did not depend only on the respectability of the subject matter. It depended heavily on the lecturing to which audiences were accustomed. In 1829 Combe delivered sixteen lectures in Dublin to an audience averaging 150 "including clergymen, physicians, barristers, fellows of Trinity, and ladies." His biographer noted, "The first lecture was free, and was attended by 500 persons." Combe noted that "Government gave the Royal Dublin Society £7000 annually to provide lectures on science gratis to the public. This had created a taste for lectures, but a great distaste to pay for them" (Gibbon 1878, 1:214). For some years, Combe remarked, his annual lectures in Edinburgh "could not subsist, if the hall were not my own property, and other expenses trifling" (Gibbon 1878, 1:279).

Other lectures were designed specifically for "the benefit of the industrial classes" by having the science lectures at times convenient for "those who were working in shops and factories" and charging only a moderate fee. In 1832 Combe, in compliance with a requisition from a large number of mechanics, shopkeepers, and clerks, delivered a course of lectures on the evenings of Mondays and Thursdays. . . . He had an attendance of over two hundred people, and eighty-four of these availed themselves of the permission granted by the Phrenological Society to examine the casts and skulls in their Museum. The languor into which phrenology had apparently fallen during the winter of 1831–32 was dispelled, and great enthusiasm for the science was displayed by the audiences in the Clyde Street Hall. . . . The immediate effect of these lectures was the formation of an association for the arrangement of annual courses of evening lectures for the working classes on chemistry, natural history, and phrenology, combined with physiology; afterwards botany, astronomy, and moral philosophy were included in the subjects of study. (Gibbon 1878, 1:255)

This association, the Philosophical Association for procuring Instruction in Useful and Entertaining Science, was followed by another inspired by Combe's lectures, the Edinburgh Ethical Society, which was also "formed by the young men who had attended his lectures" (Combe 1848). So lectures could often inspire their listeners to create something new based on what they had experienced—even if not specifically phrenological. A visit by a well-known phrenologist to a local cell could also generate a lot of enthusiasm. Combe's visit to Manchester in 1837 was said to be "like a constant jubilee" among the local phrenologists (Gibbon 1878, 2:10).

Although the letters and other writings of phrenologists refer to many anatomists and other academic lecturers who mentioned phrenology in their lectures after attending phrenological lectures themselves, there was only one appointed lecturer on phrenology in Britain during the nineteenth century: Dr William Weir, a physician and a clinical lecturer in the Glasgow Infirmary who won the lectureship at the Andersonian University at Glasgow in 1846. The post was subsidized by the Henderson Trust. However, after only two sessions, "the number of students being so small that perseverance could only bring ridicule upon the teaching, the Henderson trustees withdrew their grant, and the class was closed" (Gibbon 1878, 2:221). By the end of the 1840s, phrenology was no longer as novel or as interesting as it had been a decade before (Cooter 1984).

The Diffusion of Innovations

The diffusion of the practice of phrenology throughout Britain is in every feature a typical example of what is called the *diffusion of an innovation*

(Rogers 1995). Unfortunately, historians of science not only are unfamiliar with the research on diffusion of innovations but often consider the term *diffusion* a dirty word implying all sorts of unacceptable assumptions (e.g., Shuttleworth and Cantor 2004, 3; Dawson 2004, 182; Winter 1994, 319; Jacyna 1984, 81; J. Secord 2000, 138). There are, of course, exceptions (e.g., Golinski 1998, 169; Clarke and Jacyna 1987, 219; Adams 1979; Gaw 1999).

The passive absorption of scientific knowledge by audiences at lectures or readers of scientific works is now widely rejected as a superficial understanding of the reception of scientific knowledge. Increased focus on the audiences for science—from gentlemanly elites, through lawyers, ministers, and their families, to clerks, artisans, and mechanics—have revealed a whole new world of science history. For example, in a frequently cited paper, Anne Secord described how scientific knowledge spread through working-class communities via the important network of local pubs (A. Secord 1994). Secord argued that a significant body of natural knowledge evolved among working people themselves and was not handed down to them from on high. Such studies of science outside the scholarly elites have revealed vast tracts of historical culture that would still be invisible and unimaginable if we assumed that all scientific knowledge had its origins in the practices of those elites.

Two articles are frequently cited as rejecting diffusion in the history of science (Hilgartner 1990; Cooter and Pumfrey 1994). In fact, neither specifically criticizes diffusion. Rather, they (rightly) critique the older view that science or other knowledge is created only by elites and then spread *passively* to lesser elites and thence to commoners—a trickle-down process of passive reception and unreworked regurgitation. Problems have arisen, however, because this old-fashioned rendition of science popularization has been widely dubbed "the diffusion model," and those who reject it claim to be rejecting "diffusion" (see esp. Fyfe 2000, introduction; Shapin 1983, 151; Winter 1998, 132; Desmond 1998, 636; Topham this volume). Calling the trickle-down process "the diffusion model" has encouraged many historians to equate passive popularization with the word *diffusion* (Shuttleworth and Cantor 2004, 3; Dawson 2004, 194). In using the term *diffusion* in this way, historians of science seem to be drawing upon a very specific definition. The *OED*'s fifth meaning of *diffusion* is: "*Physics.* The permeation of a gas or liquid between the molecules of another fluid placed in contact with it; the spontaneous molecular mixing or interpenetration of two fluids without chemical combination" (*Oxford English Dictionary* 1994). Thus, in physics, *diffusion* does imply the inevitable and predictable spread of a gas or liquid through another gas or liquid.

However, the more general definition of *diffusion*, used in the study of culture or society, is: "3b. *Anthropol*. The spread of elements of a culture or language from one region or people to another" or "Spreading abroad, dispersion, dissemination (of abstract things, as knowledge)" (*Oxford English Dictionary* 1994). *Diffusion* is thus nothing more than the description of the de facto dispersion of cultural elements. How, why, by whom, and how quickly or slowly and by what means something is diffused, or not diffused, are not assumed or implied by the use of this word. It is used correctly by every discipline that studies culture except the history of popular science. To read inevitability or passivity into the word in a cultural context is to use the wrong definition.

There is a vast body of interdisciplinary research known as the diffusion of innovations. After more than seventy years of concerted scholarship, it is a sophisticated and powerful tradition of study for scholars from sociology, anthropology, archaeology, economics, social and historical linguistics, public health, and marketing research. Historians of science need to be aware of this because it is directly applicable to much of their work.

Fortunately, there is a superb overview: Everett Rogers's *Diffusion of Innovations* (1995). First published in 1962, revised editions appeared up to 1995 (the fourth edition of 1995 is vastly rewritten and improved and is the edition to be consulted). Rogers's book is very impressive and authoritative—not just because it is widely respected and cited nor only because it is the continued work of many years on the same issues and has benefited from much criticism, but especially because it draws on almost four thousand studies of the diffusion of many kinds of cultural innovations from numerous cultures and from various historical periods. As Rogers remarked, "No other field of behavior science research represents more effort by more scholars in more disciplines in more nations" (Rogers 1995, xv; see also Valente 1993; Henrich 2001). One cannot be aware of this work and maintain that the word *diffusion* is inappropriate for the history of science. The assessment by a reviewer of the first edition of Rogers in *Isis* is as apt today as it was in 1963: "Scholars are not always aware of the overlap of their activities, even the identicalness of some of these activities, with those of men in different disciplines. This seems to be the case with the study of the diffusion of innovations. Most scholars follow their own tradition of work, heedless of new concepts, theories, and methodologies that may have been developed in other scholarly fields for problems similar to their own" (Barber 1963, 296).

Rogers defined *diffusion* as "the process by which an innovation is communicated through certain channels over time among the members of a social system" (Rogers 1995, 5). Some examples of diffusion research include the long and intermittent acceptance of using citrus juice to prevent scurvy in the Royal Navy; the failure of "Miracle Rice" to displace traditional farming among Bali farmers in the 1970s; the diffusion of modern mathematics among administrators in Pennsylvania and West Virginia high schools between 1958 and 1963; the diffusion of birth control in Taiwan in the 1960s; and the communication of major public events such as the 1963 assassination of the American president John F. Kennedy or the 1986 explosion of the American space shuttle *Challenger* (Mosteller 1981; Lansing 1987; Carlson 1965; Freedman and Takeshita 1969; Rogers 1995).

Diffusion-of-innovations scholars study the same kinds of things many historians of science study: how some particular idea or practice spread, or did not spread, through a population and precisely why (Crane 1972). Key differences are that diffusion-of-innovations scholars study a wider variety of culture than do historians of science, and diffusion scholars use a general language, thus they are better prepared to understand and compare different, unique cases.

Some historians object to comparing different cases because, they believe, there are too many unique variables in each historical event to allow cross-case comparisons. This objection is contradicted by the fact that thousands upon thousands of cases have actually been compared by diffusion scholars. Not only have they discovered general features that occur in most diffusion events but they also have uncovered how particular differences alter the rate or course or can hinder or prevent an innovation's diffusion (Rogers 1995). Of course every case is composed of unique particulars that must be understood in their own context—this is essentially a tautology. What is particularly interesting is that there are clear and measurable parallels between the diffusion of utterly different cultural things in different times, places, and contexts. The parallels reveal something that can never be found by assessing or emphasizing the uniqueness of any particular case, namely, that there are general features found in the diffusion of cultural things in any time or place (Rogers 1995).

Rogers introduced diffusion studies in his book with a famous early study from the United States involving the use of a new hybrid corn in Iowa between the late 1920s and the 1940s. Several basic features were identified, which are found in countless later studies. Rogers used this example precisely because the general features are not unique to the diffusion of hybrid corn. Hundreds of farmers were interviewed by sociologists about which corn they used, when and how and from whom they had first heard of the new hybrid corn, and if or when they had adopted it and why. The investigators found that 99 percent of the farmers who did adopt the new hybrid corn adopted it between 1928 and 1941, a rather rapid rate of adoption considering the amount of behavioral change required because the new corn seed had to be purchased each year. With traditional seed, farmers had kept some of their best seed from the previous season.

It was also found that the rate of adoption of hybrid corn, if plotted on a graph, formed an S-shaped curve over time. In the first five years only 10 percent of the farmers adopted the new seed. Then the curve "took off," shooting up to 40 percent adoption in the following three years. The rate of adoption then leveled off as fewer and fewer farmers who would or could adopt it remained to do so. The diffusion of most innovations forms an S-shaped curve, though some do not. Many variables have been identified that correlate with these effects (Rogers 1995). The time course of the propagation of linguistic changes also typically follows an S-shaped curve (Enfield 2003; Croft 2000; Chambers 1992; Kroch 1989, 203).

The investigators in the hybrid corn study also found clear social differences between early and late adopters. Compared to later adopters, the early adopters had larger farms, higher incomes, and more formal education and were more cosmopolitan. Diffusion-of-innovations scholars thus tend to distinguish between a simple awareness of the existence of an innovation and active adoption. In the case of hybrid corn, many people heard about the new corn through advertisements but were far more likely actually to adopt it if they received a personal recommendation or saw their neighbors using it. In contrast, the earliest adopters were persuaded by salesmen. In the case of phrenology, Spurzheim might be likened to its first traveling salesman.

Rogers divided the diffusion of innovations into four main elements: (1) an innovation, (2) communication channels, (3) time, and (4) the social system or population. Diffusion tends to begin between individuals with links—what we might call personal networks. It has been consistently found that diffusion occurs relative to the degree of similarity between individuals; that is, the more culturally alike two people are, the more likely that an innovation will be communicated between them. Other general features that affect the diffusion of innovations have been identified, such as the relative perceived advantage of an innovation for a potential adopter, compatibility with existing belief or practice, the complexity or simplicity of the innovation, the observability or visibility of the innovation, and the "trialability" or testability of the innovation.

Although diffusionists eschew abstract groups and study real individuals and their actions, there are some problems with their usual approach. Despite Rogers's use of the term "re-invention" to refer to the degree to which an innovation is changed or modified in the process of adoption or use, many diffusionists tend to treat innovations as essentially static entities. I think historians of science have something to teach other scholars here because of our particular sympathies for complexifying received stories, context dependence, and sensitivity to the ever-changing character of scientific and other knowledge.

The Diffusion of Phrenology

This chapter addresses how phrenology became so widespread. In 1813 there were no phrenologists at all in Britain. By 1840 there were thousands, and phrenology was known to practically every person in Britain. How did that happen? It has been argued here that the *practice* of phrenology was diffused by traveling lecturers, in contrast to *knowledge* of phrenology, which was diffused more quickly and widely by print. The science had been known of in Britain in medical, intellectual, and well-to-do circles since 1800 from numerous references in the periodical literature and the publications of Gall and Spurzheim (Anon. 1800; Bojanus 1802; [Yelloly] 1802; [Brown] 1803; Arneman 1805, 327–36; Anon. 1806; Bischoff 1807; Blöde 1807; Gall and Spurzheim 1810–19). Yet not a single practicing phrenologist was created by the publications in fourteen years. Spurzheim, in just two years, created many practicing converts throughout the country.

This explanation of diffusion through lectures makes sense only if we adopt an expanded understanding of the concept of lectures in order to take into account all the activity otherwise abbreviated by the word *lectures* alone. Lectures became, over the course of the first decades of the nineteenth century, one of the major sources of public education and entertainment in Britain. The majority of activity actually engendered by lectures was clustered around them: all of the anticipation and discussion beforehand and personal contacts and intimate opportunities for persuasion and encouragement in the hours and days surrounding actual lectures. Lecturers traveled to cities or towns where they were invited or expected to find a sizable audience. Usually they had local contacts or letters of introduction. Hours of conversation were spent while touring local sights and institutions. These events gave the lecturer ample opportunity to personally demonstrate the marvels of his science and to convince some individuals to adopt it. Many more people, who did not have the opportunity of conversing with the lecturer, would have discussed the science currently being presented in public lectures. The lecturer was often invited to the homes of local families for dinners, and these private occasions gave the lecturer the opportunity to demonstrate his skills in character reading. George Combe is only the most prominent convert met in this way. Combe's converts also tended to be friends, relatives, or colleagues.

The naturalist Alfred Russel Wallace left a fascinating description of his conversion to phrenology (Wallace 1905, vol. 1). Wallace first learned about phrenology from books, but he attended the lectures by two itinerant lecturers, E. T. Hicks and J. Q. Rumball, who visited Neath in the mid-1840s. After the lectures Wallace queued up to have his head read and later recalled that "in both cases I had my character delineated with such accuracy as to render it certain that the positions of all the mental organs had been very precisely determined [by phrenology]" (Wallace 1905, 1:257). It was the performance of the phrenologist in front of him, and on his own head, which convinced Wallace that phrenology must be true. Wallace continued, "On the whole, it appears to me that these two expositions of my character, the result of a very rapid examination of the form of my head by two perfect strangers, made in public among, perhaps, a dozen others, all waiting at the end of an evening lecture, are so curiously exact in so many distinct points as to demonstrate a large amount of truth—both in the principle and in the details—of the method by which they were produced" (Wallace 1905, 1:262). Wallace remained a professed believer in phrenology until the end of his long life in 1913.

If we describe the spread of phrenology in terms of Rogers's sociology of diffusion, the equivalents of his innovation, communication channels, time, and the population are readily apparent. In this case the innovation is phrenology, the communication channels were periodicals and books, then lecture tours and word of mouth. As with the sociologists' studies, here too we find that diffusion occurred between individuals who were most alike. Initially Spurzheim convinced fellow medical men such as Edward Barlow (1779–1844), senior physician at Bath Hospital and Bath United Hospital from 1807 who attended Spurzheim's 1814 lectures in Bath; Samuel Scilly (1805–71), member of the Royal College of Surgeons; John Elliotson (1791–1868) of St. Thomas's Hospital in London; and Andrew Combe, a medical student and brother of George Combe (Cooter 1989, 20, 308; Cooter 1984, 314; Combe 1850). Spurzheim, again through personal contacts, made other converts who were not medical men such as Thomas Forster (1789–1860), whom Spurzheim met in London through the surgeon William Lawrence in 1814 (Forster 1837, 16–18).

Similarly, George Combe, not a medical man, made mostly nonmedical converts. There was, for example, his fellow Edinburgh middle-class amateur naturalist Robert Chambers (1802–71) who adopted phrenology during Combe's lectures in 1834 (Combe 1836; de Giustino 1975, 52). Certainly very interested, and at least to some degree convinced, was his friend the astronomer John Pringle Nichol (1804–59; see J. Secord 2000; Combe 1836). Similarly, Combe's demonstration visits to New Lanark in 1820 converted fellow visitors from the *London Magazine* to a belief in phrenology, and the socialist reformer Robert Owen (1771–1858) was sufficiently impressed to purchase books and busts to look into phrenology further (Cooter 1984, 232–33; Anon. 1823, 541).

In contrast to the diffusion of the practice of phrenology, which was communicated by personal instruction and therefore stayed little altered for many years, the natural philosophy of George Combe's *Constitution of Man* (1828) diffused predominately through print. This difference may explain why Combe's natural philosophy varied and changed much more and eventually melded indistinguishably into surrounding late nineteenth-century naturalism (van Wyhe 2004). The reason seems to be that much more tacit knowledge and information is passed on through personal communication than in texts, where the learner is far freer in interpreting or generating new knowledge based on the cues provided.

Another important feature named by Rogers is compatibility with existing belief or practice. For many people in early nineteenth-century Britain, phrenology was not incompatible with their existing beliefs because they had no existing popular or systematic psychology. For devout Christians and dualists, phrenology was also compatible as all the major practitioners from Gall and Spurzheim onward declared that the brain was merely the material interface between mind and body. They did not teach that the mind was simply the function of the brain. In these respects phrenology might be expected to be fairly compatible with what many people believed. An obvious exception would be those who already had an alternative set of beliefs covering the same territory as phrenology: academic philosophers and brain anatomists. Notoriously, it was precisely these sorts of people, such as philosopher Thomas Brown (1778–1820) or anatomist John Gordon (1786–1818), who were the greatest opponents in the early years of phrenology's introduction to Britain (see Shapin 1975; Cooter 1984).

The observability of an innovation is also a particularly relevant factor. Certainly phrenological busts or diagrams are instantly recognizable and unmistakable for anything else. Less obvious would be the gaze of one in the know at the heads of others. More visible, and more convincing by far, were public head readings. These demonstrations might be part of a lecturer's stage performance or the attempt of a phrenologist to convince companions at a school or asylum. Witnessing the apparently correct divination (achieved by cold reading and vague predictions) of hidden character traits from sight or touch was as moving to people in the nineteenth century as horoscopes or psychic readings are to countless people today.

This observability leads to the other feature named by Rogers, the "trialability" or testability of the innovation. Virtually all the phrenologists who left sufficient record for us to examine show that they first tried phrenology to see if it worked. They either had their own heads read or would compare the head of a companion or servant to the phrenological charts. Even one apparent confirmation usually has more psychological impact than a number of misses.

The irascible botanist and phrenologist Hewett Cottrell Watson wrote a book entitled *Statistics of Phrenology: being a sketch of the progress and present state of that science in the British islands* (1836) for which he sent out questionnaires to phrenologists throughout the United Kingdom (Egerton 2003). The responses, in addition to other evidence, allowed Watson to create an accurate overview of the diffusion of phrenology to date. Watson was convinced that phrenology was on the eve of becoming universally accepted because he could see that it had become very widely known and adopted. Watson was writing at what we can now see was the peak of phrenology's first wave of popularity and status. After a slow start, phrenology became increasingly popular in the 1820s and 1830s. Active practice of phrenology remained widespread through the mid-1840s and then began to decrease for some years.

Watson argued in 1836 that the reason phrenology was not more widely known was because it had spread quickly through a small network of devotees and was not adequately spread beyond them. This assessment was probably accurate. Phrenology had already spread to all those who would adopt the science in its current form. From that point onward phrenology would slowly decline in popularity.

Combining Watson's statistics with other evidence (especially from Cooter 1989), it is possible to compile some tables that by showing the year-by-year numbers of phrenologists and newly founded phrenological societies indicate the rate of diffusion of the practice of phrenology in Britain (see tables 3.1, 3.2). Watson also estimated that between 1815 and 1825 about twenty courses of lectures on phrenology were offered and that between 1826 and 1836 there were nearly one hundred (Watson 1836, 231).

Of course "the phrenology" that went from Gall to Spurzheim and from Spurzheim to Forster, Combe, Mackenzie et al.; from Combe to William

TABLE 3.1 Estimated number of phrenologists in Britain

Year	Number of phrenologists
1813	0
1814	1
1816	3
1817	20
1826	500
1836	5,000

Sources: Data from Watson 1836; Cooter 1989.

TABLE 3.2 Estimated number of phrenological societies founded

Year(s)	Number of societies founded
1815	0
1816–20	1
1821–25	3
1826–30	9
1831–35	15
1836–40	10
1841–45	9

Sources: Data from Watson 1836; Cooter 1989.

Scott; from Crook to Goyder; and to thousands of others was not an unchanging static body of knowledge. The diffusion of knowledge is not like the diffusion of such static objects as stones or skulls, which remain the same in every context and unaltered as they pass through the hands, or rather the heads, of many people. A popular science such as phrenology is a kind of knowledge—it is something that people *know*—and knowledge, or culture, is an altogether different beast from objects in that knowledge is cognitive. As modern cognitive anthropology and psychology have shown, knowledge "transmission" is not like digital data transfer but more like Chinese whispers (see Boyer 1994; Sperber 1996, 2000). A person had to pass through a process of learning phrenology and only then could subsequently influence others to generate very similar knowledge of their own.

Spurzheim was the junior assistant of Gall for many years. Yet when Spurzheim set out on his own in 1813 to teach what he tendentiously called "our doctrine," he taught something quite different from Gall. Spurzheim not only added several cerebral organs to bring the total from twenty-seven to thirty-three but also arranged them in a classification of orders and genera. Spurzheim changed the names of the faculties such as Gall's *Mord* (murder) which became the organ of Combativeness, an organ which Spurzheim contended had proper and improper functions. Gall had no notion of organs having improper functions. Spurzheim's doctrine was much more ambitious in this sense—almost a religion of scientific certainty.

Just as Spurzheim's version of Gall's *Schädellehre* was unique, so was George Combe's version of Spurzheim's "physiognomical system." Combe added yet another faculty and a host of other changes to make his own distinctive brand of "phrenology," which was moralistic and more like a religion of science than even Spurzheim's version. This process of change was repeated in every case of the diffusion of phrenology. In the hands of William Scott, David Goyder, H. Lundie, or any of thousands of other phrenologists, we see very similar but always unique versions of phrenology (Lundie 1844). By the 1880s, for example, phrenology was still around, but it was quite a different strain after it had spread to America, evolved further, and then been reintroduced to Britain with the Fowler family (Stern 1971). In the United States Charles Caldwell, Orson Fowler, and others followed the examples of Spurzheim and Combe. The same practice by Combe, Robert Noel, and Gustav Scheve took an Anglicized phrenology back to Germany in the late 1830s and 1840s, which is why the comparatively low-key German phrenology of the 1840s to the 1860s was not at all Gallian but Combean. A comparison of British, German, and American phrenological busts makes this difference clear (see figures 3.5, 3.6, and 3.7). There is no doubt that it was all recognizably "phrenology," but it is equally apparent upon close inspection that the phrenology of this later time had changed in many ways. Thus what was diffused was not static. The people to whom phrenology spread did not have their heads stuffed with "the science" by their teachers. Each person actively learned or constructed his or her own version from what was experienced.

The particular historical data for the diffusion of phrenology throughout Britain derived from Watson's and Cooter's indispensable biobibliographies represent a classic innovation diffusion (Watson 1836; Cooter 1989). Initial converts were made by personal contact with an early adopter (Spurzheim)—those who first took up the science already had much in common, such as "middle-class" profession (often medical, then legal in Edinburgh) or social status (Cooter 1984). The rate of active adoption of phrenology describes an S-shaped curve. This shape occurs if one measures publications, public lectures, estimated believers, or phrenological societies founded.

The same features of slow initial spread among a group of similar individuals, usually with personal links, followed by a takeoff phase and then a slow decrease as saturation approached is common to countless other

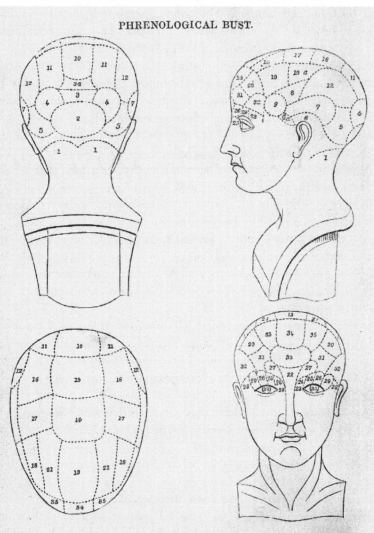

PHRENOLOGICAL BUST.

NAMES OF THE MENTAL FACULTIES, THE POSITIONS OF THE ORGANS OF WHICH ARE MARKED UPON THE BUST.

AFFECTIVE.

I. PROPENSITIES.

1. Amativeness, vol. i. p. 183
2. Philoprogenitiveness, 193
3. Concentrativeness, 211
3. a Inhabitiveness, ib.
4. Adhesiveness, 237
5. Combativeness, 243
6. Destructiveness, 255
6. a Alimentiveness, 277
7. Secretiveness, 294
8. Acquisitiveness, 311
9. Constructiveness, 326

II. SENTIMENTS.

10. Self-Esteem, vol. i. p. 341
11. Love of Approbation, 357
12. Cautiousness, 369
13. Benevolence, 382
14. Veneration, 399
15. Firmness, 413
16. Conscientiousness, 418
17. Hope, 443
18. Wonder, 449
19. Ideality, 469
19. a Unascertained, 477
20. Wit or Mirthfulness, 490
21. Imitation 511

INTELLECTUAL.

I. PERCEPTIVE.

22. Individuality, vol. ii. p. 28
23. Form, 35
24. Size, 41
25. Weight, 46
26. Colouring, 58
27. Locality, 72
28. Number, 83
29. Order, 90
30. Eventuality, 92
31. Time, 104
32. Tune, 110
33. Language, 124

II. REFLECTIVE.

34. Comparison, vol. ii. p. 151
35. Causality, 163

Figure 3.5. British phrenological bust illustrated in frontispiece to Combe, *A System of Phrenology*. Reprinted from Combe 1853; in author's collection.

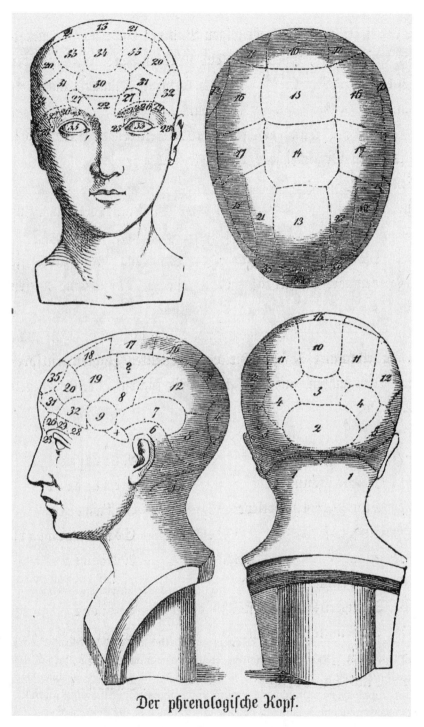

Figure 3.6. German phrenological bust illustrated in engraving "Der phrenologische Kopf" (The phrenological head). Reprinted from Scheve 1863; in author's collection.

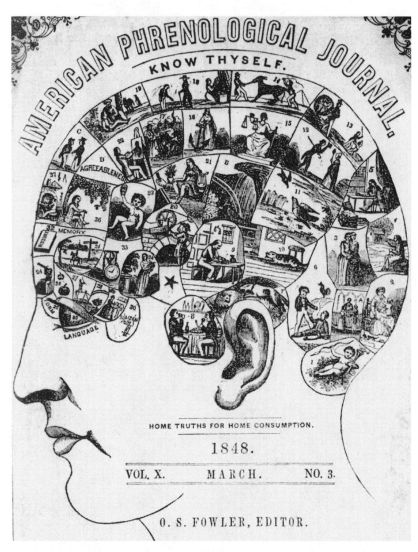

Figure 3.7. American phrenological bust illustrated in frontispiece to *American Phrenological Journal* 10 (1848); in author's collection.

innovations in the history of science, such as the spread of Darwin's theory of evolution or the use of the telescope in astronomy. The neglect of diffusion by historians of science is unfortunate. In sociology, archaeology, anthropology, economics, marketing, and linguistics, diffusion is an indispensable concept. It is readily applicable to much of the history of science. We should welcome some interdisciplinary diffusion. It promises much for understanding not only the history of science but cultural change in general.

᭡

NOTES

I would like to thank Aileen Fyfe for helpful conversation, feedback, and inviting me to speak on phrenology, out of retirement, as it were. I am also grateful to Bernard Lightman for helpful comments; Anne Secord, Jim Secord, and Richard Noakes for encouraging discussion; and the participants at the Toronto conference where an earlier version of this chapter was presented. This chapter has also benefited from the encouraging and constructive comments of two anonymous referees. I am also grateful to the participants at the Institutkollokvium ved Institut for Videnskabshistorie at Aarhus University in 2003 where an initial version was read. I am grateful to Harvard Medical Library in the Countway Library of Medicine; the National Library of Scotland; and the Bodleian Library, Oxford University, for permission to quote from unpublished manuscripts in their possession.

REFERENCES

Manuscripts

George Combe Collection. National Library of Scotland, MS.7218 ff.93–98.
Forster Papers. MS Eng. Letters f. 181, Bodleian Library, Oxford University.
Minute Book of the Phrenological Society, 1, 1820–41, Edinburgh University Library, Gen 608/2.
Spurzheim Letters. Johann Spurzheim to Honorine Pothier. Harvard Medical Library in the Countway Library of Medicine.

Published Works

Adams, Mark. 1979. "From Gene Fund to Gene Pool: On the Evolution of Evolutionary Language." *Studies in the History of Biology* 3:241–85.
Anon. 1800. *Medical and Physical Journal* 4:50.
Anon. 1806. "An Account of Dr. Gall's system of Craniology." *Medical and Physical Journal* 15:201–13.
Anon. 1823. "Phrenology." *London Magazine* 8: 41–44.
Anon. 1824. "List of Members." *Transactions of the Phrenological Society* 1:ix–xii.
Anon. 1826. "List of Members." *Phrenological Journal* 3:476–81.
Anon. 1865. *Der Fliegenden Blätter* 42.
Arneman, J. 1805. "A Concise Account of Dr. Gall's New Doctrine of the Brain, and the Faculties of the Mind." *Monthly Magazine*, October, 327–36.

Arnold, J. T. F. K. 1805. *Dr. Joseph Gall's System der Gehirn- und Schädelbaues nach den bis jetzt über seine Theorie erschienenen Schriften: Als Leitfaden bei akademischen Vorlesungen dargestellt.* Erfurt, Ger.: Henning'schen Buchhandlung.

Barber, Bernard. 1963. "Rogers, *The Diffusion of Innovations* (1962)." *Isis* 54 (2): 296–97.

Bischoff, C. H. E. 1807. *Some Account of Dr. Gall's New Theory of Physiognomy Founded upon the Anatomy and Physiology of the Brain and the Form of the Skull with the Critical Strictures of C. W. Hufeland, M.D.* Trans H. C. Robinson. London: n.p.

Blöde, K. A. 1807. *Dr. F. J. Gall's System of the Functions of the Brain extracted from Charles Augustus Blöde's, Account of Dr. Gall's Lectures, held on the abore [sic] subject at Dresden.* n.p.

Bojanus, Ludwig Heinrich. 1802. "A Short View of the Craniographic System of Dr. Gall, of Vienna." *Philosophical Magazine* 14:77–84, 131–38.

Boyer, Pascal. 1994. *The Naturalness of Religious Ideas: A Cognitive Theory of Religion.* Berkeley and Los Angeles: University of California Press.

[Brown, William.] 1803. "Villers, sur une Nouvelle Theorie de Cerveau." *Edinburgh Review*, 2:147–60.

Cantor, G. N. 1975. "The Edinburgh Phrenology Debate: 1803–1828." *Annals of Science* 32:195–212.

Carlson, Richard O. 1965. *Adoption of Educational Innovations.* Eugene: University of Oregon, Center for the Advanced Study of Educational Administration.

Carmichael, Andrew. 1833. *A Memoir of the Life and Philosophy of Spurzheim.* Dublin: Wakeman.

Chambers, J. K. 1992. "Dialect Acquisition." *Language* 68:673–705.

Clarke, Edwin, and L. S. Jacyna. 1987. *Nineteenth-Century Origins of Neuroscientific Concepts.* Berkeley and Los Angeles: University of California Press.

Combe, George. 1819. *Essays on Phrenology, or an Inquiry into the Principles and Utility of the System of Drs. Gall and Spurzheim, and into the Objections Made Against It.* Edinburgh: Bell and Bradfute.

———. 1828. *The Constitution of Man and its Relation to External Objects.* Edinburgh: John Anderson; London: Longman.

———. 1836. *Testimonials on Behalf of George Combe, : As a Candidate for the Chair of Logic in the University of Edinburgh.* Edinburgh: John Anderson.

———. 1839. *Lectures on phrenology.* Edinburgh: Maclachlan and Stewart; London: Longman; London: Simpkin, Marshall.

———. 1848. *Lectures on Popular Education, delivered to the Edinburgh Philosophical Association for Procuring Instruction in Useful and Entertaining Science, in April and November, 1833.* Edinburgh: Anderson.

———. 1850 *The Life and Correspondence of Andrew Combe, M.D.* Edinburgh: Maclachlan and Stewart; London: Longman; London: Simpkin, Marshall.

———. 1853. *A System of Phrenology*. Edinburgh: Maclachlan and Stewart; London: Longman; London: Simpkin, Marshall.

Cooter, R. 1984. *The Cultural Meaning of Popular Science: Phrenology and the Organization of Consent in Nineteenth-Century Britain*. Cambridge: Cambridge University Press.

———. 1989. *Phrenology in the British Isles: An Annotated, Historical Bibliography and Index*. Metuchen, NJ: Scarecrow Press.

Cooter, R., and S. Pumfrey. 1994. "Separate Spheres and Public Places: Reflections on the History of Science Popularisation and Science in Popular Culture." *History of Science* 22:237–67.

Crane, Diana. 1972. *Invisible Colleges: Diffusion of Knowledge in Scientific Communities*. Chicago: University of Chicago Press.

Croft, William. 2000. *Explaining Language Change: An Evolutionary Approach*. Harlow, Essex, UK: Pearson Education.

Crook, W. H. 1827. *Syllabus at a Course of Lectures on Phrenology, by Mr. Crook*. n.p.

Dawson, Gowan. 2004. "Victorian Periodicals and the Making of William Kingdon Clifford's Posthumous Reputation." In *Science in the Nineteenth-Century Periodical: Reading the Magazine of Nature*, ed. Geoffrey Cantor, Graeme Gooday, Gowan Dawson, Richard Noakes, Sally Shuttleworth, and Jonathan R. Topham, 259–84. London: MIT Press.

de Giustino, D. 1975. *Conquest of Mind: Phrenology and Victorian Social Thought*. London: Croom Helm.

Desmond, Adrian. 1998. *Huxley: From Devil's Disciple to Evolution's High Priest*. London: Penguin.

[D'Oyly, George.] 1815. "The Physiognomical System of Drs. Gall & Spurzheim." *Quarterly Review* 13: 159–78.

Egerton, F. 2003. *Hewett Cottrell Watson: Victorian Plant Ecologist and Evolutionist*. Aldershot, UK: Ashgate.

Enfield, N. J. 2003. *Linguistic Epidemiology: Semantics and Grammar of Language Contact in Mainland Southeast Asia*. London: Routledge.

Follen, C. 1834. "Funeral Oration: Delivered before the Citizens of Boston assembled at the Old South Church, November 17, 1832, at the Burial of Gaspar Spurzheim, M.D." *Phrenological Journal* 8:317–31.

Forster, T. I. M. 1837. *Recueil de ma vie, mes ouvrages et mes pensées: Opuscule philosophique*. 3rd ed. Brussels: Ve A. Stapleaux.

Freedman, Ronald, and John Y. Takeshita. 1969. *Family Planning in Taiwan*. Princeton, NJ: Princeton University Press.

Froriep, L. F. 1799. *Darstellung der ganzen auf Untersuchungen der Verrichtungen des Gehirnes gegründeten Theorie der Physiognomik des Dr Gall in Wien*. Weimar: n.p.

Fyfe, Aileen. 2000. "Industrialised Conversion: The Religious Tract Society and

Popular Science Publishing in Victorian Britain." PhD thesis, Cambridge
University.

Gall, Franz Joseph. 1822–1825. *Sur les fonctions du cerveau et sur celles de cha-
cune de ses parties, avec des observations sur la possibilité de reconnaitre les
instincts, les penchans, les talens, ou les dispositions morales et intellectu-
elles des hommes et des animaux, par la configuration de leur cerveau et de
leur tête.* 6 vols. Paris: J. B. Baillière.

Gall, Franz Joseph, and Johann Gaspar Spurzheim. 1810–19. *Anatomie et physi-
ologie du système nerveux en général, et du cerveau en particulier, Avec des
observations sur la possibilité de reconnoître plusieurs dispositions intellec-
tuelles et morales de l'homme et des animaux, par la configuration de leurs
têtes.* 4 vols. Paris: F. Schoell. (first two vols 1810, 1812 only with Spurzheim,
plus one-vol. Atlas of 100 engravings (1810) supervised by Spurzheim).

Gaw, J. 1999. *"A Time to Heal": The Diffusion of Listerism in Victorian Britain.*
Philadelphia: American Philosophical Society.

Gibbon, Charles. 1878. *The Life of George Combe: Author of "The Constitution of
Man."* 2 vols. London: Macmillan.

Golinski, J. 1998. *Making Natural Knowledge: Constructivism and the History of
Science.* Cambridge: Cambridge University Press.

[Gordon, John.] 1815. "The Doctrines of Gall and Spurzheim." *Edinburgh Review*
25:227–68.

Goyder, D. 1857. *My Battle for Life: The Autobiography of a Phrenologist.* London:
Simpkin, Marshall.

Henrich, J. 2001. "Cultural Transmission and the Diffusion of Innovations: Adoption Dynamics Indicate That Biased Cultural Transmission Is the Predominate Force in Behavioral Change." *American Anthropologist* 103 (4): 992–1013.

Hilgartner, Stephen. 1990. "The Dominant View of Popularization: Conceptual
Problems, Political Uses." *Social Studies of Science* 20:519–39.

Jacyna, L. S. 1984. "Principles of General and Comparative Physiology: The Comparative Dimension to British Neuroscience in the 1830s and 1840s." *Studies
in the History of Biology* 8:47–93.

Jena, Detlef. 2003. *Katharina Pawlowna.* Regensburg, Ger.: Pustet.

Kaufman, M. H. 1999. "Phrenology—Confrontation between Spurzheim and
Gordon—1816." *Proceedings of the Royal College of Physicians of Edinburgh*
29:159–70.

Kroch, Anthony S. 1989. "Reflexes of Grammar in Patterns of Language Change."
Language Variation and Change 1:199–244.

Lansing, Stephen. 1987. "Balinese 'Water Temples' and the Management of Irriga-
tion," *American Anthropologist* 89:326–41.

Lundie, H. 1844. *The Phrenological Mirror or Delineation Book.* Leeds, UK:
C. Crowshaw.

Mosteller, Frederick. 1981. "Innovation and Evaluation." *Science* 211:881–86.

Oehler-Klein, Sigrid. 1990. *Die Schädellehre Franz Joseph Galls in Literatur und Kritik des 19. Jahrhunderts: zur Rezeptionsgeschichte einer biologisch-medizinischen Theorie der Physiognomik und Psychologie.* Stuttgart: Fisher.

Oxford English Dictionary. 1994. 2nd ed. Oxford: Oxford University Press. Also available on CD-ROM, version 1.10.

Phrenological Journal. 1823–47. Cited in text as *PJ* with volume, year, and page number.

Rogers, E. 1995. *Diffusion of Innovations.* 4th ed. New York: Free Press.

Scheve, Gustav. 1863. *Phrenologische Reisebilder.* Cöthen, Ger.: Paul Schettler.

Secord, Anne. 1994. "Science in the Pub: Artisan Botanists in Early Nineteenth-Century Lancashire." *History of Science* 32:269–315.

Secord, James A. 2000. *Victorian Sensation: The Extraordinary Publication, Reception, and Secret Authorship of "Vestiges of the Natural History of Creation."* Chicago: University of Chicago Press.

Shapin, Steven. 1975. "Phrenological Knowledge and the Social Structure of Early Nineteenth-Century Edinburgh." *Annals of Science* 32:219–43.

———. 1983. "'Nibbling at the Teats of Science': Edinburgh and the Diffusion of Science in the 1830s." In *Metropolis and Province: Science in British Culture 1780–1850,* ed. I. Inkster and J. Morrell, 151–78. London: Hutchinson.

Shuttleworth, Sally, and Geoffrey Cantor. 2004. Introduction to *Science in the Nineteenth-Century Periodical: Reading the Magazine of Nature,* ed. Geoffrey Cantor, Graeme Gooday, Gowan Dawson, Richard Noakes, Sally Shuttleworth, and Jonathan R. Topham, 1–16. London: MIT Press.

Sperber, Dan. 1996. *Explaining Culture: A Naturalistic Approach.* Oxford: Blackwell.

———. 2000. "An Objection to the Memetic Approach to Culture." In *Darwinizing Culture: The Status of Memetics as a Science,* ed. R. Aunger, 163–74. Oxford: Oxford University Press.

Spurzheim, J. G. 1814. Lecture advertisement. *Times* (London), October 10.

———. 1815. *The Physiognomical System of Drs. Gall and Spurzheim; founded on an Anatomical and Physiological Examination of the Nervous System in General and of the Brain in Particular; and indicating the Dispositions and Manifestations of the Mind. Being at the same time a book of reference for Dr. Spurzheim's demonstrative lectures.* London: Baldwin, Cradock and Joy.

Stern, M. 1971. *Heads and Headlines: The Phrenological Fowlers.* Norman: University of Oklahoma Press.

Transactions of the [Edinburgh] Phrenological Society. 1823. Edinburgh: John Anderson.

Valente, T. W. 1993. "Diffusion of Innovations and Policy Decision-Making." *Journal of Communication* 43 (1): 30–45.

van Wyhe, John. 1999–. Web site, "The History of Phrenology on the Web." http://pages.britishlibrary.net/phrenology/.

———. 2002. "The Authority of Human Nature: The Schädellehre of Franz Joseph Gall." *British Journal for the History of Science* 35 (March): 17–42.

————. 2004. *Phrenology and the Origins of Victorian Scientific Naturalism.* Aldershot, UK: Ashgate.

Wallace, Alfred Russel. 1905. *My Life: A Record of Events and Opinions.* 2 vols. London: Chapman and Hall.

Watson, H. C. 1836. *Statistics of Phrenology: Being a Sketch of the Progress and Present State of that Science in the British Islands.* London: Longman.

Wells, Charlotte Fowler. 1896. *Some Account of the Life and Labours of Dr. François Joseph Gall: Founder of Phrenology and his disciple Dr. John Gaspar Spurzheim.* New York: Fowler and Wells.

Winter, Alison. 1994. "Mesmerism and Popular Culture in Early Victorian England." *History of Science* 32:317–43.

————. 1998. *Mesmerized: Powers of Mind in Victorian Britain.* Chicago: University of Chicago Press.

[Yelloly, John.] 1802. "Lettre de Charles Villers, etc. i.e., A letter from Charles Villers to George Cuvier, of the National Institute of France, on a new theory of the brain by Dr. Gall, in which that viscus is considered as the immediate organ of the moral faculties." *Monthly Review* 39:487–90.

CHAPTER FOUR

Lecturing in the Spatial Economy of Science

BERNARD LIGHTMAN

When he arrived in Boston in 1884 to present the prestigious Lowell Lectures, Robert Ball, then astronomer royal of Ireland, was already a seasoned science lecturer with ten years of experience under his belt.[1] But a reporter for the *Boston Herald* observed that Ball "suffered from a slight impediment in his speech" and that "he has a smooth, clear voice, with a use of it, at times, quite clergymanic" ("Modern Astronomy" 1884). Ball's response to the report is revealing. "The Boston Herald says that I have a hesitation in my speech and that my style is sometimes clergymanic," he wrote. "I must try and correct these trifles" (as quoted in Wayman 1986, 192). Ball resolved to eliminate any trace of the sermon in his oral style, and not just because he was a confirmed scientific naturalist. He realized that a successful scientific lecturer needed to entertain as well as to instruct. Three years earlier he had given his first lecture at the illustrious Royal Institution on "The Distance of the Stars." Honored by the invitation, he "took great pains with the lecture which was to be delivered in such a place and before such an audience" (Ball 1915, 203–4). The Royal Institution was famous for its lecturers, among whom were Humphry Davy, Michael Faraday, and John Tyndall. Ball had frequently heard Tyndall lecture at the Royal Institution, acknowledged his special genius as an entertaining public speaker, and learned a great deal from him about communicating effectively with an audience. As he prepared for his lecture, he recollected how James Clerk Maxwell, when about to speak in public, was in the habit of "'Tyndallising' his imagination up to the point of being able to devise picturesque phraseology and to accompany it with effective experiments" (Ball 1915, 203–4).

Ball delivered his lectures during a period when there was a veritable explosion of books and periodicals devoted to conveying science to a broad audience of interested readers. The growth of an educated middle class and of literacy among members of the working class combined with the invention of new printing technologies gave birth to an unprecedented mass market that provided new opportunities for careers in science journalism and writing. Dawson, Noakes, and Topham have pointed out that new ways of presenting the sciences to general audiences were being developed in established and new serial forms in periodicals during the middle of the century (Dawson, Noakes, and Topham 2004, 16–17). Similarly, Gates and Shteir have argued that during this period, science writers invented a new literary narrative to meet the needs of the mass reading public eager to understand the larger significance of recent scientific discoveries (Gates and Shteir 1997, 12). But the increased interest in science was not limited to the realm of print culture, as other public spaces for the exchange of knowledge, such as museums, exhibitions, zoos, aquaria, gardens, lecture halls, and conversaziones, multiplied and diversified. Just as scientific authors sought in their books to forge a new style of writing, those who engaged in public speaking tried to create innovative lecturing styles that would attract an audience with high expectations when it came to entertainment. The popularity of the Crystal Palace and the spectacular shows of London had raised the bar.

David Livingstone has stressed how place is not a neutral container but rather "constitutive of systems of human interaction" (Livingstone 2003, 7). He distinguishes public from elite spaces and asserts that a recognition of how science has been "practiced in a variety of popular arenas" should help "widen our awareness of the range of spaces in which scientific knowledge has been produced and propagated" (Livingstone 2003, 85). If place is to be considered as an important factor in our understanding of the communication of scientific ideas to the public, then we must think about how lecturing was experienced differently by audiences depending on the sites of delivery. Here I will examine how science lecturers refashioned the sites in which they lectured and how they hoped thereby to provide their audiences with a different experience—one that was entertaining as well as instructive. I will begin with a discussion of the widespread demand for science lecturers in the second half of the nineteenth century, concentrating on the most active speakers, their extensive tours, their earning power, their mode of delivery, and their use of visual aids. Like Ball, they avoided using the traditional sermon, didactic and serious in nature, as a model for lecturing as in their efforts to draw large audiences they were competing with a wide range of mass-cultural forms.[2] Then I will focus on one representative from each important type of science lecturer, those like John Henry Pepper who operated from within institutions devoted to exhibiting science in an established public space, and those like Frank Buckland who had to fashion a scientific space for themselves every time they lectured. Humphry Davy had been applauded for blending entertainment with instruction, through his reproduction of experiments during lectures, but Pepper and Buckland went a step further. They drew upon cultural forms connected to the world of entertainment to refashion sites of knowledge-production and learning and by doing so provided their audiences with a new experience of public science more akin to that of visitors to the museum or the theater. While Pepper is perhaps best remembered for his optical illusions that featured ghostly apparitions and Buckland for his contributions to British fisheries, they both drew on cultural forms connected to the world of entertainment. Their distinctive blend of instruction and amusement became a hallmark of lecturing in this period. They were indicative of the expansive force of public science in the second half of the nineteenth century, as its sites multiplied endlessly and it transformed the spatial economy of science.[3]

The Scientific Lecturing Scene

In his will, the wealthy philologist John Borthwick Gilchrist (1759–1841) had left a considerable amount of money in an educational trust. During his life he had been involved in various projects to boost popular education. In 1823, he had helped George Birkbeck found the London Mechanics' Institution, and he had also been involved in the establishment of the University of London. Part of the Gilchrist Trust was to go toward funding an annual series of public lectures in British industrial centers. But the money was tied up in litigation for more than twenty-five years, and the lectures were put on hold until the issue was resolved (Prior 2004, 219). When the Gilchrist Trust began to plan to send science lecturers throughout the country, there were doubts that they would be a success. Ball, who began his twenty-year connection with the trust in 1880, recalled one of his earliest experiences lecturing at Blackburn. The heavy rains had led the mayor to worry that attendance would be low. As they reached the hall they saw no crowds, except for the policemen standing outside the door. The mayor was about to declare the lecture a "total failure" when the policemen informed him that the hall was filled to capacity and that they had turned away two hundred people half an hour ago. Ball asserted that the general experience of those who gave the Gilchrist lectures was "house packed and every inch of standing room occupied" (Ball 1915, 217).

Ball and the other successful science lectures of the latter half of the nineteenth century were able to take advantage of the institutionalization of scientific lecturing earlier in the century. Hays has argued that a scientific lecturing empire was established in London by the 1820s and 1830s that contributed to the support and professionalization of "men of science." During this period scientific lecturing had become formalized when scientific societies and institutions began to concentrate their activities into a season between November and June. The Royal Institution and the London Institution presented a series of lecture courses through the season. Lectures were also offered by the Russell Institution in Bloomsbury, by the Surrey Institution near Blackfriars Bridge, and by a number of Mechanics' Institutes centered in London. University College and King's College offered scientific instruction as well. Hays asserts that by the 1830s it was Michael Faraday who was seen as the model lecturer due to his clarity, neatness, arrangement, and concentration on the subject under discussion, while Humphry Davy's appeal to moral elevation, poetic inspiration, mental cultivation, business profit, and amusement was no longer in vogue.[4] Among the important lecturers were John Millington, Charles Frederick Partington, Edward William Brayley, William Thomas Brande, William Ritchie, and Dionysius Lardner. Some of the London lecturers, such as Ritchie and Lardner, were in great demand in the provinces (Hays 1983, 91–97, 101–2).[5]

In the latter half of the nineteenth century, those who were hired to lecture to the public on a regular basis by formal institutions were in a more secure situation than those who were independent—as long as they maintained good relations with the directors. John Henry Pepper (1821–1900) was appointed lecturer and analytical chemist at the Royal Polytechnic Institution in 1848. A colorful showman who could draw in the crowds, he lectured daily at the Polytechnic until 1872 on such topics as electricity, chemistry, and optical illusions. But those without institutional positions were also much in demand as public lecturers, and they had the freedom to accept or reject whatever invitations came their way. Frank Buckland (1826–1880) and John George Wood (1827–1889) were well-known lecturers on natural history, while Robert Stawell Ball (1840–1913) specialized in astronomical topics. The number of science lectures that these men delivered during their lives is staggering, and the punishing demands of the lecturing circuit could often take a physical toll. By 1884, Ball had already delivered more than seven hundred lectures (Ball 1915, 224). Wood reportedly gave an average of ninety lectures each season (Upton [1910], 171). According to his son, his best season was that of 1881–1882, when he delivered more than 120 lectures (Wood [1890], 254). Eventually the stress of lecturing caught up to Wood. The *Times* reported in 1889 that he died "while in harness." He caught a severe chill while on a lecturing tour, "took to his bed at Coventry on Saturday, and died on Sunday" (Whitehead 1889, 15).

Many lecturers traveled extensively throughout Britain, and several undertook lecture tours of the United States and other parts of the world. Buckland's lecturing engagements were limited primarily to Britain. He spoke in London at the Royal Institution, the London Institution, and the South Kensington Museum, and he also delivered lectures at Brecon, Nottingham, Oxford, Sheffield, Windsor, and Witney. Wood began his lecturing career in the mid-1860s and by 1879 had engaged the services of a booking agent because he was traveling all over England (Wells 1990, 58). He went to the United States twice, once in 1883–84 to deliver the Lowell Lectures and a second time in the fall of 1884. Pepper ultimately became the most cosmopolitan of the group. After twenty-four years based at the Royal Polytechnic Institute in London, he resigned and took his optical illusions to Canada, the United States, and Australia. As astronomer royal of Ireland, Ball was centered in Dublin before he was appointed Lowndean Professor of Astronomy at Cambridge in 1892, but he often made winter visits to England. By 1887 he was limiting himself to two trips, one in November and another in January. He managed the November trip himself, accepting invitations that came during the summer. He planned the lecture tour so that he would begin in the north and work his way gradually southward. The Gilchrist Trust managed the January trip, and he dutifully followed the itinerary laid out for him. He visited mostly small towns. In his first set of Gilchrist lectures in 1880, he went to Rochdale, Accrington, Huddersfield, Preston, and Bury (Ball 1915, 223, 217). Ball came to the United States to deliver lectures three times, in 1884, 1887, and 1901. He apparently left a lasting impression. The *Boston Evening Transcript* obituary asserted that "since the days of the great and only Huxley, no one has put more of natural science into the minds of men through the medium of the tongue" (Collins and Smith 1915, 5). Extensive lecturing allowed these popularizers to become widely known throughout Britain and sometimes beyond.

Successful science lecturers could earn significant sums through their speaking engagements. Ball was an astute businessman, whose lecturing and books earned him a personal fortune (Wayman 1986, 187). He charged between £25 and £40 per lecture (Wayman 1987, 124). He cleared £165 when he gave the Lowell Lectures in 1884 and was convinced that he could make £100 a week if he stayed (Wayman 1986, 194–95). In 1892, when Ball was weighing whether to accept the Lowndean chair, he drew up a budgetary scheme for living in Cambridge. He estimated that he could earn £600 annually from his lecturing, which was equal to the salary he was to receive as Lowndean chair. In contrast, he expected to receive only £440 from his books and literary projects (County Record Office 1892). In a letter to a friend in 1897, he stated why he preferred lecturing to writing. "Lecturing is a more permanent source of income than writing," Ball wrote, "for the same lecture will be available scores of times, while there is (or ought to be) a limit to the number of times the same thing can be written." Ball also enjoyed lecturing—it was a rest and a change of pace from his duties as Lowndean Professor—and found writing articles to be "an awful grind" (Ball 1915, 221). Wood had also taken up extensive lecturing tours as a means of augmenting his income, prompted by the depression in the publishing trade in 1879 (Upton [1910], 165). However, he was not nearly as financially successful as Ball. His son claimed that he cleared about £300 per year, but his trips to the United States were financial failures due to mismanagement (Upton [1910], 171; Wood [1890], 170–71, 234, 249–52). When Wood died, he left his family in dire financial straits.

To be successful, science lecturers needed to exploit every possible angle, including the perfection of their own speaking skills and their use of visual aids. Though they offered instruction, scientific lecturers were also expected to be entertaining, especially in the post–Crystal Palace era. In London, they were competing for audiences with the theaters, museums, panoramas, and other spectacles. Lecturers became comfortable with diverse modes of delivery. Up until 1884, Ball had been delivering his lectures without a set text. In the middle of delivering the Lowell Lectures in October 1884—after a lecture in which he had "stammered and hesitated horribly"—he decided to experiment by writing and then reading his paper. He resolved in the future to write the lectures ahead of time and either memorize or read them, allowing for some improvisations along the way (Wayman 1986, 193–94). By contrast, Wood did not use a prepared manuscript and relied on rapid impromptu sketches to illustrate key points in his lectures (Wood [1890], 145). Although one critic implied that his use of "freehand diagrams" was actually to draw attention away from his "ineffective" delivery, Wood's sketches were usually considered the highlight of his lectures ("Rev. J. G. Wood" 1890). "One of his best," his son wrote, "was that of two ants fighting, in which jaws, limbs, and antennae were hopelessly interlocked, and yet the individuality of each insect was clearly preserved." He was also known for his drawing of a sperm whale and a male stickleback (Wood [1890], 159; Lightman 2000, 657–61). For his visual aids, Buckland relied on masses of specimens of bones and skin of exotic animals from around the world.

Visual aids were a crucial dimension of the presentation for public science lecturers, and Buckland and Wood, the natural historians, made little use of the magic lantern in comparison to most speakers. Ball was more typical in his reliance on the oxyhydrogen lantern. He was using photographs in 1881, charts and the stereopticon in 1884, and the oxyhydrogen lantern in 1890 ("Lecture at the Midland Institute on Tides" 1881; "Modern Astronomy" 1884; "Gilchrist Lectures at Goole" 1890). Pepper's use of expensive technological apparatuses to illustrate his lectures was typical of those who worked at science museums, though, as well shall see, his spectacular optical ghost illusions were far more sophisticated than most.

Successful science lecturers were well traveled, well paid, and innovative. One key to their success was their degree of sensitivity to the nature of the sites in which they delivered their lectures. For those who worked in scientific institutions that contained places designed to reach out to the public, it was not necessary to create a new space for their activities. But for those, like Frank Buckland, who labored in spaces not specifically designated as scientific, more work had to be put into fashioning an appropriate site.

Frank Buckland and the Culture of Display

On the evening of October 24, 1861, Frank Buckland presented a lecture on "Curiosities of Natural History" at the town hall in Witney, Oxfordshire. The enthusiastic account of the lecture in *Jackson's Oxford Journal* emphasized that Buckland had "by the simple force of his style and the sterling value of his matter, so fully realized all the great things that were expected of him" as the son of the famous Oxford geologist and dean of Westminster William Buckland, reputed to be one of the most popular lecturers at Oxford in his day. "That power of thought," the anonymous reviewer declared, "that faculty of observation, which so greatly distinguished the father, have descended unimpaired to the son." The report also highlighted the younger Buckland's orthodoxy. It described Buckland's remarks on how nature "always has a most cogent reason for working in a particular manner." Buckland then "compared the works of the Almighty with the mightiest of man's works—the *Great Eastern*, for example,—and designated them puny and weak." Finally, the anonymous author discussed how Buckland illustrated his main theme, the universal law pervading the whole of nature of "eat and be eaten." Buckland's lecture "consisted of a series of observations, original, clearly explained, and embodying great scientific truths. These truths were illustrated and demonstrated by numerous diagrams and specimens, forming of themselves quite a museum"

("Witney" 1861). If only for a single evening, Buckland had transformed Witney Town Hall into a museum of science.

Like Ball and Wood, men without a permanent public institution in which to present their lectures, Buckland delivered his addresses in town halls, churches, assembly halls, and coffeehouses—sites that were not specifically scientific. Buckland had the daunting task of turning diverse sites into ephemeral scientific ones. Through his use of specimens and his appeal to the tactile senses of his audience, he attempted to refashion these sites, temporarily, into museums. Buckland's lecturing style was shaped by the culture of display to be found within existing scientific traditions in museums. He was influenced by the medical tradition of using specimens to illustrate lectures, as established by John Hunter. In his fascination with "curiosities"—the bizarre and unusual—Buckland also drew on the natural history tradition and its love of the exotic, which he combined with his interest in zoos, circuses, and freak shows. Buckland's lecturing was therefore shaped by a variety of scientific and cultural forms.

After completing a Bachelor of Arts degree at Christ Church, Oxford, in 1848, and then medical training at St. George's Hospital from 1848 to 1851, Buckland combined natural history writing and lecturing with a career as a surgeon, first at St. George's Hospital (1852–1853) and then in the Second Life Guards (1854–1863). He began lecturing in 1853 and wrote numerous articles for various periodicals, including, on a regular basis beginning in 1856, the *Field* newspaper. He collected many of his articles together and published them in 1857 in his *Curiosities of Natural History,* and he later produced three more series with the same title. In 1863, he resigned from the Second Life Guards in order to pursue his interests in natural history. After a quarrel, he severed his relationship with the *Field* in 1865 and the following year started his own weekly journal, *Land and Water,* devoted to sport and natural history. But Buckland's new passion was fish hatching, which later led to his appointment as inspector of fisheries in 1867. His duties included the submission of annual reports to the Home Office based on extensive inspections of English rivers and their stock of fish. The books he produced during this time reflected his work on fish, including his *Fish Hatching,* published in 1863; *Logbook of a Fisherman and Zoologist,* published in 1875; and *Natural History of British Fishes,* published in 1881. Buckland remained, to the end of his life, a devout Christian and a determined opponent of Darwin's evolutionary theory. He carried on the natural theology tradition of his father, preparing a revised edition of William's Bridgewater Treatise in 1858.[6]

Buckland spoke on a variety of natural history topics (though he had a set repertoire like other lecturers), at a number of different institutions, and to diverse audiences. Early on, he lectured primarily to members of the working class. His first lecture on "The House We Live In," given in December 1853, was delivered at a workingmen's coffeehouse and institute in Westminster established by his mother (Bompas 1909, 79). In this lecture, he drew an analogy between the human body and a house, with its doors, windows, pumping apparatus, pipes, and telegraph wires, in order to teach elementary facts of physiology and hygiene as well as demonstrate how animal bodies were examples of divine handiwork (Bompas 1909, 80; Upton [1910], 85).[7] In 1858, he repeated this lecture at the Mechanics' Institutes of Abington, Newbury, and Wantage (Bompas 1909, 90; see also figure 4.1.) In that same year, he gave his first lecture at the South Kensington Museum on "Horn, Hair and Bristles," one of a series of six lectures addressed to workingmen (Bompas 1909, 88). The other speakers included Richard Owen, Lyon Playfair, and Thomas Huxley. A year later, he was lecturing at Windsor and Burnham to an audience of "working men, their wives and babies" (as quoted in Bompas 1909, 95).

For this audience, Buckland developed a light, entertaining style with plenty of humor. Even though Buckland's message was heavily inflected with natural theology themes, he, like Ball, avoided the sermonizing style. Buckland did not feel comfortable until he was able to draw laughter from the crowd. According to Bompas, who was his brother-in-law, he used to say, "I can't get on . . . until I make them laugh; then we are all right" (as quoted in Bompas 1909, 79). Buckland was known to have a collection of comical stories on hand for every speaking engagement (Walpole 1881, 306). A sample from a report on a lecture in 1863 was presented to the audience as a conundrum of natural history: "It is known to many of our readers that the Platipus is a something between a bird and an animal, but more of the latter; the witty speaker therefore asked why is the Platipus like a Tailor? Because it is a BEAST with a BILL" ("St. John's School" 1863). According to reports in contemporary periodicals, Buckland's audiences appreciated his sense of humor. Reporting on a lecture on "Fish and Oyster Culture" given at the Mechanics' Hall in Nottingham in 1865, the *Nottingham and Counties Daily Express* referred to his "racy, amusing style" as being "much enjoyed by the audience" ("Fish and Oyster Culture" 1865, 8).

Accounts of Buckland's lectures in the periodical press often noted that his specimens, used for illustrative purposes, left a marked impression on his audience. Buckland drew on his own private collections (he frequented auctions and kept a large menagerie of live exotic animals in his home) as well as specimens that he had obtained from friends. In 1859, Buckland was

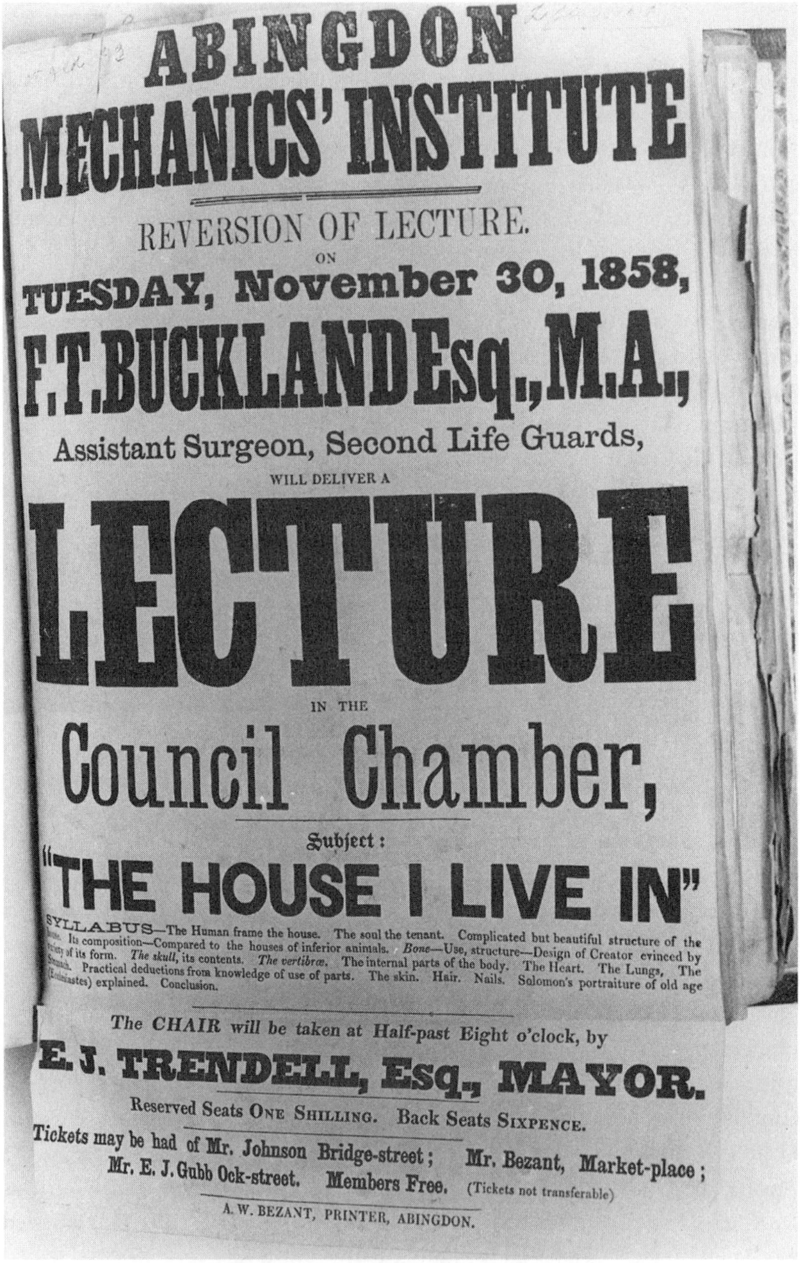

Figure 4.1. Poster for Buckland's lecture on "The House I Live In." Reprinted by kind permission of the president and council of the Royal College of Surgeons of England from *Frank Buckland, Records of My Life* [Commonplace Book], vol. 1, p. 159, Library, Royal College of Surgeons, London.

delivering his lecture on "The House I Live In" at Windsor to the Working Men's Association. He emphasized how the human body was like a movable or walking house: eyes are like windows, mouths resemble a door, and heads are comparable to a cupola. Just as a house is designed, so is the human body. The reporter from the *Windsor and Eton Express* was struck by how "the action and object of the skull, brain, teeth, lungs, stomach, hair, and skin, and all the more prominent and important members of the body were clearly explained and illustrated by the most curious and interesting specimens" ("Working Men's Association Lecture" 1859). Among Buckland's specimens were a New Zealander's tattooed head, a large shell from the China Seas, a rat with huge teeth, the vertebrae of a boa constrictor, the thighbone of a lion, and a monkey's skeleton (Burgess 1968, 73). After giving his presentation, Buckland wrote, "I lectured pretty well, but, as usual, had too many things to show" (as quoted in Bompas 1909, 95). Two years later Buckland was back in Windsor, when he spoke on "Curiosities of Natural History" in the Town Hall. The lecture was illustrated, in the words of the *John Bull* reporter, "with an exceedingly large and valuable collection of specimens, drawings, diagrams, and views." Among the drawings were life-size colored pictures of the great gorilla, magnified sketches of human hair and skin, and depictions of extinct British animals. "The specimens, however," *John Bull* declared, "were the most remarkable; they comprised exquisitely stuffed heads of the lion and bison, with parts of their articulation and structure, the eland, walrus, hippopotamus, elephant's skull and bones, rhinoceros, wild boar, beaver, polar bear, giraffe; skins of the bison, black, white and grisly bears, the platypus, hyena, gavial, and many others too numerous to mention" (*John Bull* 1861). Buckland's vast collection of specimens must have turned the lecture hall into a museum-like exhibition of stuffed animal parts, skulls, bones, and skins. Unlike other science lecturers, Buckland did not make extensive use of the magic lantern. By 1864 he was using the oxyhydrogen microscope when he lectured on "Fish Hatching," but he continued to use diagrams and specimens ("Reading Room and Library" [1864]).

At the Windsor and Eton Literary, Scientific, and Mechanics' Institute in 1861, Buckland reportedly brought with him "notes on the subject, but he preferred to lecture upon the objects and specimens before him; as if they saw a thing they could hang on other facts to it, and arrive at a great result" ("Mr. Buckland on Natural History" 1861). Buckland's rejection of the prepared script, his use of anecdotes (a common convention in written natural history) and humor, and his reliance on his specimens to move the lecture forward gave his public presentations a casual quality that allowed him to establish a strong rapport with his audience. One reporter referred to his lecture at St. John's School in Hammersmith in 1863 as "an illustrated conversation from Natural History, which he kept up for more than an hour without the slightest interval, excepting the necessary repose consequent upon cheers and laughter, produced by pointed anecdotes." The reporter was aware that the conventional term to describe Buckland's speech was *lecture*. "We have adopted the term conversation in preference to lecture," he asserted, "from the peculiarly sincere and friendly manner in which Mr. Buckland addressed his audience." The "conversation" was an illustrated one due to Buckland's use of an artist, who drew freehand sketches as Wood did, in addition to all of his other visual aids. "The word illustrated must be taken in its fullest sense," the reporter insisted, for "a great portion of the walls of the building [were] covered with diagrams and pictures, artistically representing microscopic sections of birds, animals, and fish, and [there was] a collection of preparations, which almost made the Lecture Table groan beneath its weight." This time Buckland had brought with him poisoned arrows from Central Africa, a whale harpoon used by South Sea whalers, elephant and giraffe tails, heads of poisonous snakes, the skin of a sixteen-foot-long African boa, lion skulls, and the shoes of a seven-foot-tall French giant ("St. John's School" 1863).

Due to his adoption of a culture of display coupled with a racy speaking style, Buckland was often criticized for delivering superficial lectures and writing ill-informed books. In his obituary in *Nature*, it was asserted that "he was in no sense of the word a profound naturalist; he could seize with alacrity the popular side of a scientific question, but he seldom went deeper" ("Frank Buckland" 1880, 175). But Buckland was a well-educated Oxford man and the son of an eminent geologist who had taught him extensively about natural history. He self-consciously designed his lectures and writings for a popular audience. When he lectured to a more educated audience, he was quite capable of adopting a different style. Buckland's later interest in fish culture brought him a few invitations to venues that catered to more-genteel audiences. On April 17, 1863, he lectured on fish culture at the Royal Institution, with the Duke of Northumberland in the chair and an audience that included Roderick Murchison, Michael Faraday, John Tyndall, and Edward Frankland (Burgess 1968, 99). Later published in book form as *Fish Hatching*, this work bears little resemblance to his *Curiosities of Natural History*, with its humorous, first person, anecdotal style. Instead, Buckland presents a sustained argument for the public utility of studying the breeding of fish scientifically. Buckland altered his lecturing style for such a distinguished audience. When he lectured on fish culture in other venues, such as the Mechanics' Hall in Nottingham, he continued to adopt a "racing, amusing style" and to illustrate his talk "by many diagrams and specimens of natural history" ("Fish and Oyster Culture" 1865, 8). One contemporary asserted that Buckland was at ease with both working-class and with more-educated audiences, though less successful with the latter. "If he had been a politician," the *Popular Science Monthly* obituary proclaimed, "he would have been a greater mob orator than Parliamentary debater" (Walpole 1881, 306).

Two key sources for the culture of display embodied in Buckland's extensive use of specimens in his lectures may have been the zoo and the freak show. Buckland frequented the Zoological Gardens in London and wrote extensively about zoo animals in the third series of his *Curiosities of Natural History*. Among others, articles covered giraffes, hippopotamuses, and lions at the Zoological Gardens. Human freaks and exotics also fascinated Buckland. According to Burgess, Buckland could never resist a sideshow. Buckland wrote numerous articles on freaks and carnival acts, many of them in the fourth series of *Curiosities of Natural History*, including such topics as giants, bearded ladies, fire-eaters, sword-swallowers, human cannonballs, and mermaids. As a result, promoters of unusual shows hoping for publicity often invited him to special sessions (Burgess 1968, 181–82). Buckland's interest led him to become friendly with Joseph Brice, the over eight-foot-tall French giant, whom he first met in 1862, Miss Swan the giantess, and the two-headed Mademoiselle Millie-Christine. He was known to throw parties for these friends "with the Chinamen, Aztecs, Esquimaux, Zooloos, Siamese twins, tattooed New-Zealanders, and whatever queer specimens of mankind happened to be on exhibition at the time, as fellow-guests" ("Sketch of Frank Buckland" 1885–86, 406; Upton [1910], 108; Burgess 1968, 110, 185). From the freak show and the zoo, Buckland drew on the display of the exotic in order to entertain, particularly when lecturing to his working-class audiences.

When Buckland wished to provide a genteel audience with a more learned, educational experience, or when he wished to edify a lower-class audience, he could draw upon the surgical tradition of John Hunter and its emphasis on specimens and hands-on experience. Buckland's identification with Hunter was not unusual. Jacyna has examined the various ways in which nineteenth-century British biomedical investigators could "consecrate a particular cause" if they presented themselves as belonging to the Hunterian tradition (Jacyna 1983, 102–3). Buckland would have been exposed to the Hunterian tradition through his medical training, his father, and his friendship with Richard Owen. From 1756 until his death in 1793, Hunter had been surgeon to St. George's Hospital, where the elder Buckland had been a medical student. The younger Buckland had been taught by his father to regard Hunter's memory with enormous respect (Burgess 1968, 42–43). In the preface to the first edition of his *Curiosities of Natural History*, Buckland referred to Hunter as a "great man" and considered himself one of his "followers" in stressing the "necessity of studying comparative as well as human anatomy" (Buckland 1860, vi).[8] Richard Owen, a correspondent, adviser, and one of his father's close friends, had had a long association with the Hunterian Museum at the Royal College of Surgeons, having been appointed assistant curator in 1827 and then curator in 1842, resigning in 1856 to take up a post at the British Museum. As the most eminent anatomist in Britain, Owen was widely regarded as having inherited Hunter's mantle. Yanni asserts that under Owen, the Hunterian was widely considered to be a "landmark in the display of natural history specimens" (Yanni 1999, 49). Through Owen, Buckland would have grasped the connection between the Hunterian emphasis on close observation and hands-on experience and the specimen as part of a museum exhibit (Rupke 1994, 297).

After Buckland became interested in fish culture, he set up a small fish hatching apparatus at the South Kensington Museum in 1863. Adding plaster casts of fish and other specimens over the years, the hatching apparatus evolved into a museum, which moved several times before being established in its final location at South Kensington in 1872 (Burgess 1968, opposite 116). The Museum of Economic Fish Culture was meant to educate the public on the natural history of fish, their commercial uses, and the development of English fisheries (figure 4.2). But it was not designed to impart information about systematics or anatomy. The specimens were arranged haphazardly. The visitor was meant to be entertained, not just instructed. Casts of fish were mounted on the wall. Various specimens were put on platforms on the floor. Small fish swam up a model salmon ladder. The museum attracted visitors of all kinds, including members of the royal family. In 1876, Queen Victoria herself visited it (Burgess 1968, 125; Bompas 1909, 279). According to Bompas, Buckland lectured there in the early 1870s (Bompas 1909, 257). It was fitting that a public speaker who had tried to construct ephemeral sites for science all over England by drawing upon the culture of display should go on to build his own museum and lecture in it.

Figure 4.2. Photograph of Buckland in his Fish Museum at South Kensington. Reprinted by kind permission of the president and council of the Royal College of Surgeons of England from *Frank Buckland, Records of My Life* [Commonplace Book], vol. 2, p. 282, Library, Royal College of Surgeons, London.

Pepper, Theater, and the Polytechnic

During the 1860s, a humorous song about John Henry Pepper and the Royal Polytechnic Institution began to circulate. It was titled "Laughing Gas or A Night at the Polytechnic." The song tells a story about a young man from rural England, Humphrey Brown, who has come to London to visit all of the popular London entertainments, including Pepper's own establishment. Exhausted by his busy day, Brown falls asleep in the Polytechnic lecture room and awakens to find that he is locked in for the night. Afraid of meeting one of Pepper's famous ghost illusions, he shores up his courage by swallowing the contents of a container marked "Improved Laughing Gas." He begins to feel light headed, as if he were drunk, and laughs uncontrollably. Then he explores the Polytechnic in this peculiar condition, playing with all of the exhibits (Nash [1860?]). The composer and singer of the song, John Nash (1830–1901), was a well-known music-hall artist and comedian (figure 4.3). An iron smelter as a youth, he was billed as "The

Figure 4.3. Illustration of John Nash, frontispiece, John Nash, *Laughing Gas or a Night at the Polytechnic*. Reprinted from Nash [1860?], courtesy of the Wellcome Library, London.

Laughing Blacksmith" when he became an entertainer in the Midlands. He performed at the South London Music Hall in 1860 and then at the Oxford Music Hall the following year. He toured the United States in 1874 and in 1876 and later formed his own touring company that he took across the Atlantic in 1886. He was the first music-hall artist to perform at royal command. In the *Cambridge Guide to Theatre*, Nash is described as "a specialist in silly walks" and in the *Oxford Companion to Popular Music* as one who "pioneered the laughing song" (Gammond 1991, 415–16; Banham 1995, 777–78). Nash liked to refer to himself as "Jolly John Nash," and to prove he deserved the nickname, he composed and published a song in the 1890s titled "I'm such a jolly Man."

Nash's selection of Pepper and the Polytechnic as an appropriate topic for a music-hall song is indicative of their popularity in the 1860s.[9] Though Altick, in his *Shows of London*, treats the Polytechnic Institution as an integral part of the London entertainment scene, he criticizes the inclusion of more-popular entertainment in the programs of the galleries of practical science. In the case of the Adelaide Gallery, it failed because it "became indistinguishable from the miscellaneous shows not far away in Leicester Square." Similarly, the Polytechnic, Altick claims, incorporated popular performances from the London amusement circuit in reaction to public pressure to entertain, and this action took it away from its "announced serious purpose" (Altick 1978, 382, 386). Pepper would have disagreed. As manager of the Polytechnic, he chose to include more-popular entertainment features and did not perceive them to be inconsistent with the goal of scientific instruction. Moreover, Pepper would have looked at Nash's song as evidence of the expanding force of science into popular culture and as proof of the success of his strategy to link instruction more closely to entertainment. Pepper was particularly interested in capitalizing on the theater as a vehicle for reaching out to general audiences, both in his lectures and in the Polytechnic. In his position as manager of the Polytechnic, Pepper had the opportunity to refashion an influential institution of science that could serve as a friendly home for his theatrical lecturing activities. Later in this volume, Morus discusses the Polytechnic and Pepper's role within this institution. Whereas Morus has focused on Pepper's use of technological display in his ghost illusion in this volume, I will discuss his lecturing in general and how it combined entertainment and instruction.

Born on June 17, 1821, in Westminster, to Charles Bailey Pepper, a civil engineer, John Henry Pepper (figure 4.4) was educated at King's College School and then later studied analytic chemistry at the Russell Institution with J. T. Cooper. In 1840, Pepper was appointed assistant chemical lecturer at a private school of medicine run by R. D. Grainger. He was hired at the Polytechnic as lecturer and analytic chemist in 1848 and then as manager in 1854. He continued in this role, with some short interruptions, until 1872, when he resigned for good after a quarrel with the board of directors over the extent of his autonomy. During his time at the Polytechnic, he published five science books for the public, including *The Boy's Playbook of*

Photo. Maull and Fox.
THE LATE PROFESSOR PEPPER.

Figure 4.4. Illustration of John Henry Pepper. Reprinted from "Personal,"
Illustrated London News 116 (April 14, 1900), 503.

Science in 1860, and established himself as one of the premier showmen of
science.[10] He tried to re-create his successful form of science entertainment
at the Egyptian Hall in Piccadilly but lost money on the venture, and he
went on tour in the United States, Canada, and Australia from 1874 to
1881. He accepted the post of public analyst in Brisbane, Australia, in 1881,
stayed there until 1889, and then returned to England, where he remained
until his death in 1900 (Boase 1965, 386–87; Cane 1974–75, 116–28; Secord
2002, 1648–49; Brock 2004, 1572–73).

Pepper had a reputation for being a lively speaker. Up until the early
1860s, he lectured on various topics, mostly in the physical sciences (figure 4.5). In 1850, he was lecturing on the chemistry of hydrogen, in 1851

Figure 4.5. Illustration of Pepper lecturing at the Royal Polytechnic Institute on spectrum analysis. Reprinted from J. H. Pepper, *Cyclopaedic Science Simplified* (London: Frederick Warne, 1869), frontispiece.

on the chemistry of the minerals and crystals at the Great Exhibition, in 1854 on the Crimea and "Munitions of War," and in 1857 on "A Scuttle of Coals" (Press Cuttings n.d.; "Royal Polytechnic Institution" 1854b, 335). Audiences found his lectures vastly entertaining. The *Illustrated London News* reported on a lecture given by Pepper in 1854 to "a large audience" on the chemistry of nonmetallic elements "in his usual popular and interesting style" ("Royal Polytechnic Institute" 1854, 179). Even the *Fun* reporter, who remarked that Pepper "goes into ecstacies a little too often," judged that his lecture was "well worth hearing" (Our Special Sightseer 1870, 223). According to Layton, Pepper's lectures deserve an important place in the history of science education, and they had an impact on such important scientists as Sir Henry Roscoe, Sir Ambrose Fleming, and H. E. Armstrong (Layton 1977, 538).

The Royal Polytechnic Institution, founded in 1838, was an unusual institution even before Pepper arrived on the scene (Morus 1998, 82). Though it served multiple purposes like other institutions of science—the Royal Institution, for example, contained both a laboratory and a lecture hall—the Polytechnic was even more diverse. It was equipped with industrial tools and machines, a laboratory, a lecture theater, and a large display room, known as the Great Hall, where the main exhibits were housed (figure 4.6). Among the main exhibits were the diving bell and diver, an oxyhydrogen microscope, large electrical machines, and model boats floating in a long canal. The diving bell, one of the feature attractions, provided a unique experience for visitors. Four to five persons could fit inside while it was submerged. In light of the wide range of activities, what, exactly, was the Polytechnic? Was it a museum, or a laboratory, or a lecture theater, or an exhibition hall, or even an amusement park? A contemporary guide to London placed it under the heading of "Miscellaneous Exhibitions," along with the Colosseum, Egyptian Hall, the Royal Panopticon of Science and Art, Wyld's Model of the Globe, Madame Tussaud and Son's Exhibition, and exhibits of dioramas and panoramas, rather than with the Royal Institution under "Educational and Scientific Institutions" (*London as it is to-Day* 1850, 268). As an indication of the Polytechnic's institutional novelty, Gerard Turner referred to it as "the first science center" (Turner 1987, 397). Even before Pepper took over as manager, the Polytechnic served multiple purposes and drew together in one hybrid location multiple sites of science open to the public. But Pepper transformed the Polytechnic into a hybrid site in a second sense by adding cultural spaces associated with the shows of London to the mix. When Pepper took over the reins of the Polytechnic in 1854, the big question for the manager of this institution was how to at-

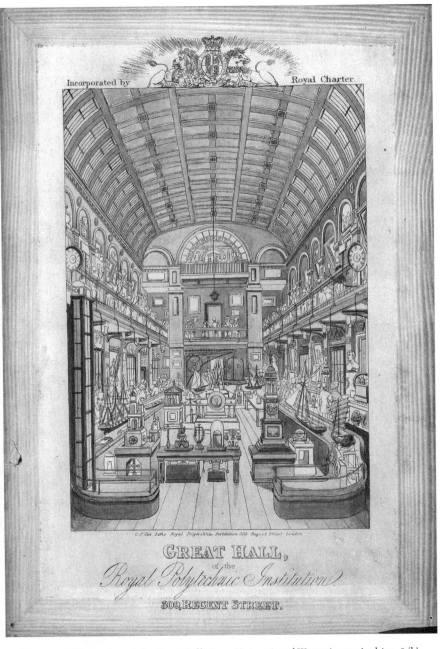

Figure 4.6. Illustration of the Great Hall. From University of Westminster Archive, 8 (b), n.d., courtesy of University of Westminster Archive Services.

tract customers whose expectations had been raised by their experiences exploring the Crystal Palace on shilling days. The Polytechnic's offerings must have seemed meager in comparison (Altick 1978, 472–73).

In 1850, *Household Words* recommended that the Polytechnic offer the public more than exhibitions of industrial machinery. "There is a range of imagination in most of us," the anonymous writer stated, "which no amount of steam-engines will satisfy." Though the Polytechnic was a "wonderful place," the author was of the opinion that "a people formed *entirely* in their hours of leisure by Polytechnic Institutions would be an uncomfortable community." Since it is probable "that nothing will ever root out from among the common people an innate love they have for dramatic entertainment in some form or other," the Polytechnic, and institutions like it, had to offer some type of dramatic amusements ("Amusements of the People" 1850, 13). Whether Pepper ever read the *Household Words* article is not clear, but shortly after he became director of the Polytechnic, he began experimenting with theatrical entertainments. In 1854, the *Athenaeum* announced that "GOOD DRAMATIC READINGS are now added to the other attractions of the Institution" ("Royal Polytechnic Institution" 1854a, 1306). Pepper stuck mainly to Shakespeare, introducing dramatic readings from *The Merchant of Venice, Hamlet,* and *Romeo and Juliet.* It was but a short step to experiment with scenes of plays and then entire plays. A jokester in *Punch* noted in 1854 that "the proprietors of the Polytechnic and Panopticon are about to introduce dramatic readings and singsongs as part of their attractions" and claimed to see no reason why plays shouldn't be used to further scientific education. But then the journalist poked fun at Pepper's innovation by offering some light-hearted suggestions, such as the creation of new plays set in the laboratory where "scenes of thrilling interest might easily be got up with the voltaic battery" ("Philosophical Drama" 1854).

In the early 1860s, Pepper increasingly exploited the relationship between the Polytechnic and the London entertainment scene (Secord 2002, 1648). Theater and, to some extent, music were the forms of popular entertainment that best suited his mix of instruction and entertainment (Brooker 2005). Panoramic and dioramic spectacle had begun to be widely used in some forms of dramatic production in London in the 1820s, contributing to the popularity of theater (Mayer 1969, 69–70; Meisel 1983, 33, 62, 380–84). Allen argues that the Victorians' insatiable appetite for novels was matched only by their voracious hunger for theatrical entertainment in all its forms, including Shakespeare, melodrama, pantomime, music hall, freak show, dancing dogs, and pyrodrama (Allen 2003, 5). Victorians flocked to the London theaters in the 1850s when Pepper first became manager of the Polytechnic, attracted by the Shakespearean revivals by Charles Kean and Samuel Phelps. Not only did they offer magnificent spectacles, which aimed to outdo the panoramas that had inspired them, but they also sought to make theater more respectable in order to draw in a higher class of clientele (Allen 2003, 20–21; Altick 1978, 473; Booth 1981, 2–3). Pepper would have realized that both the theater and public science were forms of mass-cultural entertainment, reliant on the steadily increasing number of consumers with money and leisure (Allen 2003, 6). The popularity of theater may have made it more difficult for the Polytechnic to compete in the London entertainment scene, as Altick asserts, but it also provided Pepper with a model of a successful public cultural form that inspired his redefinition of the practice of public science at the Polytechnic, including the nature of his lectures (Altick 1978, 473). Pepper's self-fashioning as a lecturer and his institution fashioning of the Polytechnic as an appropriate scene for his lecturing activities were intimately connected.

Pepper took the theatrics of science to a whole new level when he came across a new scientific principle for generating surprisingly realistic optical illusions suggested by the inventor Henry Dircks. Pepper had already been lecturing on "Optical Illusions" in 1856 and on "Remarkable Optical Illusions" in 1857 and saw this as an area that could attract a substantial audience ("Royal Polytechnic" 1856, 1612; "Royal Polytechnic Institution" 1857, 35). But just before Christmas Day in 1862, Dircks's invention, vastly improved by Pepper, was used to produce a ghost illusion that stunned a small audience of scientific friends and members of the press previewing a performance of Edward Bulwer-Lytton's "A Strange Story" at the Polytechnic. Instead of explaining how the illusion worked, as he had intended, the following day he hurriedly took out a provisional patent, sensing its almost unlimited potential (Pepper 1890, 3). Pepper prepared a companion lecture for the play, "A Strange Lecture," where he explained the wonders produced by the "Photodrome," an optical apparatus that caused phantoms to appear at will ("Polytechnic Institution" 1863a, 19). At some point, Pepper began to tell a story in his "Strange Lecture," about a student who sees the apparition of a skeleton late at night and whose sword swings right through it (Pepper 1890, 29). By February, Pepper had introduced a new lecture, "Burning to Death, and Saving from Death," followed by the still popular ghost scenes from the "Strange Lecture." The "Spectre Drama" (figure 4.7) was playing in the morning and the evening, except on Tuesdays and Wednesdays ("Polytechnic" 1863a, 218). By Easter, the play had become so popular that it was moved into the larger theater of the Polytechnic where the dissolving views were usually exhibited. "Special

Figure 4.7. Illustration of "The Spectre Drama at the Polytechnic Institution."
Reprinted from "Spectre Drama at the Polytechnic Institution" 1863.

written permission" was obtained from Dickens to mount a production of
his "The Haunted Man" as a vehicle for exhibiting the ghost illusion, and
it ran for fifteen months (Pepper 1890, 12). In the years that followed, more
plays were mounted featuring the ghost illusion, and they became a regular
part of the Polytechnic's program. In December 1864, two new spectral tab-
leaux were announced, entitled "The Indian Widow's Suttee" and "Snow
White and Rosy Red" ("Royal Polytechnic" 1864, 666). In 1865, the *Times*
announced that "Mr. Pepper's ghost is put to new uses in a dramatic entertainment, devised by Mr. Pepper himself and entitled the 'Poor Author
Tested'" ("Polytechnic Institution" 1865a, 12).

Pepper also continued to lecture on ghosts and optical illusions and
to mount new exhibits. On May 20, 1863, he delivered his ghost lecture
to the queen and the royal family ("Polytechnic" 1863c, 9). The following
month he was giving his ghost lecture twice a day ("Polytechnic" 1863b,
610). To keep the interest of his audiences, Pepper began to modify the
illusion in order to present a variety of startling effects. In 1865, visitors
to the Polytechnic could see the disembodied head of Socrates deliver a
rhymed speech and Sir Joshua Reynolds's cherubs, or at least their heads,
singing a choral song ("Polytechnic Institution" 1865b, 10). The following

year, the *Art-Journal* pointed out that Pepper, the Polytechnic, and optical illusions were inextricably connected in the public's mind. According to the *Art-Journal*, "Mr. Pepper is true to himself and to the optical phenomena which he has associated as well with his own name as with the institution that is identified with him; and so, . . . he passes . . . to fresh applications of the optical illusive impersonations that now are expected to be displayed at the Polytechnic." To the disembodied head of Socrates he had added Shakespearean creations, including the floating and speaking heads of Hamlet and Lear ("Polytechnic Institution" 1866a, 256). As a result of Pepper's innovations, the public never seemed to tire of his optical illusions.[11] In 1866, an advertisement in the *Illustrated London News* claimed that 109,000 visitors had already seen them ("Royal Polytechnic" 1866, 511). Pepper's ghost had become such a fixture in the Polytechnic that a notice in November 1870 listing the attractions at the Polytechnic announced "the explanation of the Ghost as usual" ("Royal Polytechnic" 1870, 538).

Pepper's theatrical use of the ghost in his lectures and the dramatic productions at the Polytechnic was, to him, completely in line with the scientific aims of his institution, and it was seen by the press in this light as well. The *Illustrated London News* presented an engraving of the strange optical effects in May 1863, shortly after their first appearance, "as produced on their original stage—the boards of the Polytechnic for purposes purely scientific" ("Spectre Drama at the Polytechnic Institution" 1863, 486). The *Times* of London recognized that illusions were used by Pepper to educate the public on the principles of optics. "Optics," the *Times* declared in 1866, "still predominate at the Polytechnic, reflection and refraction." Whereas the principle of reflection was illustrated in Pepper's lecture on Brewster's discoveries, in the appearance of the disembodied heads of Hamlet and Macbeth, and in the exhibition of the floating cherubs of Sir Joshua Reynolds, refraction was demonstrated in a series of dissolving views of the "Lady of the Lake" ("Polytechnic Institution" 1866b, 10). Entertaining and spectacular, his lectures were nevertheless intended to be instructive.

In lectures delivered in the late 1860s, Pepper also drew on his optical illusions to expose the fraudulence of spiritualists. The *Times* reported on a series of lectures by Pepper in 1867 and 1868 that dealt with spiritualism. On December 23, 1867, Pepper lectured on the frauds who claimed to produce spiritual manifestations ("Polytechnic Institution" 1867, 6). On March 9, 1868, he presented another lecture on spiritual manifestations, arguing that, in the words of the reporter, "effects vulgarly supposed to indicate supernatural causes are admirably produced by natural means"

("Polytechnic Institution" 1868a, 7). A *Times* article on his lecture on April 14, 1868, reported that he "denounced the utter absurdity of table-turning and spirit-rapping. He endeavoured to show that on his platform, fitted up as an ordinary drawing-room, he could exhibit the most striking of the spiritual manifestations with which Mr. Home and others a few seasons ago startled the London public." Pepper then mesmerized a female medium and seemingly levitated her, chairs, and tables ("Polytechnic Institution" 1868b, 9). Fearing that the spiritualists were seducing uninformed audiences, Pepper was willing to be as dramatic as his opponents.

Pepper's increased emphasis on entertainment in his lectures and in the Polytechnic in general transformed it into one of the leading London entertainment venues in the 1860s.[12] One periodical described the Christmas program at the Polytechnic for 1862, with particular reference to Pepper's optical illusions, as "unusually varied and interesting," so much so that the institution was now "in a position it has not enjoyed for several years at least—namely, in a highly prosperous and paying position" ("Polytechnic Institution" 1863b). The *Illustrated London News* judged in 1865 that the lineup of attractions offered during the Christmas season, including Pepper's ghost, placed the Polytechnic at the top of the list of best holiday amusements. There is "nothing similar in the metropolis that can compete," the newspaper announced ("Christmas Amusements" 1865, 19). However Pepper allowed the ghost to haunt other London theaters. The Haymarket, the Britannia, the Adelphi, and Drury Lane were among those theaters that took out licenses to use the illusion (Pepper 1890, 30). Pepper's ghost became so omnipresent in London at one point that *Punch* complained in October 1863 that "there is now a glut of ghosts everywhere" ("Ghosts without Spirit" 1863, 146). The ghost and some of Pepper's other illusions were later shown outside London, appearing at Leatherhead (just south of London) and Norwich (Magic and Mystery Box 2 n.d.). The adventures of young Humphrey Brown in "Jolly" Jack Nash's ditty, "Laughing Gas," is another reminder of Pepper's success. Nash mentions many of the feature attractions in his song, and the humor assumes that his audience is familiar with them. Under the influence of the laughing gas he has swallowed, Nash's country bumpkin wanders around the Polytechnic as if in a dream. "He saw Cherubs floating about in the air," put on the diver's suit and "made love to the diving Bell," turned on the "dissolving views," and, after bumping into the induction coil, received the shock of his life (Nash [1860?], 5). Pepper's incorporation of the theatrical into his lectures and the general program of the Polytechnic led to a remolding of this crucial site of public science. A self-fashioned theatrical lecturer par excellence, Pepper simultaneously embarked on an ambitious project of institution fashioning. It not only changed the terms on which science could be defined as "popular," it also gave science a prominent place in the London entertainment scene. In the process, Pepper created a new public space for science that, unlike the Royal Institution, was open to an audience composed of a more diverse social composition.

Pepper's bid to reenvision the spaces of public science by incorporating more theater led to a debate in the periodical press about whether the Polytechnic had gone too far in the direction of entertainment. This debate went to the heart of the questions: What kind of science should be provided for the general public? What was the appropriate mixture of entertainment and instruction in such science? Pepper's approach had led to a definition of "popular" science that was not universally accepted. Some periodicals praised the Polytechnic's measured balance of entertainment and instruction both before and during the time when Pepper was in control. In 1844, the *Pictorial Times* declared that "in no exhibition in London are amusement and instruction so thoroughly combined" ("Easter Monday and Its Amusements" 1844, 233). Later, similar expressions of satisfaction were expressed. One journal stated in 1862 that "at the Polytechnic Institution, science and fun will hold joint festival," another asserted that "the Polytechnic makes science amusing, and amusement it causes to become scientific," while a third affirmed that in combining "no little instruction with a good deal of amusement" the Polytechnic's attractions "are just suited therefore to the taste of that large section of sightseers who care little for theatre or wish to vary their amusements" (Press Cuttings n.d.; "Polytechnic Institution" 1866a, 256; "Polytechnic" 1869, 10). In 1880, eight years after Pepper had resigned from the Polytechnic for good, *Punch* compared his accomplishments with those of the new directors, who "did not seek to disguise the fact that in their opinion chemistry had been unduly sacrificed to comic entertainments, and that mechanical engineering had been altogether put on one side to make room for 'ghosts' and optical illusions." Not only did the reporter find that entertainment remained on the program, in acts such as the singing Adèson Family and a ventriloquist, but their unimpressive performances led him to recall "the past glories of the old place" ("Round About Town" 1880, 133). The ideal goal for public science institutions was to strive toward a perfect mix between instruction and entertainment, and for these periodicals, the Polytechnic under Pepper's regime had provided a model of how to achieve the proper balance.

Others agreed on the ideal but did not see the Polytechnic as having embodied it. For some, the Polytechnic needed more instruction and less entertainment. A visiting *Punch* reporter observed the confusion of two Frenchmen at the Polytechnic in 1865 when they viewed the entertainment. Under the impression that the Polytechnic "was The Literary and Scientific Institution of England," the two foreigners mistook "the comic dialogue as the lecture of some learned profession" and wondered if they should have come at all ("A Wonderful Shillingsworth" 1865, 236). When Pepper briefly retired from his position as manager of the Polytechnic in 1858, the *Illustrated London News* hoped that his successor would "not permit the desire of gain to pervert it to the more ordinary ends of mere entertainment," adding that lately "there was a tendency this way, as if amusement were about to supersede instruction" ("Royal Polytechnic Institution" 1858, 631). A few months later, with Pepper still out of the picture, the *Illustrated London News* praised the new management for banishing "everything which is not in some way connected with the purposes for which it was originally designed, and substituted in the room of exhibitions fit only for a place of mere amusement lectures on all subjects connected with popular science and natural philosophy" ("Polytechnic Institution" 1858, 241).

But where some were critical of the Polytechnic for its seeming emphasis on entertainment, others saw it as presenting too much instruction. The *Illustrated London News* had just three months earlier encouraged the new management to reduce the amount of sugarcoating needed to induce visitors to swallow the bitter pill of instruction. During the Christmas season, it pleaded for more sugar as "just now, we don't want the pill at all" ("Christmas Holidays" 1858, 608). In 1869, a reporter from *All the Year Round* recalled visiting the Polytechnic as a boy and suspected that he had been lured there under false pretenses. "There was an indefinable feeling," he remembered, "as if it were not a real, out-and-out, holiday place: as if our education were in some way going on whenever we were there. Instruction, we felt, lurked behind amusement, and it was impossible to forecast from the programme of the entertainments, exactly at what point the baleful genius of mental improvement might be expected to claim its victim." Whatever the reactions of boys of the past, the reporter thought that the boys of 1869 might find the Polytechnic somewhat dull ("Playing with Lightning" 1869, 617). The combination of instruction and entertainment in public science proved during this period to be volatile and unstable. At what point did science lecturers like Pepper go too far in incorporating entertainment? Was there a point at which they ceased to present popular science and offered mere amusement?

Refashioned Sites, Changing Experiences, and Scientific Spaces

In the post–Crystal Palace era, science lecturing was a competitive business. Wood barely eked out a living as a lecturer and scientific author, even though the popularity of his sketch lectures led to extensive tours across England and to the United States. Not only were lecturers competing with one another to draw audiences, they were also vying with the theater, the panorama, the exhibition, museums, and other forms of popular entertainment. In order to compete, both Pepper and Buckland refashioned the sites at which they lectured by bringing in elements drawn from cultural spaces associated with the world of entertainment. When Buckland went to speak at various sites around England, he transformed them by incorporating features of the freak show, the zoo, and the museum in his lectures. Pepper's Polytechnic was a new kind of hybrid scientific institution, which included elements of the exhibition hall, museum, laboratory, and lecture hall all under one roof. But Pepper refashioned this already existing site through his use of the theatrical. In the process of refashioning these sites, Pepper and Buckland also reformulated scientific lecturing and thereby provided new experiences for their audiences. Pepper's audiences could be terrified by the appearance of his ghost and then calmed and edified by his scientific explanation for apparent supernatural phenomena. When Buckland lectured, his audience was amused by his racy style, fascinated by exotic specimens, and taught to see design in the scheme of things. Instead of sermonizing, predominantly an aural experience, Pepper, Buckland, Ball, and Wood incorporated a variety of visual elements into their lecturing. More than ever, those who attended lectures experienced science as if it were part of the world of entertainment.

Though the inclusion of more entertainment in lecturing raised questions about the validity of "popular" science, it also allowed Buckland, Pepper, and other lecturers to bring science into new spaces and thereby into the center of Victorian culture. Alberti has remarked that the inclusion of lectures and displays relating to science side by side with musical performances and other entertainments in a typical conversazione program reflected "both the heterogeneity of Victorian institutional culture, and the increasing prominence of natural knowledge within it" (Alberti 2003, 215). One of the hallmarks of science from about 1830 was the proliferation of its sites. More and more science periodicals began to appear, accompanied by a huge explosion of books, periodical literature, museums, exhibitions, and many other sites that are explored in this book. In older stories about the formation of the worship of science from about 1850 to 1890, historians tended to credit elite scientists such as Darwin or Huxley for the tenacious hold that science seemed to have on the hearts and minds of Victorians. But as scientific naturalists began to cultivate the strategy of professionalization, it committed them to privileging select spaces in which to practice legitimate science, such as the laboratory above all else. They were also selective about the sites in which they would communicate the results of their research and their views on the broader implications of scientific discoveries, whether it be in the *Nineteenth Century, Nature,* or other respectable periodicals, in elite scientific institutions such as the Royal Institution or the annual meeting of the British Association for the Advancement of Science, or on carefully organized lecture tours abroad. By pursuing professionalization as a route to reforming science, scientific naturalists left huge cultural spaces open to lecturers such as Pepper and Buckland. They and individuals like them set out to fill up all of the cultural nooks and crannies they could find with science and to expand the extent and nature of the diverse sites of science. As a result, they altered the spatial economy of science.

NOTES

I am indebted to James Secord, whose work on John Henry Pepper gave me the basic idea for this piece. He also identified some of the illustrations and commented on an early draft of the essay. William Brock, Jeremy Brooker, Mark Butterworth, Jill Howard, Frank James, and Dana Rovang all supplied me with useful insights into various aspects of popular science lecturing. This piece was vastly improved as a result of the attention of Aileen Fyfe's keen editorial eye. Jean Koo obtained copies of the illustrations as well as permissions to reproduce them. I am grateful to Brenda Weeden, Elaine Penn, and the staff at the University of Westminster Archives for their crucial help in locating key sources on the Royal Polytechnic Institute. Quotations from materials held at the University of Westminster Archives are reproduced by courtesy of the University of Westminster Archive Services. Quotations from Frank Buckland's Commonplace Book are reproduced by kind permission of the President and Council of the Royal College of Surgeons of England.

1. The Lowell Institute was established after the death of John Lowell Jr. in 1839 when he left $250,000 to endow a public series of lectures by well-known intellectual figures. The physicist John Tyndall gave the lectures in 1872, and J. G. Wood delivered a series of twelve lectures in 1883 (Sopka 1981, 193, 202; Wood [1890], 189).

2. Some Christian groups were also moving away from traditional forms of worship in the latter half of the century. For a study on how evangelical Protestant groups used the theater as the basis for reenvisioning their worship space, see Kilde 2002.

3. I am using the term *spatial economy of science* as a playful parallel to the nineteenth century notions of economy, and economy of nature, that signified an interlocking, complex system in overall balance. By *spatial economy of science*, I hope to convey the idea of a complicated system of people, actions, and movements occupying diverse spaces and places in the marketplace of science.

4. For a more recent and detailed analysis of Davy's lecturing, see Golinski 1992, 188–285. Ralph O'Connor has pointed out that geologic lecturers such as Gideon Mantell and Hugh Miller still appealed to moral and poetic elevation after 1830 (O'Connor 2003). Hays's assertion that such appeals were no longer in vogue must therefore be qualified.

5. For a discussion of the public lecture as a means of scientific education in this period, see Inkster 1980.

6. There are two biographies of Frank Buckland: Bompas 1909 and Burgess 1968.

7. Buckland may have found the basic theme for his lecture in the American William Alcott's 1834 book *The House I Live In; or, The Structure and Functions of the Human Body*. First published in 1837 in England, Buckland likely encountered this work on physiology as a medical student. He used a similar title for the lecture and drew the same analogies, and he adopted the natural theology framework. In his preface, the editor, Thomas C. Girtin, presented the book more as "an appropriate introduction" to Paley and natural theology than as a physiology handbook ([Alcott] 1837, v]. I am indebted to Aileen Fyfe for pointing this out.

8. Buckland maintained his high regard for Hunter throughout his life. In *Notes and Jottings from Animal Life*, Buckland remarked that as Hunter was the "founder of the system of modern surgery, and the discoverer of many of Nature's sanitary laws, [he] may be justly regarded as one of the greatest benefactors to the human race" (Buckland 1882, 84].

9. Nash's song about the Polytechnic should not be seen as unusual because scientific themes were taken up in the performing arts, including burlesque, throughout the nineteenth century. Jane R. Goodall has explored the appearance of evolutionary themes in the performing arts during Darwin's lifetime (Goodall 2002].

10. For a current reprint edition of this book with a useful introduction by Jim Secord, see Pepper 2003.

11. For a discussion of some of Pepper's other illusions, see Lamb 1976, 43–50.

12. Altick's account of the declining fortunes of the Royal Polytechnic Institution in the 1860s, after an unfortunate accident that claimed the life of a young girl in 1858, is somewhat misleading. According to him, the Polytechnic never recovered despite being rescued temporarily by the introduction of Pepper's ghost (Altick 1978, 388–89). Although the Polytechnic was in a precarious economic position, it nevertheless managed to pull in the crowds up until Pepper's departure in 1872.

REFERENCES

Alberti, Samuel J. M. M. 2003. "Conversaziones and the Experience of Science in Victorian England." *Journal of Victorian Culture* 8:208–30.

[Alcott, William]. 1837. *The House I Live In; or, The Structure and Functions of the Human Body.* Ed. Thomas C. Girtin. London: John W. Parker.

Allen, Emily. 2003. *Theatre Figures: The Production of the Nineteenth-Century British Novel.* Columbus: Ohio State University Press.

Altick, Richard D. 1978. *The Shows of London.* Cambridge, MA: Harvard University Press, Belknap Press.

"Amusements of the People." 1850. *Household Words* 1:13–15.

Ball, W. Valentine, ed. 1915. *Reminiscences and Letters of Sir Robert Ball.* London: Cassell.

Banham, Martin. 1995. *The Cambridge Guide to Theatre.* Cambridge: Cambridge University Press.

Boase, Frederic. 1965. "Pepper, John Henry." In *Modern English Biography, Volume VI L–Z, Supplement to Volume III*, 386–7. London: Frank Cass.

Bompas, George C. 1909. *Life of Frank Buckland.* London: Thomas Nelson and Sons.

Booth, Michael R. 1981. *Victorian Spectacular Theatre 1850–1910.* Boston: Routledge and Kegan Paul.

Brock, W. H. 2004. "Pepper, John Henry." In *Dictionary of Nineteenth-Century British Scientists*, vol. 3, ed. Bernard Lightman, 1572–73. Bristol: Thoemmes Continuum.

Brooker, Jeremy. 2005. "Paganini's Ghost: Musical Resources of the Royal Polytechnical Institution." In *Realms of Light: Uses and Perceptions of the Magic Lantern from the Seventeenth to the Twenty-first Century*, ed. Richard Crangle, Mervyn Heard, and Ine van Dooren, 146–54. London: Magic Lantern Society.

Buckland, Francis T. 1860. *Curiosities of Natural History.* 4th London ed. New York: Rudd and Carleton.

———. 1882. *Notes and Jottings from Animal Life.* London: Smith, Elder.

Burgess, G. H. O. 1968. *The Eccentric Ark: The Curious World of Frank Buckland.* New York: Horizon Press.

Cane, R. F. 1974–75. "John H. Pepper—Analyst and Rainmaker." *Journal of the Royal Historical Society of Queensland* 9:116–28.

"Christmas Amusements." 1865. *Illustrated London News* 46 (January 7): 19.

"Christmas Holidays." 1858. *Illustrated London News* 33 (December 25): 608.

Collins, J. P., and J. Walter Smith. 1915. "Robert Ball, a Star Among Star Gazers." *Boston Evening Transcript*, April 10, 5.

County Record Office. 1892."Memoranda of merits and disadvantages of accepting the Lowndean Chair 1892." R32/61. Shire Hall, Cambridge.

Dawson, Gowan, Richard Noakes, and Jonathan R. Topham. 2004. Introduction to *Science in the Nineteenth-Century Periodical*, by Geoffrey Cantor, Gowan Dawson, Graeme Gooday, Richard Noakes, Sally Shuttleworth, and Jonathan R. Topham, 1–34. Cambridge: Cambridge University Press.

"Easter Monday and Its Amusements." 1844. *Pictorial Times* 13: 233, in University of Westminster Archives, R Illust 11.

"Fish and Oyster Culture." 1865. *Nottingham and Counties Daily Express,* March 4, 8.

"Frank Buckland" 1880. *Nature* 23 (December): 175.

Gammond, Peter. 1991. *The Oxford Companion to Popular Music.* Oxford: Oxford University Press.

Gates, Barbara T., and Ann B. Shteir, eds. 1997. *Natural Eloquence: Women Rein-scribe Science.* Madison: University of Wisconsin Press.

"Ghosts without Spirit." 1863. *Punch* 45 (October 10): 146.

"Gilchrist Lectures at Goole. Mr. Robert Ball on 'Other Worlds.'" 1890. *Goole Weekly Times,* January 10, 3.

Golinski, Jan. 1992. *Science as Public Culture: Chemistry and Enlightenment in Britain, 1760–1820.* Cambridge: Cambridge University Press.

Goodall, Jane R. 2002. *Performance and Evolution in the Age of Darwin: Out of the Natural Order.* London: Routledge.

Hays, J. N. 1983. "The London Lecturing Empire, 1800–50." In *Metropolis and Province: Science in British Culture, 1780–1850,* ed. Ian Inkster and Jack Morrell, 91–119. Philadelphia: University of Pennsylvania Press.

Inkster, Ian. 1980. "The Public Lecture as an Instrument of Science Education for Adults—The Case of Great Britain, c. 1750–1850." *Pedagogica Historica* 20:80–107.

Jacyna, L. S. 1983. "Images of John Hunter in the Nineteenth Century." *History of Science* 21:85–108.

John Bull. 1861. In *Frank Buckland, Records of My Life* [Commonplace Book], vol. 2, p. 38, (Nov. 30, 1861), Library, Royal College of Surgeons, London.

Kilde, Jeanne. 2002. *When Church Became Theatre: The Transformation of Evangelical Architecture and Worship in Nineteenth-Century America.* Oxford: Oxford University Press.

Lamb, Geoffrey. 1976. *Victorian Magic.* London: Routledge and Kegan Paul.

Layton, David. 1977. "Founding Fathers of Science Education (4): A Victorian Showman of Science." *New Scientist* 75 (September 1): 538–39.

"Lecture at the Midland Institute on Tides." 1881. *Birmingham Daily Gazette,* October 25, 6.

Lightman, Bernard. 2000. "The Visual Theology of Victorian Popularizers of Science: From Reverent Eye to Chemical Retina." *Isis* 91: 651–80.

Livingstone, David N. 2003. *Putting Science in Its Place: Geographies of Scientific Knowledge.* Chicago: University of Chicago Press.

London as it is To-Day: Where to Go, and What to See. 1850. London: H. G. Clarke.

Magic and Mystery Box 2. n.d. Ticket Show Places 27a and 27b, John Johnson Collection, Bodleian Library, Oxford University.

Mayer, David III. 1969. *Harlequin in His Element: The English Pantomime, 1806–1836.* Cambridge, MA: Harvard University Press.

Meisel, Martin. 1983. *Realizations: Narrative, Pictorial, and Theatrical Arts in Nineteenth-Century England.* Princeton, NJ: Princeton University Press.

"Modern Astronomy. Prof. Ball's First Lecture at the Lowell Institute." 1884. *Boston Herald,* October 15, 4.

Morus, Iwan Rhys. 1998. *Frankenstein's Children: Electricity, Exhibition, and Experiment in Early-Nineteenth-Century London.* Princeton, NJ: Princeton University Press.

"Mr. Buckland on Natural History." 1861. *Windsor, Eton, and Slough Royal Standard and Four Counties General Advertiser* (Feb. 2), in *Frank Buckland, Records of My Life* [Commonplace Book], vol. 2, p. 6, Library, Royal College of Surgeons, London.

Nash, John. [1860?]. *Laughing Gas or A Night at the Polytechnic.* London: Weippert.

O'Connor, Ralph. 2003. "Hugh Miller and Geological Spectacle." In *Celebrating the Life and Times of Hugh Miller: Scotland in the Early 19th Century,* ed. Lester Borley, 237–58. Cromarty: Cromarty Arts Trust.

Our Special Sightseer. 1870. "Monday Out." *Fun,* December 3, 223.

Pepper, J. H. 1890. *The True History of the Ghost and All About Metempsychosis.* London: Cassell.

———. 2003. *The Boy's Playbook of Science,* vol. 6 of *Science for Children,* ed. Aileen Fyfe. Bristol: Thoemmes Press.

"Philosophical Drama." 1854. *Punch* 27:179.

"Playing with Lightning." 1869. *All the Year Round,* May 29, 617, in University of Westminster Archive, Original Cutting, R Illust 21b.

"Polytechnic." 1863a. *Illustrated London News* 42 (February 28): 218.

"Polytechnic." 1863b. *Illustrated London News* 42 (June 6): 610.

"Polytechnic." 1863c. *Times* (London), May 20, 9.

"Polytechnic." 1869. *Times* (London), March 30, 10.

"Polytechnic Institution." 1858. *Illustrated London News* 33 (September 11): 241.

"Polytechnic Institution." 1863a. *Illustrated London News* 42 (January 3): 19.

"Polytechnic Institution." 1863b. [January 4], in University of Westminster Archives, R 82.

"Polytechnic Institution." 1865a. *Times* (London), October 13, 12.

"Polytechnic Institution." 1865b. *Times* (London), December 25, 10.

"Polytechnic Institution." 1866a. *Art-Journal* 28:256, in University of Westminster Archives, Copies from Contemporary Periodicals.

"Polytechnic Institution." 1866b. *Times* (London), April 10, 10.

"Polytechnic Institution." 1867. *Times* (London), December 23, 6.

"Polytechnic Institution." 1868a. *Times* (London), March 9, 7.

"Polytechnic Institution." 1868b. *Times* (London), April 14, 9.

Press Cuttings. n.d. Book of Press Cuttings Relating to R[oyal] P[olytechnic] I[nstitution]. University of Westminster Archives. R 82.

Prior, Katherine. 2004. "Gilchrist, John Borthwick." In *Oxford Dictionary of National Biography*, vol. 22, ed. H. C. G. Matthew and Brian Harrison, 217–20. Oxford: Oxford University Press.

"Reading Room and Library." [1864]. *County Times*, in *Frank Buckland, Records of My Life* [Commonplace Book], vol. 2, p. 169, Library, Royal College of Surgeons, London.

"Rev. J. G. Wood." 1890. *Saturday Review* 69:479.

"Round About Town. At the Polytechnic." 1880. *Punch*, September 25, 133, in University of Westminster Archive, R82.

"Royal Polytechnic." 1856. *Athenaeum*, December 27, 1612.

"Royal Polytechnic." 1864. *Illustrated London News* 45 (December 31): 666.

"Royal Polytechnic." 1866. *Illustrated London News* 48 (May 26): 511.

"Royal Polytechnic." 1870. *Illustrated London News* 57 (November 26): 538.

"Royal Polytechnic Institute." 1854. *Illustrated London News* 25 (August 26): 179.

"Royal Polytechnic Institution." 1854a. *Athenaeum*, October 28, 1306.

"Royal Polytechnic Institution." 1854b. *Illustrated London News* 25 (October 7): 335.

"Royal Polytechnic Institution." 1857. *Art-Journal* 19:35, in University of Westminster Archive, Copies from Contemporary Periodicals.

"Royal Polytechnic Institution." 1858. *Illustrated London News* 32 (June 26): 631.

Rupke, Nicolaas A. 1994. *Richard Owen: Victorian Naturalist.* New Haven, CT: Yale University Press.

Secord, James A. 2000. *Victorian Sensation: The Extraordinary Publication, Reception, and Secret Authorship of "Vestiges of the Natural History of Creation."* Chicago: University of Chicago Press.

———. 2002. "Quick and Magical Shaper of Science." *Science* 297 (September 6): 1648–49.

"Sketch of Frank Buckland." 1885–86. *Popular Science Monthly* 28:406.

Sopka, Katherine Russell. 1981. "John Tyndall: International Populariser of Science." In *John Tyndall: Essays on a Philosopher*, ed. W. H. Brock, N. D. McMillan, and R. C. Mollan, 193–203. Dublin: Royal Dublin Society.

"Spectre Drama at the Polytechnic Institution." 1863. *Illustrated London News* 42 (May 2): 486.

"St. John's School—Mr. Frank Buckland's Lecture on Natural History." 1863. *West London Observer*, November 7, in *Frank Buckland, Records of My Life* [Commonplace Book], vol. 2, p. 148, Library, Royal College of Surgeons, London.

Turner, Gerard L'E. 1987. "Scientific Toys." *British Journal for the History of Science* 20:377–98.

Upton, John. [1910]. *Three Great Naturalists.* London: Penguin Press.

Walpole, Spencer. 1881. "Mr. Frank Buckland." *Macmillan's Magazine* 43 (February): 303–9.

Wayman, P. A. 1986. "A Visit to Canada in 1884 by Sir Robert Ball." *Irish Astronomical Journal* 17:185–96.

———. 1987. *Dunsink Observatory, 1785–1985: A Bicentennial History.* Dublin: Dublin Institute for Advanced Studies and Royal Dublin Society.

Wells, Ellen B. 1990. "J. G. Wood: Popular Natural Historian." *Book and Magazine Collector*, no. 79 (October): 56–64.

Whitehead, Alfred. 1889. "The Late Rev. J. G. Wood." *Times* (London), March 9, 15.

"Witney. Witney Natural History Society." 1861. *Jackson's Oxford Journal*, November 2, in *Frank Buckland, Records of My Life* [Commonplace Book], vol. 2, p. 48, Library, Royal College of Surgeons, London.

"A Wonderful Shillingsworth!" 1865. *Punch* 49 (December 16): 236–37.

Wood, Rev. Theodore. [1890]. *The Rev. J. G. Wood: His Life and Work.* New York: Cassell.

"Working Men's Association Lecture." 1859. *Windsor and Eton Express*, February 12, in *Frank Buckland, Records of My Life* [Commonplace Book], vol. 1, p. 160, Library, Royal College of Surgeons, London.

Yanni, Carla. 1999. *Nature's Museums: Victorian Science and the Architecture of Display.* London: Athlone Press.

Print

CHAPTER FIVE

Publishing "Popular Science" in Early Nineteenth-Century Britain

JONATHAN R. TOPHAM

In 1849, a reviewer in the *Quarterly Review* told readers that "popular science" was "less a concession" to the spirit of the age than was generally supposed. Ever since Newton, he asserted, "popular science" had been "the humble attendant on mathematical philosophy," and he cited in evidence the Newtonian publications of David Gregory, John Keill, John Desaguliers, and Colin Maclaurin. The reviewer nevertheless quickly ran out of such examples, admitting, "There is no eventful occurrence to record till the establishment, in our own day, of Mechanics' Institutes, of which a prominent design was the propagation of elementary science among *the people*," and particularly acclaiming the publications of the Society for the Diffusion of Useful Knowledge (SDUK), which "largely promoted the spread of popular science" ([Elwin] 1849, 307, 314, 316). The failure of this attempt to give "popular science" a history before the nineteenth century is highly instructive. While numerous publications in the seventeenth and eighteenth centuries attempted to present the arcane findings of the new philosophy to nonmathematical readers, it was not until the 1820s and 1830s that publications were produced under the designation "popular science."[1] These new publications were intended for far different audiences: ones defined by the new social and intellectual divisions of the industrial age.

Historians of science are now generally agreed that "popular science" is a problematic category of analysis, not least because it has historically carried with it unsustainable assumptions about the passive diffusion of scientific knowledge from expert scientists to inexpert lay publics (see, e.g., Cooter and Pumfrey 1994; Hilgartner 1990; and van Wyhe's opposing view in chapter 3 of this volume). However, the historical emergence and development of this category, and of its attendant diffusionist assumptions, has received relatively little scholarly attention. While historians continue to struggle to find an analytic vocabulary in which to discuss the engagement of exoteric publics with science, it is striking how little historical actors' own use of the term *popular science* has been examined. By paying closer attention to the diverse, changing, and often competing ways in which the concept and terminology of "popular science" has been used, we can expect to achieve a deeper understanding of what historical actors were attempting to achieve. The value of this approach becomes particularly apparent when we consider that the phrase "popular science" originated in early nineteenth-century Britain at a moment when both of its component words were undergoing redefinition. Indeed, the origination of a new specialized and disciplinary notion of "science" in these years—what has been described as the "invention of science"—was closely associated with the development of new "popular" audiences both for science and, more generally, for printed matter (Schaffer 1986; Topham 2000).

The English adjective *popular* was from an early period applied in regard both to "the common people" and to "the people as a whole as distinguished from any particular class" (*OED*, 3rd ed., s.v. "Popular"). However, neither of these senses equated to the bourgeois public addressed by the scientific publications of the eighteenth-century Newtonians. Moreover, when, in the early nineteenth century, the term *popular* began to be applied to particular forms of publication, it reflected a growing sense of the disintegration of a unitary bourgeois public, in which the latest findings in the sciences could be discussed among enlightened gentlemen, and of the diversification of reading audiences. The new "popular" publications which were produced were "intended for or suited to ordinary people" (*OED*, 3rd ed., s.v. "Popular") in terms of ease of understanding, or in terms of price, or as measured by their success (Shiach 1989, 27). For some producers, this meant the production of literature specifically designed for a working-class readership, often with a view to controlling the growing body of potentially revolutionary or atheist workers. For others, the target was a unified mass audience, with working- and middle-class readers brought together for purposes of social harmony or to maximize a market (see, e.g., Topham 2004a, 43). In all cases, however, the notion was loaded with consciousness of the new social order.

The production of "popular" publications also had important implications for the intellectual order. The new, increasingly specialist "science" conducted within newly founded disciplinary societies and their publications reflected the growing sense of a fragmenting public sphere. As printed matter became available to a far larger proportion of the populace, the notion that knowledge was the preserve of an enlightened bourgeois public rapidly came under threat. In response, natural knowledge was increasingly defended as the preserve of disciplined experts and romantic geniuses (Schaffer 1986). In 1830, Coleridge famously complained, "You begin . . . with the attempt to popularize science: but you will only effect its plebification. It is folly to think of making all, or the many, philosophers, or even men of science and systematic knowledge" (Coleridge 1830, 82). These issues applied to all kinds of "knowledge" texts. Indeed, Morag Shiach argues that the new sense of the term *popular* generally related to those "excluded from the institutions of knowledge production" (Shiach 1989, 27). However, the parallel emergence of new senses of "popular" and "science" indicates the particular importance of this development in understanding the history of science in the period. Of course, the boundaries between the popular and the learned remained uncertain and were constantly renegotiated. Publications addressed to mechanics encouraged readers to make original contributions to knowledge (Sheets-Pyenson 1985). Indeed, by the mid-nineteenth century, the gentlemanly specialists were shocked to learn that "popular science" publications could challenge their own authority before an eager public (Yeo 1984).

The origin of self-designated "popular science" publishing in early nineteenth-century Britain is, of course, part of a much larger history. However, its striking historical specificity warrants further attention. Alan Rauch begins his study of the cultural significance of "knowledge" in early nineteenth-century Britain by noting the lack of attention devoted by scholars to what he calls the "knowledge industry" of the period (Rauch 2001, 22). His own brief survey emphasizes the development of encyclopedias, educational books for children, and the publications of the SDUK. In this chapter, my object is to situate the development of the printed commodity of "popular science" in relation to the book trade of the period, and in particular to examine the role of commercial publishers in that development. "Popular science" has often been treated as a purely intellectual or ideological production, shaped by the concerns of authors or such ideological organizations as the SDUK or the Society for the Promoting of Christian Knowledge (SPCK). Yet it was also a commodity, manufactured and sold by tradespeople, many of whom have subsequently been forgotten or been obscured by histories emphasizing the contributions of prominent ideologues.

One of the leading contemporary historians of "popular science" publishing in early nineteenth-century Britain was Charles Knight, who had a vested interest in making the role of the SDUK, with which he was for so long associated, central (see C. Knight 1854, 1864–65). To Knight, the

SDUK's determination to produce "popular science" stood in opposition to the reactionary attitudes of the leading early nineteenth-century publishers. The high price of books in the early nineteenth century was, he argued, due to a significant extent to "the determination of the great publishers not sufficiently to open their eyes to the extension in the number of readers" (C. Knight 1854, 238–39). Quoting from Harriet Martineau's *History of England During the Thirty Years' Peace*, published in 1849–50, he claimed that the SDUK "established the principle and precedent of cheap publication (cheapness including goodness), stimulated the demand for sound information, and the power and inclination to supply that demand; and marked a great æra in the history of popular enlightenment" (C. Knight 1854, 240). There had, he acknowledged, been some earlier cheap publishers of "knowledge" texts, but their many failings meant that they warranted little attention. "From the time when the Society commenced a real 'superintendence' of works for the people," he argued, "the old vague generalities of popular knowledge were exploded; and the scissars-and-paste school of authorship [i.e., hack compilers] had to seek for other occupations than Paternoster-row could once furnish" (C. Knight 1854, 241).

The importance of the SDUK in the invention of "popular science" is difficult to overestimate. Yet, ignoring the activities of others in the literary marketplace who were involved in the production of cheap "knowledge" texts obscures the diversity of motivations and activities involved in making the new commodities. The purpose of this chapter is to give an account of some of the key developments in cheap publishing in Britain that led to the development of the new printed commodities of "popular science" during the first quarter of the nineteenth century, prior to the inception of the SDUK. I begin by outlining the state of the book trade in the early nineteenth century, pointing out the implications of the substantial developments of the period for the production of "popular science." In the next two sections, I review some of the more noteworthy developments that took place in the publication of cheap scientific matter in the period between 1815 and 1825. First, I examine developments in cheap educational publishing, focusing in particular on the innovative ventures of Richard Phillips and William Pinnock. Then I consider parallel developments in periodical publishing, examining the emergence of the cheap weekly press in the early 1820s and its implications for the conceptualization of "popular science" as a printed commodity. In conclusion, I return to the more familiar territory of the late 1820s, when the SDUK and such leading literary publishers as John Murray, Archibald Constable, and Longmans began to publish works of self-conscious "popular science," briefly indicating the continuity between these and earlier ventures. The chapter is necessarily exploratory, since little work has been done on the history of cheap publishing in this period, but it brings to light a number of people and publications whose role in the history of "popular science" has hitherto been largely obscured.

The Early Nineteenth-Century Book Trade

The creation of the new commodity of "popular science" by early nineteenth-century publishers was deeply embedded in the transformations which took place in the book trade. These changes had radical effects on the emerging role and identity of the "publisher," as Samuel Smiles outlined in an epilogue to his memoir of the second John Murray (1768–1843). In the eighteenth century, he reported, the spirit of "commercial monopoly" was still evident "in the practice of co-operative publication which produced the 'Trade Books'" and in a "deep-rooted belief in the perpetuity of copyright, which only received its death-blow from the celebrated judgement of the House of Lords in the case of Donaldson v. Becket in 1774" (Smiles 1891, 2:508). By contrast, coming in the wake of the French Revolution, Murray's career was, he argued, characterized by the rise of free trade, unfettered literary genius, and political reform. To Smiles, Murray was a remarkable manifestation of his times: he was an entrepreneur with a touch of genius in his "largeness of view," who could claim, "The business of a publishing bookseller is not in his shop, or even in his connections, but in his brains" (Smiles 1891, 2:510, 511). Murray was, we are told, perhaps the first publisher to abandon the retail and country trade to specialize in publishing. His financial calculations were not those of "ordinary commercial shrewdness"; instead, he possessed a mixture of "dash and steadiness" (Smiles 1891, 2:512). With a "high sense of rectitude" and an "unfeigned love of literature," he became the hub of a literary coterie, extending to authors in a commercial age the kind of patronage once extended by aristocrats. Murray was, Smiles claimed, a transitional figure, embodying the best of the old and the new. In his time, "the old association of booksellers, with its accompaniment of trade-books, dwindled with the growth of the spirit of competition and the greater facility of communication" (Smiles 1891, 2:513, 515, 516). At the same time, the risks of his adventurous publishing were mitigated by the "high prices then paid for ordinary books" (Smiles 1891, 2:517).

This portrait of a leading early nineteenth-century publisher highlights many of the key features of the transformation in the book trade discussed in William St. Clair's magisterial new book *The Reading Nation in the Romantic Period*. St. Clair grounds his account of reading in this period in a detailed analysis of the economics of the book trade. His study substantiates and extends the notion that changes in intellectual property rights were critical to the development of the market for print and the history of reading. In particular, the 1774 House of Lords ruling against perpetual property was the "most decisive event in the history of reading in England since the arrival of printing 300 years before," breaking up "as perfect a private monopoly as economic history can show" (St. Clair 2004, 109, 101). What followed was a dramatic shake-up of business practices, as the trade came to terms with the fact that the old monopoly enjoyed by London's copyholding booksellers had been swept away. New entrants to the trade issued cheap reprints of standard works, anthologies, and abridgements, spawning an unprecedented increase in the market for printed matter. St. Clair's researches broadly substantiate James Lackington's assessment in 1791 that in twenty years the number of books sold had quadrupled. "All ranks and degrees," Lackington exclaimed, "now READ" (Lackington 1791, 255; St. Clair 2004, 118.). Of course, as this finding implies, the transformation in the book trade was accompanied by social, cultural, educational, and infrastructural changes—most notably the development of provincial middle-class culture, the commodification of leisure, and the extension of elementary education among an increasingly urban working class—but the magnitude of the changes in the trade itself should not be underestimated.

The newly competitive trade conditions led, as Smiles identified, to changes in the role of the copyholding bookseller. As the example of Murray nicely demonstrates, the emergence of the nineteenth-century "publisher" involved the gradual separation of publishing from the retail and wholesale trade, the breakup of the system of publishing "congers" for "trade" books, and the rise to dominance of the literary entrepreneur. However, as Smiles observed, entrepreneurs such as Murray made the headway they did partly because they were able to command high prices for the new literary products they financed. This increase in prices for new works was fostered by the provisions of the new Copyright Acts of 1808 and 1814. Following the 1774 ruling, copyright was restricted to the period of fourteen years laid down by the Copyright Act of 1710, with the possibility of an extension to twenty-eight if the author was still alive at the end of the first period. However, copyright was extended to a statutory twenty-eight years in 1808 and to the life of the author, if longer, in 1814. William St. Clair argues that this legislation effectively closed the "brief copyright window" that the 1774 ruling had opened, restoring power to the publishers and effectively hypostatizing an "old canon" of out-of-copyright works which could be reproduced cheaply, while the prices of new books soared, protected as they were by the longer copyright term.

The rise in price of in-copyright books during the early nineteenth century was in part due to rises in wages and the cost of paper, both exacerbated by the war with France (C. Knight 1854, 138; St. Clair 2004, 116). However, the role of a still-conservative book trade in maintaining and protecting the practice of issuing new books in small editions at high prices was crucial. James Lackington, who made his fortune by developing a trade in cheap remainders, inscribed on his carriage the motto "SMALL PROFITS DO GREAT THINGS," but this was a motto few in the book trade were prepared to follow, preferring instead to destroy the bulk of any remainders and to sell those they retained at full price (Lackington 1791, 234, 224). In Charles Knight's assessment, "The very notion of cheap books stank in the nostrils, not only of the ancient magnates of the East, but of the new potentates of the West. For a new work which involved the purchase of copyright, it was the established rule that the wealthy few, to whom price was not a consideration, were alone to be depended upon for the remuneration of the author and the first profit of the publisher" (C. Knight 1854, 1:276). The consequence was that, for the growing numbers of lower-middle- and working-class people equipped for reading by the rapid expansion of elementary education, the bulk of newly copyrighted works were beyond reach. Instead, they had to rely on reworkings of older books. The working class had access to a chapbook literature now updated from a premodern to a more recent canon, and those in slightly less straightened circumstances could afford cheap reprints of "old canon" texts. However, relatively little of this material dealt with scientific subjects (St. Clair 2004, 339–56). In addition, as St. Clair points out, out-of-copyright books were plundered extensively in the compilation of new textbooks and schoolbooks, and more generally the burgeoning market in books for children provided some of the cheapest new works in the period (St. Clair 2004, 137). Nevertheless, by the end of the war in 1815, the opportunities for the bulk of readers to gain access to scientific reading matter were severely limited.

Ten years later, contemporaries were agreed that the position had radically changed. In 1824, Thomas Dibdin, writing in the preface to his latest guide to gentlemanly book collecting, made a conspectus of the state of the book trade. For all that his own interest was in what he called the "bibliomania" attaching to fine works, he waxed lyrical about the rapidly accelerating "growth of knowledge," which, he claimed, was swiftly and widely "diffused" through "popular and useful" works (Dibdin 1824, xviii, xii).

First among these works were books for children, ranging from the educational books of John Aikin, Anna Barbauld, and John Bonnycastle to
William Pinnock's sixty-four educational catechisms. Looking through
"the CATALOGUES of SCHOOL BOOKS circulated by the two greatest Publishers in England"—by which he meant Longmans and the newcomers
G. and W. B. Whittaker—he observed:

> The most superficial view of the contents of these Catalogues, shews
> the extraordinary and advantageous variety of instruction which they
> contain. Science, Arts, Trade, Manners, Customs—something of every
> thing, and of the very best kind—will be found in each. . . . Let the re-
> flecting reader consider, from the data here laid down, what is the quan-
> tity of instruction which is daily in circulation among the infantile
> world; or among those who have scarcely reached their sixth year? Fifty
> years ago there was hardly any *pabulum* of the kind; or that pabulum
> was exclusively distributed, from the repository of our old acquaintance
> Dan Newbury, in St. Paul's Church-yard. (Dibdin 1824, xii, xiii)

This transformation was also, he reported, to be found in publications for
adults. The "diffusion of knowledge" for such readers had "of late years,
or rather very recently" become "equally rapid and efficient." Dibdin had
particularly in mind here the inception of the new weekly journals: both
the relatively expensive literary journals—the shilling *Literary Gazette*
established in 1817 and the sixpenny *Literary Chronicle* in 1819—and the
new twopenny and threepenny weeklies including the *Mirror of Literature* started in 1822, the *Mechanic's Magazine*, and the *Chemist* in 1824.
Collectively, he reported, the new cheap weeklies sold in excess of one hundred thousand copies. Moreover, similar cheap forms of publication were
beginning to be used for books. In such a climate, he asked, was it "chi-
merical to suppose that *Bacon's Abridgement* (of the Law) and *Comyn's
Digest*" would soon be "produced in the same manner?" (Dibdin 1824,
xv, xvii).

Dibdin's report provides a valuable appraisal of the development of
"popular," particularly knowledge-based, publishing prior to the inception
of the SDUK. In the remainder of this chapter, I examine the activities
and motivations of a number of the publishers who were involved in the
development of the cheap "knowledge" industry in this key period. In particular, I draw on Dibdin's account to explore two key nodes of activity:
the production of cheap educational books for children and cheap journalism. My argument is not that these ventures were exercises in "popular

science" publishing avant la lettre but that they were important in serving to develop and define "popular" publishing in general and "popular science" in particular. As this observation implies, "popular science" did not develop in isolation from other "popular" forms. However, as indicated in the introduction, the implications of the development of "popular" publishing for the development of specialized disciplinary "science" makes it of particular importance for historians of science.

Cheap Educational Publishing

The development of a market for instructive books for children in late eighteenth- and early nineteenth-century Britain, many dealing with scientific subjects, has been a growing focus of historical scholarship in recent years. From the 1750s, the book trade outsider John Newbery pioneered the publication of children's leisure books to exploit the growing market of the eighteenth-century middle classes, and, as Jim Secord has shown, he included such scientific books as the *Newtonian System of Philosophy* in his output (Secord 1985). By the end of the century, the market for children's science books was becoming increasingly large and sophisticated. Secord observes that the decades around 1800 "saw a veritable explosion of new titles in the previously restricted field of juvenile science" (Secord 1985, 140). This increase in output was not least a reflection of the growing demand for educational works among the dissenting middle classes, and, as Aileen Fyfe and John Issitt have shown, Unitarian authors (among them John Aikin, Anna Barbauld, and Jeremiah Joyce) and publishers (such as Joseph Johnson) were prominent in the production of highly successful educational books including *Evenings at Home,* published in 1792–96, and *Scientific Dialogues,* published in 1800–1803 (Fyfe 1999a, 1999b, 2000; Issitt 2000).

What has not attracted so much attention is the development in the early decades of the nineteenth century of the cheaper and often much more crudely didactic books for children described by Dibdin, many of which also dealt with scientific subjects. While these publications were often explicitly aimed at a juvenile school market, the period was one in which divisions between formal and informal education, and between juvenile and adult education, were less strongly marked. The new cheap educational works of the early nineteenth century thus made scientific material available to a large public consisting not only of children but also of adults. In much the same way, cheap reprint editions later made the more-expensive children's books on science of the early years of the nineteenth century available to poorer adult learners of successive generations (see Fyfe 1999a, 1999b). Moreover, as we shall see, the new cheap educational publications of the 1810s and 1820s began to use the vocabulary of the "popular" in ways which suggest continuity between cheap educational works for children and later forms of "popular science" for explicitly adult audiences. As David Knight has observed, the frontiers between textbooks and more explicitly "popular" works were "hazy, particularly at first" (D. Knight 2000, 188).

While much still remains to be done to chart the development of the market for children's educational books in the early nineteenth century, one feature which is clear is that it was an important platform on which a number of leading publishers established their businesses. Some were new-comers to the trade, as Newbery had been; others were established busi-nesses, which used the market for educational books to launch themselves into the new era. Perhaps the most outstanding of these was the house of Longmans, whose dominance Dibdin emphasized (see Wallis 1974; Cox and Chandler 1925; Blagden 1949; Briggs 1974). The Longman family had been leading members of the book trade since the first Thomas Longman bought the bookshop of his former master in 1724, and it had always had some involvement with the trade in schoolbooks. However, under the con-trol of the third successive Thomas (Thomas Norton Longman, 1771–1842), the firm underwent the characteristic changes described in regard to Mur-ray earlier—reducing its reliance on publishing congers, offering handsome rewards to high-profile authors, holding regular literary meetings and din-ners, and developing specialized departments to handle various aspects of the business (so many partners were taken on to manage these aspects that Walter Scott dubbed Longmans the "long firm"). In addition to high-profile literary publishing, the firm came to dominate the market in stan-dard trade works such as textbooks, becoming, in Thomas Moore's phrase, the "establishment publisher" (quoted in St. Clair 2004, 159). Other pub-lications were somewhere between the fashionable and the educational. Jane Marcet was paid handsomely for her various scientific *Conversations*, earning 2,329 pounds from her *Conversations on Chemistry* alone between 1806 and 1865 (Horrocks 1987). On the back of such publications, Long-mans became phenomenally large and powerful. In 1775, the firm had been insured for 5,000 pounds, but by 1822, this figure had risen to a stupendous 71,000 pounds, and in that year the firm was reported to have sold five mil-lion volumes (Barnett 1998, 53; St. Clair 2004, 172).

In addition to such established publishers as Longmans, there were many new entrants to the trade, such as Sir Richard Phillips (1767–1840) (see Timbs 1872; Timbs 1865, 94–123; Issitt 1998; Axon 1888, 238–65; Herne 1892–95). The son of a Leicestershire farmer, Phillips was brought up by an uncle in London, a brewer to whom he was heir. His uncle sent him to several schools, where he showed ability in mathematics and English composition, and with a "passion for literature and experimental philosophy" he returned to Leicester to set up his own school in 1788 (Timbs 1872, 222). By 1790, however, he had become a bookseller, and his shop became a depository of radical reform literature. He soon added a printing press, and in 1792 he commenced the *Leicester Herald,* which attracted contributions from Joseph Priestley and which Phillips continued to edit despite a period of imprisonment in Leicester Gaol for selling Paine's *Rights of Man.* Following a devastating fire, he moved to London in 1796, where in July he started the radical *Monthly Magazine,* under the literary editorship of John Aikin. By its inclusion of political and "philosophical" content, this magazine achieved significant success—the first number selling three thousand copies, which "astounded the booksellers" (Timbs 1865, 98, 121). Phillips soon also became a general publisher, developing a particularly strong educational list (Timbs 1865, 101–2). By 1807, he was a sufficiently prominent tradesman to achieve election as high sheriff of London, and, during his shrievalty, he was knighted. Two years later, in circumstances that are not clear, he became bankrupt, although with help from friends he managed to rescue some of his most successful copyrights and reestablish himself. Phillips retired to Brighton in 1823, selling a third of his literary property to the rising educational publishers G. and W. B. Whittaker and leaving the whole of it (which he claimed had originally cost him seventy thousand pounds) under their management. The Whittakers were particularly severely affected by the financial crisis of 1825, in which Phillips lost between forty thousand and fifty thousand pounds (Timbs 1871, 470; Timbs 1872, 222; Axon 1888, 256; Topham 2004b).

Phillips was a singular man—a radical jailbird who was knighted by George III and a man whose vegetarianism, abstinence from alcohol, and anti-Newtonian views were the butt of some humor. Yet, his background as a provincial schoolmaster and radical bookseller with connections to leading figures in rational dissent was not atypical among successful entrants to the book trade at this period. It was a background that fostered a belief both in the importance of education and in the commercial potential of educational literature. The development of Phillips's educational output deserves extended study, but an outline can be pieced together. Among his first authors were some, such as William Mavor and Jeremiah Joyce, who had written for the *Monthly Magazine.* In the course of time, however,

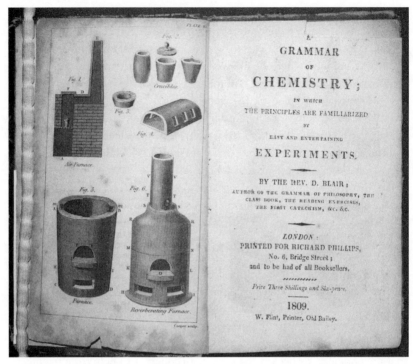

Figure 5.1. Title page of *A Grammar of Chemistry*. The pseudonym "Rev. D. Blair" served to efface Richard Phillips's radicalism, evoking instead the conservatism of Hugh Blair's enormously popular *Sermons* (St. Clair 2004, 272). Reprinted from Blair 1809 by kind permission of Leeds University Library.

Phillips himself took up the pen, writing numerous works under a range of pseudonyms (see, e.g., figure 5.1), carefully chosen to give his publications a wholesome, establishment air (Timbs 1865, 105; St. Clair 2004, 272).[2] As his publishing output grew, he began to shape it into a coherent program, issuing groups of related products and giving the whole a pedagogical rationale by inventing what he termed the "interrogative system" in 1816 (Timbs 1865, 105). His procedure was to publish an elementary textbook on a particular subject (these he often termed "grammars"), then a quarto copybook containing questions on the textbooks for students to fill in (hence "interrogative"), and finally a tutor's "key" giving the answers (Phillips [1823?], reproduced in Issitt 1998, 30–45; see also Issitt 2000, 229–30). Ranging widely over many fields of learning, Phillips inevitably found himself in need of help from hack writers, and he became notorious as a publisher who was not afraid to bend the necks of authors to his own perceptions of market demand. In his semiautobiographical novel *Lavengro*, for instance,

George Borrow, who worked as one of Phillips's hacks around 1823, gave a fictionalized account of the publisher, portraying him as "highly clever and sagacious" but cynically opportunist in what he commissioned and capriciously interfering with his compilers' endeavors (Borrow 1905, 226; Boyle 1951).

One of the most striking features of Phillips's capricious interference was in the maintenance of his anti-Newtonian views. Augustus De Morgan noted of James Mitchell's 1823 *Dictionary of the Mathematical and Physical Sciences* that it contained "a temperate description of [Phillips's] doctrines, which one may almost swear was one of his conditions previous to undertaking the work" (De Morgan 1915, 1:242). Yet, while Phillips's scientific interests were somewhat unorthodox in this respect, they were nonetheless deeply rooted. As a young teacher in Leicester, he ran evening classes in mathematics and philosophy and was president of the Adelphi philosophical society (Herne 1892–95, 65–66). The *Monthly Magazine,* too, was notable for its scientific contents and contributors, and when Phillips began in the educational market, many of his books were on scientific subjects. Puffing his books in the 1820s, he recalled:

> Until the appearance of the first of this series of Elementary Books, there existed no really practical means of extending scholastic instruction, beyond the Grammars of Languages, the Elements of Arithmetic and Algebra, and some branches of mixed Mathematics. . . . The first essay towards an improvement, was made in the publication of Dr. MAVOR's British Nepos, and his Natural History, drawn up by that gentleman, on the suggestion of the publisher. The success of these led to the creation of other useful Books, in the various branches of elegant and useful knowledge. (Phillips [1823?], 1)

While there is an element of hubris about this claim, there can be no doubt that Phillips's activities in the first two decades of the nineteenth century significantly broadened the available range of elementary books on the sciences. Moreover, his publications were frequently described as being on a "popular plan," giving a popular "account," "view," "introduction," or "descriptions," or being written in a "popular manner." Such books were certainly not cheap in absolute terms. Usually selling for several shillings, or at least half a crown (two shillings and sixpence), they were the preserve of the middle class.[3] Nevertheless, many were very successful, suggesting that they were affordable to the lower middle class. For instance, John Issitt reports that between 1813 and the mid-1850s, ten thousand to twenty thousand copies were printed annually of the 1803 *Easy Grammar of General Geography* (Issitt 2002, 100). Around 1809, Phillips claimed that William Mavor's 1801 *English Spelling Book*, priced at one shilling and sixpence, was selling eighty thousand copies per year, and by 1836, it was claimed to be in its 423rd edition (Phillips [1809?], 30).[4]

While Phillips's publications thus provided elementary scientific reading to a wide readership, it was not until the "propitious epoch" after the end of the war with France that elementary scientific works became widely available at prices accessible to working-class readers. Contrasting these new publications with those discussed earlier in this chapter, the journalist William Jerdan observed:

> No doubt we had or have, our Watts, Goldsmiths, Blairs [i.e., Phillips],
> Mavors, Murrays, Mangnalls, Enfields, and others, whose productions
> contributed much towards promoting the education of the *many* in
> their time; but they were rather insulated in their valuable efforts, and
> it was not till the period was probably riper for the wished-for progress,
> and Pinnock and Maunder startled us with their Catechisms and con-
> densed Histories, that the aim was brought to bear upon the millions,
> now better prepared and more ready and apt for the reception of infor-
> mation. (Jerdan 1866, 336–37)

The reference here is to William Pinnock (1782–1843), another newcomer to the book trade who combined spectacular success with spectacular failure (Curtis 1896, 159; Hawkins 1973, 158). Pinnock was the son of a laborer in the Hampshire market town of Alton, but, like Phillips, he became first a school usher (at Eggar's Grammar School) and then a printer and stationer, writing and publishing his own sets of educational works (probably in emulation of Phillips's), including a *Universal Explanatory Reader*, published in 1809, with *Sequel*, the second edition of which appeared in 1810, and *Introduction*, the fifth edition of which appeared in 1811, and similar suites of books on punctuation and spelling. These volumes clearly achieved some success and were sold by agents in London and numerous provincial towns and cities. By 1812, Pinnock had moved twenty miles west to the cathedral city of Winchester and had begun to publish his later much-famed educational "catechisms."[5] They were issued in a cardboard wrapper at an extraordinarily cheap nine pence and consisted of two sheets of demy printing paper, each folded into eighteen leaves, making a small seventy-two-page book (figure 5.2). Titles issued in 1812 included a *Catechism of Ancient History*, a *Catechism of Mythology*, and a *Catechism of*

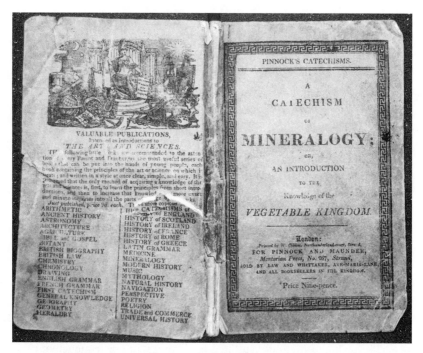

Figure 5.2. Front and back covers of *Catechism of Mineralogy*. Britannia presides over a panoply of Christian learning on this paper-covered book measuring just 135 by 90 millimeters (5.3 by 3.5 inches). Reprinted from [Pinnock 1819?], in author's collection.

the History of England, but others soon followed, and by 1816, twenty-two were in production, including catechisms on arithmetic, chemistry, and astronomy.[6] By 1814, Pinnock had settled twenty miles to the north in the Berkshire market town of Newbury, but before he left Winchester, he began to employ (later entering into partnership with) the printer, Samuel Maunder (1785–1849). Though possibly originating in Winchester, Maunder came from a Devonshire family and had apparently been a bookseller and printer in Kent for a number of years—from 1805 to 1808 in Bromley and from 1808 onward in Tonbridge, where Pinnock married Maunder's sister Ann in December 1814.[7] From that time onward, Maunder became Pinnock's great ally and collaborator in the growing catechism business.

Instructional catechisms on secular subjects were, of course, far from new, with their long, authoritative lists of questions and answers. Moreover, with the expansion of the market for middle-class educational literature from the late eighteenth century, the catechismal form had been repeatedly used in the preparation of books on the sciences, such as Samuel

Parkes's 1809 *Chemical Catechism* (D. Knight 1986). Yet, at prices running to several shillings, works such as Parkes's were well beyond the means of most readers. Even in their cheapness and format, however, Pinnock's catechisms had important precedents. By 1809, Richard Phillips had issued for nine pence a *First Catechism for Children* in the same small seventy-two-page format later used by Pinnock, of which thirty thousand copies were produced in the first six months (Phillips [1809?], 30). Phillips later claimed that his was "the first of those varied Series of Catechisms with which, since its appearance, the public have been overwhelmed" (Phillips 1821, 2). There were certainly many imitators, including Phillips's collaborator William Mavor, who claimed to have harbored the idea for ten years (Mavor 1809). Mavor certainly took the formula further, issuing from 1809 a number of similar seventy-two-page catechisms on a range of subjects including botany, "animated nature," health, and "general knowledge" with the cheap publishers Lackington, Allen, and Company. Yet, as Dibdin and Jerdan testified, Pinnock and Maunder developed this approach on an altogether larger scale. Before they could do so, however, they had to overcome the problems of trading from a small provincial town. Not only was their access to national markets restricted by their location, but at the start of 1816, Pinnock and Maunder suffered bankruptcy in consequence of their financial involvement with the Newbury Bank, which collapsed after being robbed in December 1815.[8] By then, however, they had already made plans to move to London.

Before the Newbury Bank failed, Pinnock and Maunder bought prominent premises at the heart of journalistic London—267 Strand, opposite St. Clement's Church—from the journalist William Jerdan, and by 1816, they were using these premises to rebuild their business.[9] In the summer of 1817, the shop became the office of the recently founded *Literary Gazette*, of which Pinnock and Maunder now became the publishers, entering into a partnership as coproprietors with Jerdan (its editor) and Henry Colburn (its instigator). While it lasted, the collaboration was highly beneficial to all parties. The *Literary Gazette* had initially been issued as a stamped paper, allowing it to pass all over the country through the ordinary mail. However, in the face of competition from cheaper imitators, an unstamped edition began to be issued in June 1818 for distribution by news vendors in the capital and through agents around the country. The business thus increasingly came to rely on the development of a provincial distribution network, independent of the wholesale agents of Paternoster Row. In this respect, it was mutually supportive of Pinnock and Maunder's "connection with Education," since that also involved "travelling throughout the country" to develop "provincial agencies" for their cheap textbooks (Jerdan 1852, 2:182; Jerdan 1866, 339). In addition, Pinnock and Maunder were responsible for changing the *Gazette*'s printer to Thomas Bensley, with whom they were for a period in partnership (Jerdan 1852, 3:62). Bensley, who had funded Friedrich Koenig's development of steam presses in the years after 1807, soon applied the new technology to the printing of the *Literary Gazette*, making it the first weekly periodical to be printed in this manner.[10] Bensley also printed many of the catechisms, and, while there is no evidence of their having been steam-printed, the print-runs of some were such that it is not inconceivable.

Once in London, Pinnock and Maunder's educational publications created a "sensation," and their premises in the Strand became, for four short years, "famed" as their "catechism-bookselling-shop" (Jerdan 1866, 339; Jerdan 1852, 2:181). They issued catechisms on a wide range of subjects, with forty-nine titles in production by 1819. Many of them were scientific, including algebra, arithmetic, astronomy, agriculture, botany, chemistry, electricity, entomology, geometry, medicine, mineralogy, natural history, and navigation.[11] According to Jerdan, they were adopted by "half the schoolmasters in England," and certainly their sales were on a grand scale (Jerdan 1866, 339). The business brought in several thousand pounds a year, and one of Pinnock and Maunder's titles alone—their edition of Goldsmith's *History of England*—reputedly earned a profit of more than two thousand pounds in one year (Jerdan 1852, 2:181; Jerdan 1866, 340). By 1824, Dibdin reported that one impression of the *Catechism of the Bible and Gospel Histories* had reached seventy thousand copies (Dibdin 1824, xiv). Such success brought its own problems, and in 1819, Pinnock and Maunder had to obtain injunctions to restrain the production of piratical editions of the catechisms.[12] Moreover, Jerdan reported that Pinnock had "no idea of details" and kept poor accounts (Jerdan 1852, 2:182). Very quickly the whole empire came crashing down. Bensley's printing office burned down in June 1819, and by November of that year, Pinnock had ruined himself and his partner by an unwise speculation in wood veneer (Jerdan 1852, 2:181, 3:62; Jerdan 1866, 341). They were forced to sell the copyright of the catechisms to G. and W. B. Whittaker. The value of the catechisms, reputed to be in excess of thirty thousand pounds, was a clear indication of the success of the series (Dibdin 1824, xiv).

To William Jerdan, Pinnock and Maunder were individuals who "came to initiate, or newly systematize a species of popular literature which exerted a wide and beneficial national influence, especially upon the lower and middle classes of the people, and showed the way to the extension

and improvement of similar works" (Jerdan 1866, 336). As this statement
implies, the cheap scientific publications they produced were of relevance
not only to children. Granting an injunction in 1819 to protect Pinnock's
catechisms from piracy, Lord Eldon observed, "It appears to me that adults
might be greatly benefited by the instruction these books contain, as well
as the younger branches of society."[13] Pinnock and Maunder's publishing
activities had demonstrated the existence of a large public interested in
acquiring cheap elementary "knowledge" texts, including adults as well as
children, and others were not slow in following where they led.

Cheap Journalism and "Popular Science"

In addition to the growth of educational literature, Dibdin was much taken
in 1824 with "how varied and wonderful [was] the periodical diffusion of
knowledge, of every description," as he surveyed the growing range of
weeklies, monthlies, and quarterlies that characterized the period (Dibdin
1824, xviii). Jerdan, too, considered that the "prodigious power and influence of the press" dated "much of their growth and increase" from the
period after the end of the war with France (Jerdan 1866, 338). A significant
feature of this development, as Charles Knight pointed out, was the inception of *Blackwood's Edinburgh Magazine* in 1817, which marked the start
of a "new era of Magazines." Formerly, the magazines had been "the vehi-
cles for the communication to the world of all sorts of opinion, theological,
moral, political, and antiquarian. They were the tablets upon which the
retired scholar or the active citizen might equally inscribe their theories or
their observations, in a familiar and unpretending style; and they at once
kept alive the intelligence of their own generation, and formed valuable
records for succeeding eras" (Knight 1864–65, 1:265). Where magazines like
the *Gentleman's* and Phillips's *Monthly* had contained articles on all fields
of learning, including natural philosophy, natural history, and the practical arts, often contributed gratuitously or for a small remuneration, *Blackwood's* and the magazines that soon followed its example (the *London*, the
New Monthly, and the *European*) were self-consciously literary, offering increasingly rich rewards to talented writers (Dawson, Noakes, and Topham
2004, 12–14). Literature was "metamorphosed," Jerdan claimed, "into the
condition of a trade" or profession, so that it was no longer "stragglers, vol-
unteers, insulated knights of the pen, or original thinkers" who wrote for
the newly literary magazines but professional authors (Jerdan 1866, 338).
At the same time, as I have shown elsewhere, there was an extremely rapid
increase in the number and range of specialized scientific periodicals—

both commercial magazines and society transactions—as technical scientific discourse became increasingly removed from a hitherto notionally unified public sphere (Dawson, Noakes, and Topham 2004, 13–14; Topham forthcoming).

Jerdan was well placed to appreciate the postwar development of periodical literature, being involved from an early stage with the first weekly literary journal to achieve lasting success. The *Literary Gazette* was begun in January 1817 by the publisher of the *New Monthly Magazine*, Henry Colburn, as a sixteen-page quarto weekly, selling for a shilling and containing synoptic reviews of new books, literary and scientific news, and other essays and articles. With the growing output of the press, the high prices of new books, and the increasing selectiveness of the magazines and reviews, the *Literary Gazette* offered what the established periodicals could not: rapid access to the contents of a large proportion of contemporary literature as well as up-to-date news of the activities of the learned and scientific societies. Undercut by the sixpenny *Literary Journal* (1818–19) and *Literary Chronicle* (1819–28), the *Gazette* soon began to issue an eightpenny unstamped edition, for those not requiring postal delivery. By 1823, the formula was an established success, with the *Gazette* selling four thousand copies a week. In 1824, Dibdin observed: "There is, at present, such an hunger and thirst after information, that the reading man looks towards his *weekly* Journal, or Register, or Chronicle, with the same eagerness and certainty that he used to anticipate his monthly supplies of mental food" (Sullivan 1983, 243; Dibdin 1824, xiv–xv).

These transformations in periodical literature in the years after 1815—with periodicals increasingly addressed to reading publics fragmented by intellectual specialisms—undoubtedly contributed to the demise of the Enlightenment notion of a bourgeois public sphere. A contemporary development that had a much more dramatic impact, however, was the re-emergence of the radical press after the end of the war. The emblem of this development was William Cobbett's "two-penny trash"—the abridged two-penny edition of his weekly *Political Register* issued from November 1816, which reportedly sold forty, fifty, or even seventy thousand copies per week (Altick 1998, 324–26). The success of Cobbett's venture brought numerous other radical publishers into the field, including William Hone, Richard Carlile, and Thomas Wooler. Against a background of radical protest and unrest, such ventures struck terror into the heart of government, leading, after the "Peterloo massacre" of 1819, to the repressive "Six Acts" (figure 5.3). These measures effectually gagged the radical press for a dozen years, but Cobbett's short-lived success raised the prospect of a large market for cheap

' The body of the people, I do think,
are loyal still,'
But pray, My L—ds and G—tl—n,
don't shrink
From exercising all your care
and skill,
Here, and at home,
TO CHECK THE CIRCULATION

OF LITTLE BOOKS,

Whose very looks—
Vile ' *two-p'nny trash,*'
bespeak abomination.
Oh! they are full of blasphemies
and libels,
And people read them
oftener than their bibles

Figure 5.3. William Hone's satirical verse and George Cruikshank's engraving encapsulate the government panic over the circulation of cheap journals in the aftermath of the Peterloo Massacre of 1819. Reprinted from [Hone] 1820, [14], in author's collection.

publications among the growing urban working class rendered literate by the religiously motivated schools of the preceding decades. The situation was recognized by Robert Southey, who in February 1822 observed, "One effect of general education (such as that education is) is beginning to manifest itself. The two-penny journals of sedition and blasphemy lost their attraction when they no longer found hunger and discontent to work upon. But they had produced an appetite for reading" (Southey 1849–50, 5:116–17). This appetite, Southey noted, was soon addressed by cheap apolitical periodicals.

One of those who were quick to observe the emerging market was Charles Knight. The son of a Windsor bookseller, and from 1812 coproprietor with his father of a weekly provincial newspaper, Knight had plans as early as January 1814 for a "cheap work. . . in weekly numbers, for the use of the industrious part of the community," which was to include "plain Essays on points of duty," "the Evidences of Christianity," and "Abstracts of the Laws and Constitution," but also "Information on useful Arts and Sciences" (Knight 1864–65, 1:226). This project was not realized, but in December 1819, Knight wrote in his *Windsor and Eton Express:*

> The mass of useful books are not accessible to the poor; newspapers, with their admixture of good and evil, seldom find their way into the domestic circle of the labourer or artizan; the tracts which pious persons distribute are exclusively religious, and the tone of these is often either fanatical or puerile. The "two-penny trash," as it is called, has seen farther, with the quick perception of avarice or ambition, into the intellectual wants of the working-classes. It was just because there was no healthful food for their newly-created appetite, that sedition and infidelity have been so widely disseminated. The writers employed in this work, and their leader and prototype, Cobbett, in particular, show us pretty accurately the sort of talent which is required to provide this healthful food. . . . They state an argument with great clearness and precision; they divest knowledge of all its pedantic incumbrances; they make powerful appeals to the deepest passions of the human heart. (C. Knight 1864–65, 1:236)

This was the exemplar, if accompanied by suitable reverence for God and the state, for those who wished to create "a more popular literature than we possess." Knight's own early attempts in this respect were a failure, however. His shilling monthly *Plain Englishman,* published from 1820 to 1822, was far wide of the mark. To begin with, as Knight subsequently conceded, the postwar political crisis had rendered him somewhat reactionary. He had "felt that one must be content for a while to shut one's eyes to the necessity for some salutary reforms, in the dread that any decided movement towards innovation" would damage the constitution, and his magazine consequently conveyed a "benevolent optimism" (C. Knight

1864–65, 1:225, 246). In addition, however, he thought his failure was due to his not exploiting the successful twopenny weekly format pioneered by Cobbett, which he had earlier intended to use (C. Knight 1864–65, 1:244).

Others were quick to see the possibilities raised by Cobbett's "twopenny trash." In Edinburgh, the young printer William Chambers and his bookseller brother Robert began a fortnightly periodical called the *Kaleidoscope; or, Edinburgh Literary Amusement* in October 1821. Like the cheap edition of Cobbett's *Register*, it consisted of sixteen octavo pages, though priced at threepence. Nevertheless, it failed to make money and was abandoned the following January (Chambers 1872, 162–71). In London, however, there was soon a dramatic eruption of such publications, mostly produced by small-time publishers and printers in the literary underworld of the Strand and surrounding streets. One of the first of these small-timers was Joseph Onwhyn, the piratical publisher of Byron's *Don Juan* (St. Clair 2004, 322, 327) and other disreputable and radical literature, who in the autumn of 1822 published the *Hive; or, Weekly Entertaining Register*. The *Hive* offered on tightly packed pages a combination of new articles and extracts from contemporary and time-honored sources, promising on its title-page "popular reviews of new works calculated to promote the diffusion of useful knowledge, with large extracts." Within weeks of the *Hive* appearing, the publisher of the *Literary Chronicle*, John Limbird, was offering similar fare in his *Mirror of Literature, Amusement, and Instruction*. Limbird, too, had radical connections, having started out as a close associate of Thomas Dolby, one of the publishers of Cobbett's *Register*. Over the next few years, the much-vaunted success of the *Mirror*—some numbers of which sold eighty thousand copies within two years—encouraged a torrent of new apolitical twopenny weeklies in the 1820s. Many of them were produced by publishers who, like Dolby, had been left stranded by the suppression of the radical press—"journeyman printers . . . out of work," as Southey put it (Southey 1849–50, 5:117). They also began to offer cheap reprints of out-of-copyright works in the same twopenny number form, which in St. Clair's opinion probably reduced prices to "the lowest levels in relation to incomes since the arrival of printing" (St. Clair 2004, 537). Certainly, members of the established trade were outraged and protested to the likes of Limbird, but, like Pinnock's catechisms and the literary weeklies, the new cheap literature was distributed through agency networks distinct from the wholesale trade and was thus outside their immediate control (Topham 2005).

By confirming the existence of a new and profitable market for cheap publications, the twopenny weekly miscellanies of the early 1820s ushered in a new era. As the editor of the *Mirror* observed, the "men of trade" had not previously understood that "small profits on an extensive sale might be as productive as exorbitant charges where the sale was limited."[14] According to a correspondent in the *Gentleman's Magazine* in 1822, the appearance of the new weekly miscellanies heralded a "Golden age of literary and commercial enterprise," in which the "diffusion of every species of information" was put on a new footing (ΠAN 1825, 483). These publications were self-consciously designed to maximize a market—to attempt to create for the first time a "mass" market. This much is clear from the example of Limbird's *Mirror*, which boasted that its low price enabled it to circulate among all social classes, from the peasant to the manufacturer and from the merchant to the nobility.[15] Indeed, Limbird's earlier sixpenny weeklies, the *Literary Journal* and the *Literary Chronicle*, had made similar claims to draw together all social classes by appearing at the cheapest possible price (Topham 2005).

In addressing this notionally all-inclusive audience, the cheap miscellanies became increasingly involved in the burgeoning discourse of the "popular." This involvement is strikingly demonstrated in the case of the *Mirror of Literature*, as I have shown elsewhere (Topham 2004a). The first two editors of the *Mirror* were hack writers who had learned their trade with Richard Phillips. The first, Thomas Byerley, followed the practice of traditional magazines, like Phillips's *Monthly Magazine*, for which he had written, in accepting miscellaneous articles from correspondents on scientific and other subjects. However, when Phillips's former amanuensis John Timbs took over the editorship in 1827, he printed far fewer scientific contributions from correspondents, preferring instead to reprint paragraphs from leading scientific journals. To these gobbets, which he placed in a new section called the "Arcana of Science," he gave the designation "popular science." The distance that was here produced between the true "arcana" or mysteries of science and what was now called "popular science" was twofold. First, the demands of the mass audience, with its socially and culturally diverse readers, required that no specialist knowledge be presupposed. This effect was redoubled, however, by the separation of literary and scientific discourses in the gentlemanly magazines of the postwar years. By the 1820s, even the *Gentleman's Magazine* was increasingly offering scientific extracts in place of original contributions, and in 1828 the new literary weekly, the *Athenaeum*, began to run a regular column of extracts headed "Popular Science" (Holland and Miller 1997). The new phrase reflected this dual dislocation: the attempt to address an imperfectly educated "mass" audience and the simultane-

Figure 5.4. Title page of *Mechanic's Magazine*. "Ours and for us": Hodgskin and Robertson's vision was of a journal in which mechanics could themselves apprehend knowledge and its attendant power. Reprinted from *Mechanic's Magazine* 1 (1823–24): [i], by kind permission of Leeds University Library.

ous emergence of specialist scientific disciplines with discrete organs of publication.

By no means all of the new cheap journals were so accepting of cultural hierarchy. One of the earliest imitators of the *Mirror* was the *Mechanic's Magazine* (figure 5.4). Begun as a threepenny weekly in August 1823, the journal was produced by the journalist and Benthamite radical Thomas Hodgskin and the patent agent Joseph Clinton Robertson, who was a friend and literary collaborator of the *Mirror*'s editor Byerley. From the start, however, the magazine sought to develop a socially distinct audience, removed from the "mass" audience of the *Mirror*. "As yet," the editors wrote in the first number, "there is no periodical Publication, of which that numerous and important portion of the community, the Mechanics or Artisans, including all who are operatively employed in our Arts and Manufactures, can say, 'This is ours, and for us'" (Anon. 1823–24a). The *Mechanic's Magazine* was designed to meet such a market, and from the beginning solicited "Communications from intelligent Mechanics, and from all others who may take an interest in the diffusion of useful information" (Anon. 1823–24a). The magazine thus provided an open forum for the scientific and technical contributions of its artisanal readers, embodying what Susan Sheets-Pyenson calls "low" scientific culture (Sheets-Pyenson 1985). Moreover, it was in the pages of the *Mechanic's Magazine*, in October 1823, that the proposal which led to the initiation of the London Mechanics' Institute appeared, emphasizing that mechanics should take the matter of their education "into their own hands" (Anon. 1823–24b, 102). While the institute was quickly appropriated by middle-class "supporters," it is significant that its origins lay in the cheap journalism of the postwar literary underworld (Stack 1998; Mussell 2006).

Conclusion

As the example of the *Mechanic's Magazine* clearly illustrates, there was not a unified and uncontested "popular science" that emerged at this period. I do not mean to claim that "popular science" had a unitary origin or, indeed, that its meaning was subsequently static. John Timbs's use of the phrase "popular science" in the *Mirror of Literature* was certainly an early usage, but it was not the first. Moreover, as we have seen, it was part of a larger discourse of the "popular" in the period that applied not only to scientific publications such as those produced by Richard Phillips but also to other knowledge publications and to fiction too. However, acknowledging that "popular science" has a history in its own terms is crucial to understanding the competing and changing visions of it that have under-pinned the activities of historical actors. My object in this account has been to sketch out some of the main developments in the book trade in the 1810s and 1820s that helped foster and sustain new notions of popular publication in general and "popular science" publications in particular. As we have seen, contemporaries such as Dibdin recognized the transformative importance of these ventures, but their impact was also clearly demonstrated by the reaction of the Society for the Diffusion of Useful Knowledge and established publishers. In this concluding section, I briefly review the attempts of such publishers to enter the field of self-consciously "popular" publication, showing how they drew upon earlier developments but were also defined in opposition to them.

The formation of the Society for the Diffusion of Useful Knowledge was clearly reactive to the commercial "popular science" ventures of small-time publishers. When Henry Brougham wrote his manifesto on the "Scientific Education of the People" in the October 1824 number of the *Edinburgh Review*, he cited the cheap publications of Limbird as the prototype for his own plans ([Brougham] 1824, 100). By the time he had expanded this article into a pamphlet, published in January 1825 with the title *Practical Observations on the Education of the People*, there were more cheap publications to report on, including the *Mechanic's Magazine*, Hodgskin's *Chemist* (1824–25), the *London Mechanic's Register* (1824–26), and Dolby's cheap histories (Brougham 1825, 3–4). Brougham's pamphlet, which passed through at least nineteen editions within the year, proposed the formation of a society to undertake a program of cheap educational publishing like the existing commercial ventures. An initial meeting of interested parties was held in April 1825, but the financial crash affecting the book trade in the winter of 1825–26 caused further action to be postponed until November 1826 ([Brougham] 1827, 235).

In the interim, numerous publishers had begun to react to the new market signaled by the success of the cheap miscellanies and by Brougham's pamphlet. In the summer of 1825, Charles Knight, by this time a publisher in Pall Mall East, planned a cheap "National Library," to consist of "about a hundred volumes, in History, Science and Art, and Miscellaneous Literature." In the prospectus, Knight was dismissive of the existing cheap publications, claiming that "with some few striking exceptions, the general Literature of our country is either addressed to men of leisure and research, and is, therefore, bulky and diffuse; or it is frittered down into meagre and spiritless outlines, adapted only for juvenile capacities." His new library was to be published in conjunction with an established literary publisher

(Henry Colburn) and a leading trade and wholesale house (G. B. Whittaker), but the financial crash again intervened (C. Knight 1864–65, 2:37). Richard Phillips, too, was up and doing. He later claimed that the idea of "sixpenny-worths of knowledge, and of a Society to give them sanction, and, above all, as a security from the craft of trade" had been all his own, and that he had communicated it to Brougham in July 1825.[16] In December 1825, he out-raged Brougham by issuing a prospectus for a "Library of the People" to be published by a "Society for Diffusing Useful Knowledge," apparently with support from Birkbeck and the scientific agriculturalist Sir John Sinclair ([Brougham] 1827, 235–36; Timbs 1865, 115). Like the *Mirror*, and the later *Library of Useful Knowledge*, Phillips's library was issued (by Knight and Lacey) in serialized octavo numbers, starting in March 1827, though appar-ently without great success.[17] Opportunity also beckoned in Edinburgh. In the months before the crash in which he was so desperately embroiled, Archibald Constable planned a cheap monthly *Miscellany* of original publications by high-profile authors. He told Walter Scott that he would revolutionize "the art and traffic of bookselling" by offering at the rate of "sixpence a-week" volumes "so good that millions must wish to have them, and so cheap that every butcher's callant [i.e., lad] may have them." As a result, he concluded, he would sell them "not by thousands or tens of thousands, but by hundreds of thousands—ay, by millions!" Such volume of sales would, he thought, make him "richer than the possession of all the copyrights of all the quartos that ever were, or will be, hot-pressed" (Lockhart 1837–38, 6:28, 31).

On the eve of the 1825–26 financial crisis, a number of the big com-mercial houses had thus clearly grasped the demand for cheap publica-tions. Once the trade began to recover from the crash, these schemes came to fruition. When the SDUK met in November 1826, Knight's friend the reformist barrister Matthew Hill mentioned his "National Library" plan to Brougham, and for a time it seemed likely to be adopted by the new society. The scheme was ultimately deemed insufficiently scientific, but the SDUK began to issue its "Library of Useful Knowledge" in sixpenny parts—closely modeled on Limbird and Dolby's publications—the follow-ing March (figure 5.5). John Murray promptly adopted Knight's "National Library," issuing a prospectus in December 1826 for a series of "Original Treatises" condensing "the information which is scattered through vo-luminous and expensive works" and encompassing all the "divisions of Popular Knowledge" so as to form a kind of encyclopedia (C. Knight 1864–65, 2:47–48; [Brougham] 1827, 230–31.). While Murray soon abandoned the scheme, he returned to it in 1829, commencing his own "Family Library"

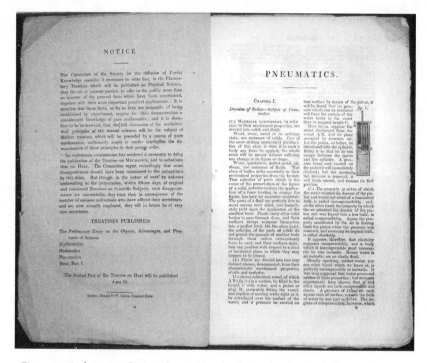

Figure 5.5. In format and price, the treatises in the Society for the Diffusion of Useful
Knowledge's Library of Useful Knowledge emulated the twopenny and threepenny
productions of Limbird, Dolby, and Robertson. Reprinted from [Lardner] 1828, in
author's collection.

of five-shilling nonfiction volumes (Smiles 1891, 2:250–51, 295–96; Bennett
1976). Constable recovered sufficiently to begin a series of monthly volumes at three shillings and sixpence under the title *Constable's Miscellany of Original and Selected Works in Literature, Art, and Science,* and
Longmans entered the market with Lardner's *Cabinet Cyclopaedia* in six-
shilling volumes in 1829 (Peckham 1951).

According to Charles Knight, it was these ventures that marked the
inception of the "modern epoch of cheap literature" (C. Knight 1854, 243).
As I have shown in this account, however, the origin of cheap literature
in general, and of "popular science" publications in particular, in the first
quarter of the nineteenth century owed at least as much to the activities of commercial publishers whose activities have since been obscured.
"Popular science" publishing did not emerge de novo in the work of the
SDUK. Instead, it emerged in the interplay of a series of largely commercial ventures in cheap educational publishing and cheap journalism in the

radically altered conditions of the early nineteenth-century book trade. The publishers involved, ranging from the radical Phillips to the Tory Jerdan, did not share a unified ideology. On the contrary, their activities and approaches were widely divergent and sometimes incompatible. Moreover, they did not share a unified vision of "popular science." Nevertheless, the designation of publications as "popular," and the use of the term *popular science*, indicated the emergence of a range of new commodities that were to have wide implications for the conception of both science and its relationship to a burgeoning range of publics. The inception of the SDUK, with its ideological concerns about managing a potentially unruly working class, was a significant intervention in this ongoing debate. However, the reconceptualization of science and its publics had already begun.

NOTES

1. There are almost no examples of the phrase "popular science" being used in book titles before 1820; after that date, the phrase rapidly came into regular usage.

2. On the use of these names to create a recognizable brand, see also Issitt 1998, 257. Phillips later claimed that "all the elementary books under the names of Blair, Goldsmith, Barrow, Pelham and Bossut" were his own productions between 1798 and 1815, and, while this claim is not entirely accurate, he was also the author of works under other pseudonyms (Phillips [1839], col. 648; Herne 1892–95, 70–71).

3. Prices of his books in the 1800s are given in Phillips [1809?], a copy of which is sewed into the copy of Blair 1809 in Leeds University Library; prices in the 1820s are given in Phillips [1823?], 12–16.

4. Numbering of editions was often spurious during this period; see St. Clair 2004, 180–81.

5. Eugenia Roldán-Vera suggests that the catechisms were begun around 1810, but I have not found any earlier than 1812 (Roldán-Vera 2003, 76 n. 77).

6. *Times* (London), August 3, 1816, 2a.

7. The *DNB* states from personal information that Maunder came from a family settled near Barnstaple, Devon. However, the occurrence of Ann Maunder's marriage in Tonbridge and of a Samuel Maunder trading as a bookseller and printer there (British Book Trade Index Project 2004) gives presumptive if inconclusive evidence of the history outlined here. Another possibility is that the Samuel Maunder trading in Kent was the father of the Samuel and Ann discussed here; if so, he may also have been the bookseller of this name trading in Winchester in 1791 (Piper 1920).

8. On the bankruptcy, see *Times* (London), January 1, 1816, 4a. On the Newbury Bank, see *Times* (London), December 16, 1815, 3g; *Times* (London), December 7,

1825, 3e; and Jerdan 1852, 2:182. In the notice of their bankruptcy, Pinnock and Maunder were listed as partners with "W. Vincent," and a man of that surname was also a partner in the Newbury Bank.

9. That the purchase took place before the collapse of the Newbury Bank is made clear in Jerdan 1852, 2:182. On January 1, 1816, Pinnock and Maunder were still listed in the *Times* as booksellers of Newbury, but they had certainly moved to the Strand by August of that year. See *Times* (London), August 3, 1816, 2a.

10. Brock and Meadows 1998, 74; *Literary Gazette* 2 (1818): 1, 23–24; *Literary Gazette* 6 (1822): 672.

11. *Times* (London), August 7, 1819, 4a.

12. *Times* (London), July 12, 1819, 3e; *Times* (London), July 28, 1819, 3c; Roldán-Vera 2003, 77 n. 79.

13. The ruling was subsequently quoted on the cover of Pinnock's catechisms.

14. *Mirror of Literature* 5 (1825): [iii].

15. *Mirror of Literature,* 5 (1825): iii–iv.

16. Timbs 1865, 122; Anon. 1833, 326; Phillips to Francis Place, 7 July 1829, Add. Ms. 37950 f. 15, British Library, London.

17. *Times* (London), March 1, 1827; [Brougham] 1827, 235–36. A second edition of some of the numbers, dated 1827, is in the Bodleian Library at Oxford University.

REFERENCES

Altick, Richard D. 1998. *The English Common Reader: A Social History of the Mass Reading Public, 1800–1900.* 2nd ed. Columbus: Ohio State University Press.

Anon. 1823–24a. "To the Mechanics of the British Empire." *Mechanic's Magazine* 1:16.

Anon. 1823–24b. "Institutions for Instruction of Mechanics: Proposals for a London Mechanics Institute." *Mechanic's Magazine* 1:99–102.

Anon. 1833. "Peterborough-Court Conversazione. July 31, 1833." *Mechanic's Magazine* 19:324–33.

Axon, William E. A. 1888. *Stray Chapters in Literature, Folk-Lore, and Archaeology.* Manchester: John Heywood.

Barnett, David. 1998. *London: Hub of the Industrial Revolution; A Revisionary History 1775–1825.* London: Tauris Academic Studies.

Bennett, Scott. 1976. "John Murray's Family Library and the Cheapening of Books in Early Nineteenth Century Britain." *Studies in Bibliography* 29:139–66.

Blagden, Cyprian. 1949. *Fire More Than Water: Notes for the Story of a Ship.* London: Longmans, Green.

Blair, David [Richard Phillips]. 1809. *A Grammar of Chemistry; in which the Principles are Familiarized by Easy and Entertaining Experiments.* London: Richard Phillips.

Borrow, George. 1905. *Lavengro: The Scholar—The Gypsy—The Priest.* London: George Routledge and Sons.

Boyle, Andrew. 1951. "Portraiture in *Lavengro:* II. The Publisher—Sir Richard Phillips." *Notes and Queries* 196:361–66.

Briggs, Asa, ed. 1974. *Essays in the History of Publishing: In Celebration of the 250th Anniversary of the House of Longman, 1724–1974.* London: Longman.

British Book Trade Index Project. 2004. Web site, "British Book Trade Index," http://www.bbti.bham.ac.uk (accessed September 30, 2004).

Brock, W. H., and A. J. Meadows. 1998. *The Lamp of Learning: Two Centuries of Publishing at Taylor and Francis.* 2nd ed. London: Taylor and Francis.

[Brougham, Henry]. 1824. "Scientific Education of the People." *Edinburgh Review* 41:96–122.

———. 1825. *Practical Observations upon the Education of the People, Addressed to the Working Classes and their Employers.* 15th ed. London: Longman, Hurst, Rees, Orme, Brown, and Green.

[———]. 1827. "Society for the Diffusion of Useful Knowledge." *Edinburgh Review* 46:225–44.

Chambers, William. 1872. *Memoir of Robert Chambers with Autobiographic Reminiscences.* Edinburgh and London: W. and R. Chambers.

Coleridge, Samuel Taylor. 1830. *On the Constitution of the Church and State, According to the Idea of Each; with Aids Towards a Right Judgment on the Late Catholic Bill.* 2nd ed. London: Hurst, Chance.

Cooter, Roger, and Stephen Pumfrey. 1994. "Separate Spheres and Public Places: Reflections on the History of Science Popularization and Science in Popular Culture." *History of Science* 32:237–67.

Cox, Harold, and John E. Chandler. 1925. *The House of Longman: With a Record of their Bicentenary Celebrations, 1724–1924.* London: Longmans, Green.

Curtis, William. 1896. *A Short History and Description of the Town of Alton in the County of Southampton.* Winchester: Warren and Son.

Dawson, Gowan, Richard Noakes, and Jonathan R. Topham. 2004. "Introduction: Reading the Magazine of Nature." In *Science in the Nineteenth-Century Periodical: Reading the Magazine of Nature,* ed. Geoffrey Cantor, Gowan Dawson, Graeme Gooday, Richard Noakes, Sally Shuttleworth, and Jonathan R. Topham, 1–34. Cambridge: Cambridge University Press.

De Morgan, Augustus. 1915. *A Budget of Paradoxes.* 2nd ed. 2 vols. Chicago: Open Court.

Dibdin, Thomas Frognall. 1824. *The Library Companion; or, The Young Man's Guide, and the Old Man's Comfort, in the Choice of a Library.* London: Harding, Triphook, and Lepard; and J. Major.

[Elwin, Whitwall]. 1849. "Popular Science." *Quarterly Review* 84:307–44.

Fyfe, Aileen. 1999a. "Copyrights and Competition: Producing and Protecting Children's Books in the Nineteenth Century." *Publishing History* 45:35–59.

————.1999b. "How the Squirrel Became a Squgg: The Long History of a Children's Book." *Paradigm* 27:25–37.

————. 2000. "Reading Children's Books in Late Eighteenth-Century Dissenting Families." *Historical Journal* 43:453–73.

Hawkins, C. W. 1973. *The Story of Alton in Hampshire*. [Alton]: Alton Urban District Council.

Herne, Frank S. 1892–95. "An Old Leicester Bookseller (Sir Richard Phillips)." *Transactions of the Leicester Literary and Philosophical Society*, n.s., 3:65–73.

Hilgartner, Stephen. 1990. "The Dominant View of Popularization: Conceptual Problems, Political Uses." *Social Studies of Science* 20: 519–39.

Holland, Susan, and Steven Miller. 1997. "Science in the Early *Athenaeum:* A Mirror of Crystallization." *Public Understanding of Science* 6:111–30.

[Hone, William]. 1820. *The Man in the Moon, &c. &c. &c*. London: William Hone.

Horrocks, Sally. 1987. "Audiences for Chemistry in Regency Britain: Mrs. Marcet's *Conversations on Chemistry*." BA diss., University of Cambridge.

Issitt, John. 1998. "Introducing Sir Richard Phillips." *Paradigm* 26:25–45.

————. 2000. "The Life and Works of Jeremiah Joyce." PhD diss., Open University.

————. 2002. "Jeremiah Joyce: Science Educationist." *Endeavour* 26:97–101.

Jerdan, William. 1852. *The Autobiography of William Jerdan, With His Literary, Political, and Social Reminiscences and Correspondence During the Last Fifty Years*. 3 vols. London: Arthur Hall, Virtue.

————. 1866. *Men I Have Known*. London: G. Routledge.

Knight, Charles. 1854. *The Old Printer and the Modern Press*. London: John Murray.

————. 1864–65. *Passages from a Working Life During Half a Century, With a Prelude of Early Reminiscences*. 3 vols. London: Bradbury and Evans.

Knight, David. 1986. "Accomplishment or Dogma: Chemistry in the Introductory Works of Jane Marcet and Samuel Parkes." *Ambix* 33:94–98.

————. 2000. "Communicating Chemistry: The Frontier between Popular Books and Textbooks in Britain during the First Half of the Nineteenth Century." In *Communicating Chemistry: Textbooks and Their Audiences, 1789–1939*, ed. A. Lundgren and B. Bensaude-Vincent, 187–205. Canton, MA: Science History Publications.

Lackington, James. 1791. *Memoirs of the First Forty-Five Years of the Life of James Lackington, The Present Bookseller in Chiswell-Street, Moorfields, London. Written by Himself. In a Series of Letters to a Friend*. London: Printed for and sold by the author.

[Lardner, Dionysius.] 1828. *Pneumatics*. 4th ed. London: Baldwin and Craddock.

Lockhart, John Gibson. 1837–38. *Memoirs of the Life of Sir Walter Scott*. 7 vols. Edinburgh: Robert Cadell; London: John Murray and Whittaker.

Mavor, William. 1809. *Catechism of Health*. London: Lackington, Allen.

Mussell, James. 2006. "'This is Ours and For Us': The *Mechanic's Magazine* and Low Scientific Culture in Regency London." In *Repositioning Victorian Sci-*

ences: Shifting Centres in Nineteenth-Century Scientific Thinking, ed. David Clifford, Elisabeth Wadge, Alex Warwick, and Martin Willis, 107–17. London: Anthem Press.

IIAN. 1825. "On Cheap Periodical Literature." *Gentleman's Magazine* 95 (1): 483–86.

Peckham, Morse. 1951. "Dr. Lardner's *Cabinet Cyclopedia*." *Papers of the Bibliographical Society of America* 45:37–58.

Phillips, Richard. [1809?]. *Useful Books Recently Published, or in the Course of Publication, by Richard Phillips*. London: [Richard Phillips].

———. 1821. *The First, or Mother's Catechism; Containing Common Things Necessary to be Known by Children at an Early Age*. London: Richard Phillips.

———. [1823?]. *Illustrations of the Interrogative System*. London: G. and W. B. Whittaker.

———. [1839]. *A Million of Facts*. Stereotype ed. London: Ward, Lock.

[Pinnock, William]. [1819?]. *Catechism of Mineralogy*. London: Pinnock and Maunder.

Piper, A. Cecil. 1920. "The Early Printers and Booksellers of Winchester." *Library*, 4th ser., 1:103–10.

Rauch, Alan. 2001. *Useful Knowledge: The Victorians, Morality, and the March of Intellect*. Durham,NC: Duke University Press.

Roldán-Vera, Eugenia. 2003. *The British Book Trade and Spanish American Independence: Education and Knowledge Transmission in Transcontinental Perspective*. Aldershot: Ashgate.

Schaffer, Simon. 1986. "Scientific Discoveries and the End of Natural Philosophy." *Social Studies of Science* 16:387–420.

Secord, James A. 1985. "Newton in the Nursery: Tom Telescope and the Philosophy of Tops and Balls, 1761–1838." *History of Science* 23:127–51.

Sheets-Pyenson, Susan. 1985 "Popular Science Periodicals in Paris and London: The Emergence of a Low Scientific Culture 1820–1875." *Annals of Science* 42:549–72.

Shiach, Morag. 1989. *Discourse on Popular Culture: Class, Gender and History in Cultural Analysis, 1730 to the Present*. Cambridge: Polity Press.

Smiles, Samuel. 1891. *A Publisher and His Friends: Memoir and Correspondence of the Late John Murray, with an Account of the Origin and Progress of the House, 1768–1843*. 2nd ed. 2 vols. London: John Murray.

Southey, Charles Cuthbert. 1849–50. *The Life and Correspondence of Robert Southey*. 6 vols. London: Longman, Brown, Green, and Longmans.

Stack, David. 1998. *Nature and Artifice: The Life and Thought of Thomas Hodgskin (1787–1869)*. London: Royal Historical Society.

St. Clair, William. 2004. *The Reading Nation in the Romantic Period*. Cambridge: Cambridge University Press.

Sullivan, Alvin, ed. 1983. *British Literary Magazines: The Romantic Age, 1789–1836*. Westport, CT: Greenwood Press.

Timbs, John. 1865. *Walks and Talks About London*. London: Lockwood.

———. 1871. "My Autobiography: Incidental Notes and Personal Recollections." *Leisure Hour*, July 29, 469–72.

———. 1872. "Thirty Years of the Reign of Victoria: Personal Recollections." *Leisure Hour*, April 6, 221–23.

Topham, Jonathan R. 2000. "Scientific Publishing and the Reading of Science in Nineteenth-Century Britain: A Historiographical Survey and Guide to Sources." *Studies in History and Philosophy of Science* 31A:559–612.

———. 2004a. "The *Mirror of Literature, Amusement and Instruction* and Cheap Miscellanies in Early Nineteenth-Century Britain." In *Science in the Nineteenth-Century Periodical: Reading the Magazine of Nature*, ed. Geoffrey Cantor, Gowan Dawson, Graeme Gooday, Richard Noakes, Sally Shuttleworth, and Jonathan R. Topham, 37–66, 261–65. Cambridge: Cambridge University Press.

———. 2004b. "Whittaker, George Byrom (1793–1847)." In *Oxford Dictionary of National Biography*. Oxford: Oxford University Press.

———. 2005. "John Limbird, Thomas Byerley, and the Production of Cheap Periodicals in the 1820s." *Book History* 8:75–106.

———. Forthcoming. "Scientific Books, 1800–1830." In *Cambridge History of the Book in Britain*, Vol. 5, *1695–1830*, ed. Michael Turner and Michael Suarez. Cambridge: Cambridge University Press.

Wallis, Philip. 1974. *At the Sign of the Ship: Notes on the House of Longman, 1724–1974*. London: Longman.

Yeo, Richard. 1984. "Science and Intellectual Authority in Mid-Nineteenth-Century Britain: Robert Chambers and *Vestiges of the Natural History of Creation*." *Victorian Studies* 28:5–31.

CHAPTER SIX

Sensitive, Bashful, and Chaste?
Articulating the *Mimosa* in Science

ANN B. SHTEIR

In the opening section of *Conversations on Vegetable Physiology* (1829), the well-known popularizer Jane Marcet focused on one plant, the *Mimosa pudica*, as a way to arouse the curiosity of her audience. This novel and exotic botanical specimen was a good example to feature in a book that aimed to introduce young readers to the excitement of new work about the structure of plants (figure 6.1). Commonly known as the Sensitive Plant, the *Mimosa pudica* has long been an object of wonder and curiosity because its leaflets close up and droop when touched, as if recoiling from the intrusion of a hand or a breath. Marcet introduces the Sensitive Plant into the conversation between a teacher named Mrs. B. and her two young female pupils about differences between the animal and vegetable creation. When Mrs. B. explains that plants are unlike animals in that they are "not endowed with sensibility," young Emily refers to the *Mimosa* as an example of a plant which seems to blur that distinction.

EMILY: But the mimosa or sensitive plant, Mrs. B., when it shrinks from the touch, wears a strong appearance of sensibility.

MRS. B.: Yet I should doubt whether it is any thing more than appearance. Some ingenious experiments have, indeed, been recently made, which tend to favour the opinion that plants may be endowed with a species of sensibility; and seem to render it not improbable that there may exist in plants something corresponding with the nervous system in animals.

CAROLINE: The sensitive plant would then, no doubt, be a nervous fine lady at the court of Flora. (Marcet 1829, 9)

Mrs. B. does not accept the personification of the *Mimosa* as "a nervous fine lady at the court of Flora" and has no interest in invoking Flora, the

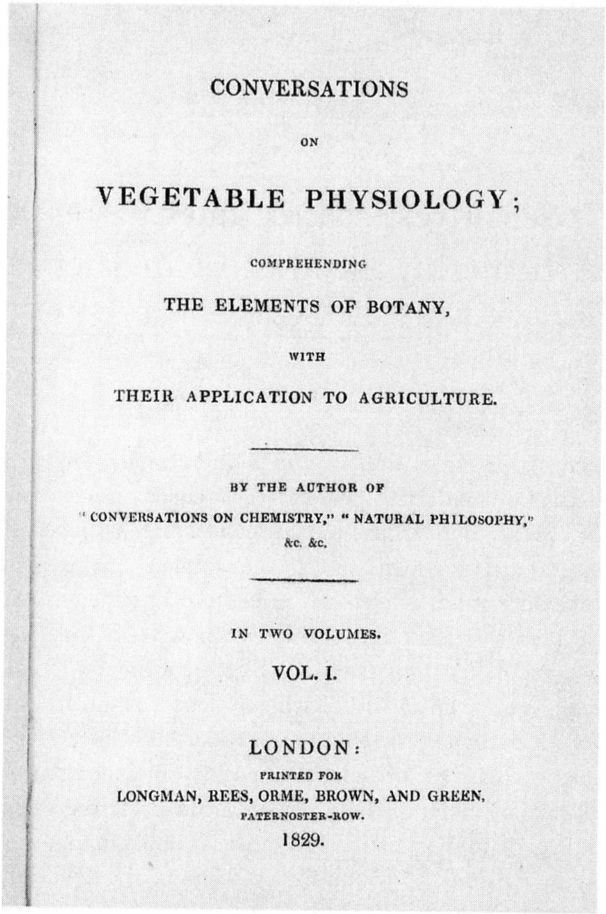

CONVERSATIONS

ON

VEGETABLE PHYSIOLOGY;

COMPREHENDING

THE ELEMENTS OF BOTANY,

WITH

THEIR APPLICATION TO AGRICULTURE.

BY THE AUTHOR OF

" CONVERSATIONS ON CHEMISTRY," " NATURAL PHILOSOPHY,"
&c. &c.

IN TWO VOLUMES.

VOL. I.

LONDON:

PRINTED FOR
LONGMAN, REES, ORME, BROWN, AND GREEN,
PATERNOSTER-ROW.
1829.

Figure 6.1. Title page of *Conversations on Vegetable Physiology*. Reprinted from [Marcet] 1829. Courtesy of the Osborne Collection of Early Children's Books, Toronto Public Library.

mythological goddess of flowers. Her orientation is toward mechanistic renderings of natural processes rather than iconographic references and literary languages of nature. Stepping away from any notion of plant volition or any figural analogy, she proceeds to teach Caroline and Emily about plant physiology instead. Mrs. B. describes in some detail a series of contemporary experiments that involved placing vegetable poisons on plants, including on the Sensitive Plant. These experiments showed exclusively physical causes for motion, providing no evidence that a plant could feel pleasure or pain or could have volition: "A plant rooted in the earth is a poor, patient, passive being: its habits, its irritability, and its contractibility, all depending on mere physical causes" (Marcet 1829, 14). Mrs. B. goes on to explain the cellular and vascular systems of plants, thus giving a technical botanical tone and direction to the rest of the book.

Jane Marcet's central placement of the *Mimosa pudica* in *Conversations on Vegetable Physiology* illustrates the fascination that the Sensitive Plant has long held for communities of readers, writers, and students of nature. Going back into sixteenth-century accounts by plant hunters and across the years into speculations by eighteenth-century systematizers and experimentalists, the Sensitive Plant captured the interest of many because of its physical responsiveness. During the early nineteenth century, accounts of the *Mimosa pudica* proliferated within print culture. Writers on scientific topics offered quite varied accounts of the movements of the Sensitive Plant as they sought to explain its curious qualities to readers of diverse ages, education levels, and experience. Three separate but overlapping *Mimosa* discourses are apparent in the books and periodicals that are the "sites" of this chapter. The Sensitive Plant appeared in scientific works, with their technicalities and styles of presentation of material. It flourished in literary writing, where it was personified, treated playfully and philosophically, and used as an object lesson for sexed and gendered human behavior. As well, it featured prominently in popularizing books and periodicals, where vocabularies for describing the *Mimosa* were often in tension. By considering accounts of the *Mimosa* as a group, we become aware of the cross talk and can more easily identify commonalities and differences among how writers presented this wondrous and mysterious botanical specimen. The lability in renderings of the Sensitive Plant makes it useful for studying the articulations in languages of nature during the years 1800 through 1830. In particular, the story of the *Mimosa pudica* within the history of science writing is a register of how and why literary and figurative elements were excised from accounts that sought scientific authority.

At the time when Jane Marcet was preparing her book about plant physiology, all three discourses about the *Mimosa*—scientific, literary, and popularizing—were in play in British writings about plants. Hence she and her publishers could make deliberate choices about the content, vocabulary, and narrative styles they would use to craft books for new audiences and to disseminate introductory material about plant physiology. Marcet chose a narrative form shared by other didactic writers of her generation. The conversational "familiar" style of *Conversations on Vegetable Physiology* links it to introductory works of informal education that had been a backbone of Enlightenment publishing for decades. Following a model established with her *Conversations on Chemistry* twenty years earlier, and through the personae of Mrs. B. and the two girls, Marcet engages the intellectual attention of young readers, particularly girls and young women. The conversational form is meant as a way to fix the attention of this audience, and by the late 1820s, when the physiological turn within botany was consolidating, Marcet wisely chose to feature topics such as roots, stems, leaves, sap, and atmosphere for readers who wanted to keep up with topics of current interest. *Conversations in Vegetable Physiology* also has an applied dimension in relation to agriculture and provides information about soil, the structure of seeds, and diseases in plants.

The interlocutors in Marcet's books illustrate the "scientific mothers" feature of the history of popular science writing (Gates and Shteir 1997, 9). The conversations between Mrs. B. and her pupils are meant to record moments of transformation, awakening to new scientific knowledge. In the opening conversation, Mrs. B. acknowledges that she herself had been bored at an earlier time by a style of botany that had only to do with names and classification. Her "conversion" came from hearing the lectures by the botanist Augustin de Candolle, and young Emily acknowledges the particular power that comes from learning from someone who herself has been "converted" (Marcet 1829, 14). The narrative form is therefore neither backdrop nor incidental to the work of this book, and the personalities and responses of the girls are part of the process. For Marcet and other writers of scientific dialogues, the scene setting and interplay of teacher and students manifest the content of the intellectual drama in ways that echo the narrative brilliance of Platonic dialogues.

Jane Marcet's *Conversations on Vegetable Physiology* participates in the history of the "familiar format" within science writing by its use of a domestic setting, characterization, and some ancillary teachings. But the real nexus of Marcet's book uses literary features precisely to take science and science teaching away from a literary approach to nature that has to do with personification, mythology, and the anthropomorphic—away, in other words, from young Caroline's figure of the *Mimosa* as "a nervous fine lady at the court of Flora." Gesturing to an older tradition of experimental work on the *Mimosa*, Marcet crafts an introductory and popularizing science book in 1829 that is of a serious kind. She joins the host of female authors who had used their science writing to expand the knowledge base for women and worked hard to model informed and serious encounters with plants. This was a polite science for its day, but her style of science was neither romantic, poetical, nor feminized. Languages of nature blend here, as

Marcet melds technical approaches to writing about science with narrative devices that help make her subject matter enticing and accessible.

The variety of discourses around the *Mimosa* makes this plant specimen an intriguing focus for studying how science is fashioned in sites within print culture. In particular, the figure of the *Mimosa* foregrounds literary strategies in botanical writing and brings choices about readership and science culture into high relief. But the *Mimosa* is not solely about crafting catchy representations of an incidental plant across many kinds of writing. The physicality of the Sensitive Plant gives it an additional dimension. As we shall see, observers of the *Mimosa*, and those writing and reading about it, often linked the sentient and corporeal character of the plant to ideas about male sexuality, beliefs about female chastity, and norms of virtue and female conduct. Such historical associations between the *Mimosa* and the human sexed and gendered body resonated within literary and scientific cultures. Depictions of the *Mimosa* serve, therefore, as a window onto social and cultural ideas circulating in audiences being cultivated for the marketplace of nineteenth-century science.

Scientific Accounts

Jane Marcet's discussion of the *Mimosa* from the 1820s is one of numerous nineteenth-century accounts of this plant by scientists preoccupied with plant movement and mechanisms. To this day, botanists still seek to understand the processes at work in the tropisms of the *Mimosa* and plants such as the Venus flytrap (Ueda and Yamamura 1999). Across earlier centuries, natural philosophers, naturalists, and physiologists found the *Mimosa* alluring in its technical and symbolic aspects. They featured the *Mimosa* in debates about the gradation of being and analogies between plants and animals, and also about whether plants move because of mechanistic "irritability" or because of some kind of "sensitivity" and volition. Poets, playing with the figurative possibilities of analogy, claimed the *Mimosa* for a range of purposes. Erotic writers too had a field day with the Sensitive Plant, linking its physical responses to sexuality as a phallic icon of male sexuality, for example, or as a figure of female chastity and self-protection. Writers of conduct literature in turn seized the potential of the *Mimosa* for moral lessons, particularly in cautionary tales for women. Moving into the nineteenth century, writers who set out to popularize the study of botany at a time of high cultural interest in natural history and science adapted and recast the *Mimosa* in a variety of vocabularies and narrative frames.

Early records of the Sensitive Plant date from the late sixteenth century when New World specimens were introduced into English and European gardens (Webster 1966, 6–9). The herbalist John Parkinson writing in *The-atrum Botanicum* (1640) described two kinds of plants that close up when touched or breathed upon. He labeled one of them "Herba Viva" (or "Herba amoris"), with the power "to procure love between man and woman," and the other "Herba Mimosa" (or "the Mimicke herbe"), "which upon the touch of any man, or his breathing onely, and not of any thing else would shrinke and seeme as withered" (Parkinson 1640, 1617). Parkinson's illustration of "Herba Mimosa" is shown in figure 6.2. Parkinson also details an experiment on a plant that grew in a Chelsea garden in 1638–39, showing that young leaves respond to touch in a way different from older leaves. He wrote that whereas older leaves "[fold] themselves upward close to the middle ribbe, upon any touch thereof," on younger leaves "the whole stalke with leaves would fall downe and shrinke up the leaves" (Parkinson 1640, 1618). The curiosity value of these plants drew the attention of early horticulturalists and experimentalists, who debated ways of explaining their motion. Does the *Mimosa* move because of animating spirits and animal-like volition, or is plant movement a mechanical response to stimuli? During the seventeenth and eighteenth centuries, key thinkers waded into discussions about whether plants have animal souls or vegetative souls and about where demarcation lines should be placed in gradations of being. The Royal Society took on the topic of the Sensitive Plant in July 1661 at the request of King Charles II and conducted experiments that explained the motions of the plant by analogy to human anatomy and physiology. By contrast, the botanist John Ray was skeptical about plant sensitivity and analogical explanation, and he argued instead in ensuing decades that movement in plants had only to do with changes in temperature and moisture (Webster 1966, 10–23). Writing in *Curiosities of nature and art in husbandry and gardening* (1707), the Abbé de Vallemont declares that the Sensitive Plant "has no more sense than a Cabbage," and he explains its movements in terms of its "juices" "retreating" in response to cold (Vallemont 1707, 87–91). At mid-eighteenth century, the botanist John Hill took a different view, based on experiments that he performed on the "sleep" of plants and cause of motion in the Sensitive Plant. He reported on his plant dissections and use of microscopes in a letter to Linnaeus in 1757 and claimed that light, rather than heat or moisture, causes the leaves to close and droop. Choosing mechanistic explanations, he argued that light puts the "fibres" into "an incessant vibration," exciting motion (Hill 1757, 38).

When Linnaeus stepped onto the stage of European botany, Hill's mechanistic type of explanation for the behavior of the *Mimosa* took a backseat

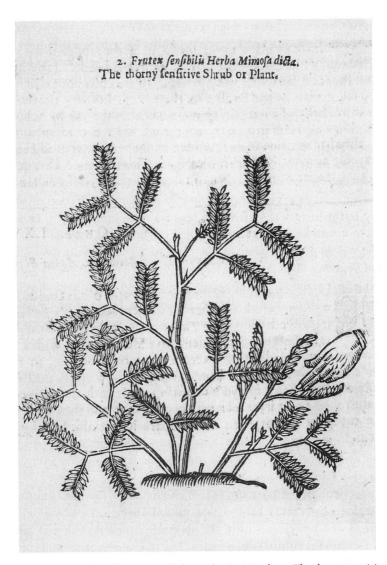

2. *Frutex sensibilis Herba Mimosa dicta.*
The thorny sensitive Shrub or Plant.

Figure 6.2. Illustration "Frutex sensibilis Herba Mimosa dicta. The thorny sensitive Shrub or Plant." Reprinted from Parkinson 1640, 1617. Courtesy of Hunt Institute, for Botanical Documentation, Carnegie Mellon University, Pittsburgh, PA.

to other types of botanical concerns for much of the eighteenth century. Linnaeus's sexual system of plant classification and his binomial system of nomenclature changed the face of botany into the early nineteenth century and indeed beyond that as well. Linnaeus based his working taxonomy of plants in assumptions about analogies between plant processes and human processes, especially beliefs about sexuality and distinctions between male and female. His theories represent what Philip Ritterbush has termed "the triumph of botanical analogy" (Ritterbush 1964, chap. 4). Linnaeus located *Mimosa* in class 23, order 1 of his sexual system. Using the reproductive parts of flowers to identify and distinguish types of plants, he classified *Mimosa* as Polyandria Monogynia, his category for plants with male, female, and hermaphroditic flowers on the same plant, which have many stamens and one pistil. Like other botanists before him, Linnaeus distinguished among different groupings of *Mimosa*, listing thirteen types in his *Hortus Cliffortianus* (1737) and expanding that number to thirty-nine in his *Species plantarum* of 1753. Linnaeus's *Species plantarum* is of special significance in the history of systematics because there Linnaeus, the father of modern botanical nomenclature, coined the binomial *"Mimosa pudica"* as the shorthand designation for the plant we conventionally now refer to as the Sensitive Plant (Linnaeus [1753] 1957, 518).[1] Linnaeus thereby distinguished that plant within the genus *Mimosa* from others, including from *Mimosa sensitiva*, a botanically separate species. By selecting the species term *pudica*, Linnaeus imported culture into botanical nomenclature on a large scale, for the Latin adjective *pudicus* means "modest" or "bashful" and derives from the verb *pudeo*, meaning "to be ashamed." Associated with "shame," "shyness," and "shameful," its antonym *impudicus* is conventionally translated as "shameless," "lewd," and "unchaste." Linnaeus's penchant for the analogical and anthropomorphic is well known, as is the imprint of his own values and cultural norms on Linnaean ideas about sexual difference and botanical systematics (Schiebinger 1993, chap. 1). Certainly, analogical assumptions and practices are at work in Linnaeus's coinage of *Mimosa pudica*.

The popularity of Linnaean botany in England across the later part of the eighteenth century gave metaphors and analogical explanations for practices and processes within the vegetable kingdom a special prominence within the cultural imagination. The playfulness of accounts of Linnaeus's ideas in publications after 1760 helped disseminate his theories to wider audiences, and this wider dissemination in turn contributed to the fashion of Linnaean botany among polite audiences for the rest of the eighteenth century. Moreover, analogical ideas about plants and animals that underlie many Linnaean writings contributed to organicist theories that characterize early romantic science, including the speculations of Erasmus Darwin. Darwin's writings from the late eighteenth century into the opening years of the nineteenth century often seamlessly meld figurative and technical approaches to nature. As we shall see, an analogical approach to nature shaped his literary rendering of Linnaean botanical ideas in the 1789 expository "Loves of the Plants" (in E. Darwin 1791). The same approach to nature also infuses his more technical writing in *Phytologia; or the Philosophy of Agriculture and Gardening* (1800). There Darwin builds on experimental work that had been conducted over the previous century to understand the physiology of plant growth, but he disagrees with those who explain plant functions in terms of mechanics alone. Drawing direct links between plants and human body parts and systems, Darwin writes in the *Phytologia* about the development of plants in terms of lungs, umbilical vessels, pulmonary and aortal arteries and veins, muscles, and nerves. According to Darwin's theory of vegetation, the *Mimosa* illustrates the vitality of the vegetable kingdom, "a sensitive sensorium, or brain, existing in each individual bud or flower" (E. Darwin 1800, 135). The movement of the leaves of plants stands as proof for him that they have sensibility and volition as well as a sense of touch.

In the years following Erasmus Darwin's physiological work, researchers continued to debate the causes of motion in the Sensitive Plant, conducting experiments, challenging the interpretations of one another's findings, and making their work known. Among early nineteenth-century plant physiologists, Agnes Ibbetson, for one, argued strongly for mechanical explanations of plant movement and functions. She strenuously disagreed with the notion that plants have feeling or volition, and she expounded her theories about plant growth in periodical publications as well as in unpublished works (Shteir 1996, 120–35). In a series of essays in William Nicholson's *Journal of Natural Philosophy, Chemistry, and the Arts* during the years 1802 through 1813, she laid out detailed accounts of experiments she conducted during a long research program that involved close observation, plant dissection, and extensive study with solar microscopes. The *Mimosa* was a prominent specimen for her work, and it gave credence to her belief that plants are "machines governed wholly by light and moisture; and dependent on these causes for motion" (Ibbetson 1809, 114).

Physiological issues such as these dominated nineteenth-century botanical inquiry about the *Mimosa* and were widely featured in magazines and books. The *Gardeners' Chronicle*, for example, reported in December 1848 on experiments conducted on the Sensitive Plant in order to explore analogies between plants and animals. One botanist used chloroform on the leaf of a Sensitive Plant, and others used nitric or sulfuric acid, the results suggesting that plants have "special organs susceptible of being affected by vegetable poisons in a way not unlike that in which the nervous system of animals is affected" (*Gardeners' Chronicle* 1848, 859).

Charles Darwin was interested in these features of the *Mimosa pudica* as well, and an exchange of letters with Joseph Hooker in October 1862 alludes to experiments he did to study the effects of ether and other narcotic and poisonous substances on how plants such as this one responded (Burkhardt 1997, 483–85, 486–88). Observations and experiments that he conducted with his son Francis on circumnutation in plants led in time to their book *The Power of Movement in Plants* (C. Darwin 1880), and the *Mimosa pudica* is one of their objects of study regarding the nyctitropic or sleep movements of leaves. Using language devoid of literary allusion, Charles Darwin writes, "At night, as is well known, the opposite leaflets come into contact and point towards the apex of the leaf; they thus become neatly imbricated with their upper surfaces protected" (C. Darwin 1880, 374). Charles and Francis Darwin conducted a series of experiments to trace the movement of the Sensitive Plant under carefully controlled conditions, and they detail precisely timed observations of movement in the Sensitive Plant in diagrams that represent a way of visualizing the *Mimosa pudica* that is in keeping with late nineteenth-century scientific languages and practices (see, e.g., figure 6.3). During this time, botanical systematists also continued to study the Sensitive Plant. George Bentham published the classic taxonomic account of the *Mimosa* in 1875, based on work about the natural order Leguminosae dating back to the 1840s, and this account remains "the fundamental text" on the genus (Bentham 1875; Barneby 1991).

Literary Accounts

Among the many magazines that cultivated the scientific curiosity of early nineteenth-century readers, the *Botanist's Repository* was an illustrated publication from the years 1797 to 1814 that featured new, rare, or otherwise interesting plants. When the botanical painter and engraver Henry Andrews depicted the *Mimosa pudica* there in 1808, he used technical scientific vocabulary from the plant chemistry of his day to describe the intriguing movements of its leaves: "Its singular quality of shrinking from the touch is supposed to be owing to its being strongly saturated with oxygen gas, which it disengages upon the slightest provocation, and its place for a short time is supplied by the atmospheric air; which retiring, the leaves again resume their former appearance, and so remain expanded till the evening, unless disturbed by design or accident; for the rude approach of the common air disorganizes its foliage" (Andrews 1808, pl. 544). At

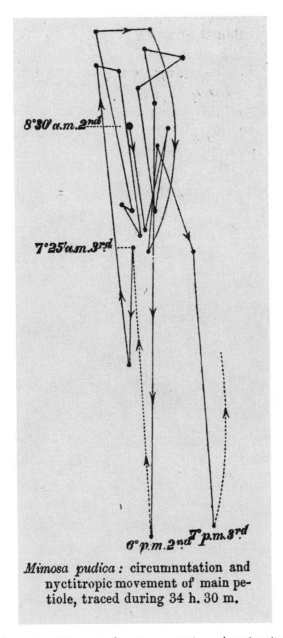

Mimosa pudica : circumnutation and
nyctitropic movement of main petiole, traced during 34 h. 30 m.

Figure 6.3. Illustration "*Mimosa pudica:* circumnutation and nyctitropic movement of main petiole, traced during 34 h. 30 m." Reprinted from C. Darwin 1880, 375, fig. 157. Courtesy of The Thomas Fisher Rare Book Library, University of Toronto.

Figure 6.4. Illustration *"Mimosa pudica/Bashful Mimosa."* Reprinted from Andrews 1808, pl. 544. Courtesy of the Hunt Institute for Botanical Documentation, Carnegie Mellon University, Pittsburgh, PA.

the same time, he also labeled the *Mimosa pudica* "Bashful *Mimosa*" (see figure 6.4) and by so doing invoked a body of figural vocabulary that carries its own cultural resonances and traditions.

These overlapping discourses of nature are less surprising than one might expect during the period under discussion here. Nevertheless, technical experimental accounts of motion in plants usually contrast in their purposes, language, and overall effect with literary accounts of the *Mimosa*. For anyone who opposes mechanistic accounts of plant motion and finds analogical patterns congenial, there was—and is—a wide world of possibilities for animating, embodying, and gendering the *Mimosa*. During the eighteenth century, ideas about "sensibility" and "sensitivity" referred to the senses and physical sensations in the body and also to feelings of an emotional or spiritual kind. In multiple genres across the print culture of that time, writers explored the *Mimosa* as metaphor. Figural uses abounded, but the Sensitive Plant had resonance in domains of the body as well. While the *Mimosa* featured in cultural representations of that time, many meanings that attached to it were about the responsiveness of the physical plant itself.

For example, the Sensitive Plant is used iconically in John Cleland's erotic classic of the mid-1700s, *Memoirs of a Woman of Pleasure*, to celebrate male genitalia. There, among the many episodes of exuberant sexuality in this novel, Fanny Hill describes a moment when she and a friend have set out to seduce a young country boy: "My fingers . . . had now got within reach of the true, the genuine, sensitive plant, which instead of shrinking from the touch, joys to meet it, and swells, and vegetates under it" (Cleland [1748–49] 1985, 162). *Fanny Hill* is but one of the erotic and pornographic books from eighteenth-century England that used plants as phallic allegories or as codes for female sexual parts. Linnaean writings about plant sexuality and the loves of plants opened literary doors for botanical erotica, and writers cultivated fertile fields of play. At a time when theories and vocabulary from botany, reproductive theories, and electricity were featured in erotic writings, female sex organs were portrayed as caves and shrubs, and male genitalia were portrayed as examples of the virile "Tree of Life," or as the Sensitive Plant (Peakman 2003, 67–76). Thus, the Sensitive Plant was figured as a male erotic icon in James Perry's "Mimosa" (1779), a parodic poem addressed to a noted aristocratic woman of fashion. The poem is full of sexual innuendo, and the fun is meant to be in the phallic imagery. Perry describes the *Mimosa* as a plant about which "ladies throb & thrill":

> This plant, enrich'd with sense and life,
> Pleases the widow, and the wife;
>   With its Articulations. . . .
> The *root*, the *pedicle*, and *stem*
> Rise up, as nature, gentle dame
>   With kind *sensation* fires.

From *drooping curves*, before our eyes,
When *perpendiculars* arise;
   What Lady but admires them.

Perry continues:

The ladies, when a *plant* they see
Exalted to a great degree,
   All long to have a *touch.* ([Perry] 1779, 8–10)

Elsewhere in the poem Perry writes:

Various, and many are the kinds,
The Botanist, and female finds
   Of this luxuriant plant:
In *hot-beds*, do they thrive and grow,
And there, to best advantage shew
   Their *motions* that inchant. ([Perry] 1779, 13)

Perry dedicated his *"Mimosa"* to the botanist Joseph Banks, who had traveled to Tahiti with Captain James Cook as a naturalist on the voyage of the *Endeavour* and was satirized as a "Botanic Macaroni" and "botanical libertine" in popular publications of that time (Fara 2003, 9; Bewell 1996, 181). Perry's erotic botanical verses illustrate, therefore, a kind of common ground between literary culture and natural history.

But there was more than just masculine hegemony in the parade of visual and figural forms of the *Mimosa* during the Linnaean years. Other kinds of gender ideology play upon the *Mimosa* as *pudica.* This species epithet offered quite different narrative possibilities to those with a more cautionary mind-set. John H. Wynne's *Fables of Flowers* (1773), for example, is a collection of verse meant to supply a female audience with lessons about conduct. A moralizing fable about "The Sensitive Plant," for example, sexes the *Mimosa* female and animates her as a "maiden" "shrunk in herself" with "modest grace" and a "spotless breast." Far from phallic and sexualized, this *Mimosa* is intent upon protecting her "virtue," and the fable ends with a warning by the plant herself that "The Nymph who slights strict virtue's guard, / Shall quickly meet a snare" ([Wynne] 1773, 42–45).

Of course, plant imagery and flower imagery are as varied as their real-life counterparts in the vegetable kingdom, and the responsive *Mimosa* is no exception. Transsexual or cross-dressed, sexual fantasy or scare tactics, the sensibility and physicality of the Sensitive Plant generate many different voices. In a "poetical essay" entitled "The SENSITIVE PLANT and The ROSE-BUD" in the popular *Lady's Magazine* for November 1802, for example, a mother uses the *Mimosa* as an object lesson for her daughter about modesty and danger: "What Nature has taught to this delicate flower, / Let a mother's fond counsel impart" ("SENSITIVE PLANT" 1802, 605). But fair young Celia is lured instead by the promise of a shepherd that his kiss will transform the rosebud into a rose. The "blooming girl," being carefully raised and cultivated, is one prominent trope in what Amy M. King recently identified as a "botanical vernacular" across English novels of the eighteenth and nineteenth centuries (King 2003). Examples abound of emblematic references to women as specific flowers and also as flowers in various stages of growth. In one such illustrated book, entitled *Floral Emblems* (Phillips 1825), the *Mimosa* is figured as the embodiment of "Bashful Modesty." The verse accompanying this illustration reads: "Whence does it happen that the plant, which well / We name sensitive, should move and feel? / When know her leaves to answer her command, / And with quick horror fly the neighbouring hand?" (Phillips 1825). The modish nineteenth-century literary confection of the "language of flowers" gave wide range to symbolic interpretations of this kind. These references often were gender-linked to features such as "feminine" emotional responsiveness and were widely associated with beliefs about women's nature.

However, references to the *Mimosa* in literary texts also could resonate in ways that carried no gender associations at that time. At a level of literary high culture, Percy Shelley fashioned his own *Mimosa* in his 1820 poem "The Sensitive Plant," seizing upon analogies between the *Mimosa* and humankind that go beyond emotions or sexuality, or even sex-specific identification. Shelley's poem depicts a garden that abounds with life, cared for by an idealized "Lady," an anima mundi, who then dies and leaves her universe in desolation. A Sensitive Plant remains in the garden, standing companionless, a "leafless wreck" (Shelley [1820] 1975, 111), a physical and mortal object, full of responsiveness, representing humankind in the cosmos, and yearning for something more. Shelley imagined "a force working like sentiency . . . in all vegetative life" (Maniquis 1969, 154), and his poem is filled with plant symbols and with characteristically romantic ideas about sensitivity, analogy, and the imagination. In a similar way, two decades later, Anne Brontë referred to the *Mimosa* in her 1847 novel *The Tenant of Wildfell Hall* in order to describe a male character whose personality was chilly and reserved. "His heart," she writes, "was like a sensitive plant, that opens for a moment in the sunshine, but curls up and shrinks into itself at the slightest touch of the finger, or the lightest breath of the wind" (Brontë [1847] 1992, 36). These references illustrate the versatility of the Sensitive Plant as a figural vocabulary in step with ideas about nature that were part of romanticism and its legacies.

Popularizing Science Books

Just as the figure of the *Mimosa* challenged ideas about clear demarcations between plant and animal, so writings about the Sensitive Plant can challenge tidy generic demarcations between scientific and literary accounts of natural knowledge. There are so many instances of writings about nature and science that feature the *Mimosa* from the 1790s across the early decades of the new century that the malleability of representations seems especially characteristic.

Accounts of the *Mimosa* that were written in verse represent a particularly pointed generic challenge. For example, when Erasmus Darwin shaped an account of the Linnaean system in "The Loves of the Plants," published as the second part of his expository poem *The Botanic Garden* (1791), was he writing a botanical poem or crafting a popularizing piece of science writing? His intention was to depict a large system of organic energy across nature's kingdoms, and to this end Darwin filled the four cantos of "Loves of the Plants" with hyperbolic verses describing examples of plants within their Linnaean categories. At the same time, the fanciful imaginings go hand in hand with extensive expository botanical footnotes throughout the long poem. In his prefatory "Advertisement," Darwin made the teacherly intention of the poem explicit; he set out, he explains in now much-cited phrasing, "to inlist Imagination under the banner of Science" (E. Darwin, 1791, "Advertisement"). Darwin introduces the *Mimosa* in the opening canto of "Loves of the Plants" as an illustration of the Linnaean class Polygamy, and the depiction illustrates the spirit of his enquiry. His *Mimosa pudica* participates in a sexualized Linnaean tradition and a system of analogy and personification. She is female, not male, and she is a figure of chastity that vibrates with potential sexuality:

> Weak with nice sense, the chaste MIMOSA stands,
> From each rude touch withdraws her timid hands;
> Oft as light clouds o'erpass the Summer-glade,
> Alarm'd she trembles at the moving shade;
> And feels, alive through all her tender form,
> The whisper'd murmurs of the gathering storm;

Shuts her sweet eye-lids to approaching night;
And hails with freshen'd charms the rising light.
(E. Darwin 1791, 1:301–8)

Darwin's plotline for the story continues into an account of the *Mimosa* on the brink of marriage:

Veil'd, with gay decency and modest pride,
Slow to the mosque she moves, an eastern bride;
There her soft vows unceasing love record,
Queen of the bright seraglio of her Lord. (E. Darwin 1791, 1:309–12)

"Loves of the Plants" works as both botanical poetry and polite science. It entices with its sumptuous language and celebration of nature animated by love, and the intricacies of Linnaean language and classification seem easier to comprehend when personified and garbed in metaphor.

The combination of languages of nature in Darwin's "Loves of the Plants" reaches out to several potential audiences. While readers with interest in figure and fantasy may linger only over the poem's verses, there is more to the account of the *Mimosa pudica* than exuberant personification of a botanical "eastern bride." Those readers ready to bridge back and forth between literary renderings and technical information from the world of Linnaean botany would have ample material for both pleasure and substantive reflection in Darwin's long footnote to this stanza. There Darwin provides a different look at the Sensitive Plant as he describes his own experiments with light sensitivity and movement and speculates about why the leaflets of the plant collapse when touched. He writes, for example, "Now as their situation after being exposed to external violence resembles their sleep, but with a greater degree of collapse, may it not be owing to a numbness or paralysis consequent to too violent irritation, like the faintings of animals from pain or fatigue?" (E. Darwin 1791, 1:301 n.). His discussion recalls earlier technical accounts of John Hill and others on the "sleep" of plants.

Like Erasmus Darwin, Frances Rowden also used verse to introduce ideas from Linnaean botany. Rowden's *Poetical Introduction to the Study of Botany* (1801) is a book of elementary botany that first explains the parts of plants and the Linnaean system and then presents verse accounts of plants to illustrate the Linnaean classes and orders. She too uses footnotes as her location for more technical botanical information than would be appropriate in poems. *Mimosa pudica* is one of her examples of Linnaean class 23, *Polygamia, Monoecia.* With an eye to attracting an audience and keeping its interest, she too, like Darwin, capitalized on analogy and gendered metaphors in support of botanical teaching. But her Sensitive Plant is a very different creature from Darwin's "eastern bride" trembling in anticipation of her wedding night. Rowden intentionally crafts a counternarrative to Darwin and depicts a *Mimosa pudica* that is characterized most of all by chastity and humility:

> As some young maid, to modest feeling true,
> Shrinks from the world and veils her charms from view,
> At each slight touch her timid form receives,
> The fair *Mimosa* folds her silken leaves;
> Her trembling nerves Sensation's law obey,
> And feel th' alternate change of night and day. (Rowden 1818, 248–49)

Since her account is meant as an exposition of Linnaean features of the plant, Rowden animates the five stamens of class *Polygamia* as "Five little Sylphs" that wait on the "fair *Mimosa*" along with "Sweet Sensibility" as guardians to the "tender plant." Her poetic botanical account ends with an exhortation to female readers to think ahead to a time in their future lives when youthful folly will give way to maternal love and "scenes more suited to thy pure desires" (Rowden 1818, 251). Rowden was both a writer and a governess. Her *Poetic Introduction to the Study of Botany* promoted knowledge of botany but through it also promoted gendered ideals, particularly those having to do with domesticity and female subordination.

In Rowden's day, writers like her who wanted to promote popular botany worked in a variety of genres. While some chose verse, others took full advantage of the prose miscellany, a narrative style that suited late Enlightenment interest in information and general education, especially when the intended audience was young readers across the middle ranks of society. This was home turf for Priscilla Wakefield, a Quaker writer, who joined Jane Marcet and others in the tradition of women popularizing science topics between the 1780s and 1820s. Wakefield's book *Domestic Recreation; or Dialogues Illustrative of Natural and Scientific Subjects* (1805) is shaped as improving conversations between a mother and her two daughters, in which their "domestic recreation" is to develop knowledge about nature. The mother encourages their curiosity about plants, insects, and animals; promotes interest in the refraction of light; and explains experiments that use solar microscopes. She links these topics to seeing "the hand of a Deity in the most minute of his works" (Wakefield 1805, 19). Wakefield allies herself by the choice of the "familiar format" with narratives of natural history, and her religious (though nondoctrinal) tags align this book with narratives of natural theology (Gates and Shteir 1997, 11). However, the trajectory is toward the didactic and scientific rather than toward the literary. Wakefield uses stories and even a piece of poetry to enliven and heighten her points, but there are no literary emblems in her presentation of material, nor does she shape her information into conduct book lessons for girls about virtue.

Wakefield features the *Mimosa* when taking up the topic of analogies between plants and animals in a chapter on the "gradation of being." Does the *Mimosa* have "something like life and instinct" as well as "the power of directing the motions of their leaves and branches"? One girl describes what appears to be volition in the Sensitive Plant: "I happened to touch one of these . . . and the leaves curled as if they were afraid of me; but as soon as I went to a distance they recovered, and looked the same as before" (Wakefield 1805, 47–48). In response, the mother steps away from personification and contemporaneous symbolic accounts, and indeed from sexing or gendering the *Mimosa* altogether. Instead, she explains tropisms in the *Mimosa* and related plants as "a mechanical power, or inferior kind of instinct, that guides them to their own preservation" (Wakefield 1805, 49). Some features of plants have "a faint resemblance to animals," but the mother uses the subjunctive voice to make points such as that circulation of sap "may be compared" to circulation of blood. Analogy is less interesting to her than the superiority of animals to vegetables in the "rank of being," based in "the degree of intellect, sensibility, and animation" (Wakefield 1805, 49–51). In this work of natural knowledge, with its didactic intentions and sturdy and serious tone, Wakefield avoids technical language and again and again promotes "attentive observation" rather than books as the way to "acquire knowledge" and "enrich the mind" (Wakefield 1805, 14). The *Mimosa* seems to have embodied the visual and empirical orientation of this book because Wakefield's episode about the Sensitive Plant is featured as the frontispiece in both the first edition of the book in 1805 (see figure 6.5) and a subsequent edition published in 1806 (Wakefield 1806).

Generic variety is a feature of many early nineteenth-century writings that integrated material from the sciences and natural history. Popularizing writers cultivated many formats within print culture and with far less generic specialization than was taken account of until recently. Because natural history and science were topics of educational value and commercial potentiality, publishers welcomed books that addressed niche markets. Numerous women writers fashioned information-rich books for these

"*I happened to touch one of these called a Sensitive Plant.*" &c.

see page 47.

Figure 6.5. Illustration from frontispiece to *Domestic Recreation; or Dialogues.* Reprinted from Wakefield 1805. Courtesy of the Osborne Collection of Early Children's Books, Toronto Public Library.

markets, and their versatile popularizations went well beyond homogeneous juvenile audiences or readers of poetry (Shteir 1997a). The popularizing writer Maria Jacson, for one, began with an elementary botany book for children and then wrote another, at a somewhat higher level, for adults, both dedicated to explaining the Linnaean system. In 1811, she published *Sketches of the Physiology of Vegetable Life,* a book in an expository popular science vein, whose format most resembles an autobiographical reportage. This book is generically more like a compilation of journal entries, studded with references to her readings and experiments about roots and sap and plant circulation, interspersed with critical reflections. Jacson positions herself and her book carefully. She refers to her audience as a "youthful student," a pronominal "he," and aims to move her readers from elementary knowledge about plant physiology into something more complex. Theories of "vegetable life" are central to her inquiry, and hence the *Mimosa* appears as an example at several points. She opens her discussion by reviewing ideas about "irritability" as the explanation for plant motion, and then puts forward her own belief, based in "an attentive observation of the motions of vegetable life" (Jacson 1811, 11), that plants have sensitivity and possess volition and a faculty of sensation. At various points throughout the book she discusses motion in other plants as well as further examples of analogy between plant and animal functions. Jacson's language is strictly based in observation and experiment, and she even criticizes the Swiss botanist Augustin de Candolle for experiments on the *Mimosa* that are "deficient in that minute accuracy which ought always to accompany the report of such researches" (Jacson 1811, 48). Jacson was by her own account not a "botanist," but she had social connections to Erasmus Darwin and the Lichfield circle and was known for having independent judgment and a keen interest in performing experiments (Shteir 1996, 108–17). She wrote *Sketches on the Physiology of Vegetable Life* in the first person and with an eye to a reading audience that could or would accommodate a goodly amount of critical engagement with "authorities" along with a piling up of information, all without a conversational format to leaven the lump.

Jacson's physiological account of the *Mimosa* in 1811 illustrates shifts taking place at that time within botanical culture in England, notably, from a focus on Linnaean systematics to matters of plant structure and from leisurely and polite pursuits toward new styles of teaching and writing about botany. Within these contexts, Jane Marcet's account of the *Mimosa* in *Conversations on Vegetable Physiology*, discussed at the beginning of this chapter, indicates that more "scientific" writers who were reaching out to broad general audiences had jettisoned various literary features of earlier expository writing about botany and instead were compiling accounts that made no reference to religion, beauty, or moral teachings. Jane Marcet's *Conversations on Vegetable Physiology* from 1829 was on its way toward being a textbook, and her book was easily adapted for that purpose (Lindee 1991).

In terms of the literary history of science, it is just a small step from

Marcet's book to the introductory account of the natural system of plant classification in John Lindley's *Ladies' Botany* (1834–37). Both books were written with an eye to expanding the audiences for botany. Lindley, a particularly influential antagonist of Linnaean systematics during the early decades of the nineteenth century, worked hard to modernize botany through his writing and teaching. He gave keen attention to specific audiences, and books such as his *Introduction to the Natural System* (1830) and *The Vegetable Kingdom* (1846) were among the many tools in his campaign. As the first professor of botany in the University of London, he used the university classroom as well as other institutional formations to shape botany particularly into a discipline suited for upwardly mobile young men. To this end, he endeavored to take the study of plants out of the orbit of a polite, literary, and fashionable pursuit associated closely in the public mind with women and female activities (Shteir 1996, chap. 6). *Ladies' Botany* is one of Lindley's sites for new-style botanical science, and his account of *Mimosa* models a popular study of plants that is serious and non-Linnaean, and not literary in any way. Choosing technical vocabulary and the language of the natural system, Lindley locates "the curious *Sensitive plants*" in the third division of the Pea tribe, "whose many parted leaves shrink from the touch of the very wind that blows upon them, which close up and appear to go to sleep at night, and which seem as if struck with sudden death if they receive any rude shock" (Lindley 1834–37, 1:128). Although the similes of sleep and death in this account show that narrative is still at work in Lindley's description, there is no trace of playfulness or conduct book instruction in his account, and his language is far removed from the colorful and anthropomorphized depictions in Erasmus Darwin's and Frances Rowden's versions of the *Mimosa*.

John Lindley wrote *Ladies' Botany* in an epistolary format and addressed it to women who take on responsibility for teaching their children about the new methods in systematic botany. The first letter begins in this way: "You ask me how your children are to gain a knowledge of Botany, and whether the difficulties which are said to accompany the study of this branch of science cannot, by some little contrivance, be either removed altogether, or very much diminished" (Lindley 1834–37, 1:1). Unlike earlier writers who set instructional letters inside contexualized family narratives, Lindley's letters are unidirectional, give no names or personalities to the author or recipients, and suggest no setting or contexts. Instead, Lindley allied himself explicitly with Jean-Jacques Rousseau's *Letters on the Elements of Botany*, well known in England in the translation by Thomas

Martyn, a work written for the same maternal audience and the same pedagogical purpose (Rousseau 1785). Lindley's book differs from Jane Marcet's *Conversations on Vegetable Physiology* in this regard. There Marcet gave powerful agency to her teacher, Mrs. B. She also modeled considerable learning on the part of her two female pupils. The narrative form of *Ladies' Botany* suggests that John Lindley does not have the same agenda for women in science (Shteir 1997b). Lindley's *Ladies' Botany* was published in multiple editions into the 1840s as a text for "schools and young persons," and the omission of any references to girls contrasts markedly with narrative practices of earlier introductory botany books.

Conclusion

Mimosa discourses are like different ingredients in cultural recipe books for science—and for science in the marketplace. Stimulating interest in science, attractive and mysterious, the Sensitive Plant was portrayed over several centuries as a wondrous natural object that could instigate experimental inquiry and also entice neophytes across the threshold into botanical knowledge. Dating back into the eighteenth century, writers played with a variety of figural and technical vocabularies for representing it, and they often integrated botanical content with other cultural themes. Poems with botanical footnotes found a comfortable place in the polite cultures of science because poets and natural historians alike could assume some general knowledge of, and interest in, the vegetable kingdom. Erasmus Darwin's "Loves of the Plants" and Priscilla Wakefield's *Domestic Recreations* show that poets and writers about science and natural history also could cultivate botanical knowledge for new readers by combining languages geared to leisured audiences with material of a more utilitarian and technical kind. Moving across the years 1800 through 1830, however, practices consolidated regarding types of knowledge for specific audiences and ways of communicating this knowledge. Shifting discourses of the *Mimosa* show that flexibility for braiding languages of nature became harder to maintain as technical and popular forms of science and science writing each took their own shape and directions. Rather than incorporating literary material, verse accounts, and family-based narratives, popularizing writers moved nearer to the technical end of the spectrum. They chose to use various techniques for making botanical knowledge interesting and accessible, and these techniques included directness of vocabulary and new generic choices. In Jane Loudon's *Botany for Ladies* (1842), for example, we encounter a self-described "popular introduction to the natural system" that bears no resemblance to the format characteristic of early science writing by women and for women. Loudon figures a Sensitive Plant that stands at a remove from literary reference. It appears without poetic tags, personifications, or allusions to the "court of Flora." Instead, the *Mimosa pudica* is simply an example of an ornamental plant in the second division of the order Leguminoseae, and the genus *Mimosa* is distinguished from *Acacia* "in the corolla being funnel-shaped, and four or five cleft" (Loudon 1842, 44). By the 1840s, writers who wanted to popularize botany with an eye to a broad readership, and a paying readership, scaled way back on literary elements and dressed their *Mimosa* in technical features and botanical data instead.

Accounts of the Sensitive Plant that have been discussed in this chapter demonstrate that literary, scientific, and popularizing discourses of nature went their separate ways by the mid-nineteenth century. Seen in the context of earlier accounts of the *Mimosa*, narrative forms such as are found in Loudon's *Botany for Ladies* take only one vocabulary as normative for popularizing writing about science. It was strategic and pragmatic on Jane Loudon's part to acknowledge the marketplace and the culture of science in the 1840s and to shape her narrative accordingly. By writing about the *Mimosa* in nonliterary ways, Loudon and other women science writers could authorize themselves within a consolidating science culture. Indeed, incarnations of the *Mimosa pudica* in British science writing during the first half of the nineteenth century show how women crafted a place for themselves in the culture of the new popular science by aligning themselves with explanatory and narrative modes that demarcated the "scientific" from other kinds of inquiry into nature. At the same time, however, Jane Loudon's way of writing about botany illustrates an acceptance of an account of nature that, by its exclusion of literary vocabularies and figural languages, represents a particular version of the culture of science. It is not surprising that the Sensitive Plant encountered in the books and magazines that were sites of popular science during the mid-nineteenth century differed from the phallic icon articulated in James Perry's eighteenth-century erotic literary verses. Directions in science culture had taken personifications of "bashful" and "chaste" plants off the agenda for scientific writers and readers by that time. Nevertheless, there may well have been a gap for some audiences between accounts of the Sensitive Plant in the print culture of science, on the one hand, and the experiences, on the other hand, of those who themselves had opportunities to witness the physicality of the *Mimosa pudica*, touching it and breathing upon its leaflets, watching the rise and fall of its parts, and responding in turn to the curiosity and surprise of the *Mimosa* and its appeal to the eye and to the imagination.

NOTES

I gratefully acknowledge funding from the Social Sciences and Humanities Research Council of Canada that made possible visits to the Hunt Institute for the Botanical Documentation at Carnegie Mellon University in Pittsburgh; the Library of the Royal Botanic Gardens, Kew, UK; and the Library of the Wellcome Institute for the History and Understanding of Medicine, London, UK. Staff members at these institutions assisted my research in many ways, and I appreciate their expertise and enthusiasm when searching for the *Mimosa* in early print and visual sources.

1. Confusions abound about the designation *Mimosa*, in part because the term was used for plants that now are botanically identified as *Acacia*. (Barneby 1991) The yellow-flowered "mimosas" found in flower shops, for example, are acacias.

REFERENCES

Andrews, Henry C. 1808. *The Botanist's Repository, for New and Rare Plants.* Vol. 8. London: Privately printed by T. Bensley.

Barneby, Rupert C. 1991. *Sensitivae Censitae: A Description of the Genus Mimosa Linnaeus (Mimosaceae) in the New World.* Bronx, NY: New York Botanical Garden.

Bentham, George. 1875. "Revision of the suborder Mimoseae." *Transactions of the Linnean Society of London* 30: 335–664.

Bewell, Alan 1996. "'On the Banks of the South Sea': Botany and Sexual Controversy in the Late Eighteenth Century." In *Visions of Empire: Voyages, Botany, and Representations of Nature,* ed. David Phillip Miller and Peter Hanns Reill, 173–93. Cambridge: Cambridge University Press.

Brontë, Anne. [1847] 1992. *The Tenant of Wildfell Hall.* Oxford: Clarendon Press.

Burkhardt, Frederick, et al. 1997. *The Correspondence of Charles Darwin.* Vol. 10, *1862.* Cambridge: Cambridge University Press.

Cleland, John. [1748–49] 1985. *Memoirs of a Woman of Pleasure.* Oxford: Oxford University Press.

Darwin, Charles, with Francis Darwin. 1880. *The Power of Movement in Plants.* London: John Murray.

Darwin, Erasmus. 1791. *The Botanic Garden: A Poem, in Two Parts, with Philosophical Notes. Part II: The Loves of the Plants.* London: J. Johnson.

———. 1800. *Phytologia; or, The Philosophy of Agriculture and Gardening.* London: J. Johnson.

Fara, Patricia. 2003. *Sex, Botany and Empire: The Story of Carl Linnaeus and Joseph Banks.* Duxford, Cambridge, UK: Icon Books.

The Gardeners' Chronicle. 1848. [Vol. 8], no. 52. London: Privately printed.

Gates, Barbara T., and Ann B. Shteir, eds. 1997. *Natural Eloquence: Women Reinscribe Science.* Madison: University of Wisconsin Press.

Hill, John. 1757. *The Sleep of Plants, & Cause of Motion in the Sensitive Plant, Explain'd.* London: R. Baldwin.

Ibbetson, Agnes. 1809. "Remaining Proof of the Cause of Motion in Plants explained." *William Nicholson's Journal of Natural Philosophy, Chemistry, and the Arts* 24:114–23.

Jacson, Maria E. 1811. *Sketches of Vegetable Physiology.* London: John Hatchard.

King, Amy M. 2003. *Bloom: The Botanical Vernacular in the English Novel.* Oxford: Oxford University Press.

Lindee, M. Susan. 1991. "The American Career of Jane Marcet's *Conversations on Chemistry,* 1806–1853." *Isis* 82:8–23.

Lindley, John. 1830. *An Introduction to the Natural System of Botany.* London: Longman, Rees, Orme, Brown, and Green.

———. 1834–37. *Ladies' Botany: or, A Familiar Introduction to the Study of the Natural System of Botany.* 2 vols. London: James Ridgway and Sons.

———. 1846. *The Vegetable Kingdom; or, the Structure, Classification, and Uses of Plants, Illustrated upon the Natural System.* London: Bradbury and Evans.

Linnaeus, Carolus. 1737. *Hortus Cliffortianus.* Amsterdam: n.p.

———. [1753] 1957. *Species plantarum.* Vol. 1. A facsimile of the first ed. London: Ray Society.

Loudon, Mrs. [Jane]. 1842. *Botany for Ladies; Or, A Popular Introduction to the Natural System of Plants, According to the Classification of De Candolle.* London: John Murray.

Maniquis, Robert M. 1969. "The Puzzling *Mimosa:* Sensitivity and Plant Symbols in Romanticism." *Studies in Romanticism* 8 (3): 129–55.

[Marcet, Jane]. 1829. *Conversations on Vegetable Physiology; comprehending the elements of botany, with their application to agriculture.* Vol. 1. London: Longman, Rees, Orme, Brown, and Green.

Parkinson, John. 1640. *Theatrum Botanicum: the Theater of Plants.* London: Tho. Cotes.

Peakman, Julie. 2003. *Mighty Lewd Books: The Development of Pornography in Eighteenth-Century England.* Houndmills, Hants., UK: Palgrave.

[Perry, James]. 1779. *Mimosa: or, the Sensitive Plant.* London: W. Sandwich.

Phillips, Henry. 1825. *Floral Emblems.* London: Saunders and Otley.

Ritterbush, Philip. 1964. *Overtures to Biology: The Speculations of Eighteenth-Century Naturalists.* New Haven, CT: Yale University Press.

Rousseau, Jean-Jacques. 1785. *Letters on the Elements of Botany.* Trans. Thomas Martyn. London: B. White and Son.

Rowden, Frances. 1818. A *Poetical Introduction to the Study of Botany.* 3rd ed. London: G. and W. B. Whittaker.

Schiebinger, Londa. 1993. *Nature's Body: Gender in the Making of Modern Science.* Boston: Beacon Press.

"The SENSITIVE PLANT and The ROSE-BUD." 1802. *Lady's Magazine,* November, 605.

Shelley, Percy. [1820] 1975. "The Sensitive Plant." In *The Poetical Works of Shelley,* vol. 3, ed. Newell F. Ford. Boston: Houghton Mifflin.

Shteir, Ann B. 1996. *Cultivating Women, Cultivating Science: Flora's Daughters and Botany in England, 1760 to 1860.* Baltimore: Johns Hopkins University Press.

———. 1997a. "Elegant Recreations? Configuring Science Writing for Women." In *Victorian Science in Context,* ed. Bernard Lightman, 236–55. Chicago: University of Chicago Press.

———. 1997b. "Gender and 'Modern' Botany in Victorian England." In *Women, Gender, and Science: New Directions,* ed. Sally Gregory Kohlstedt and Helen E. Longino, *Osiris* 12, 29–38. Chicago: University of Chicago Press.

Ueda, Minoru, and Shosuke Yamamura. 1999. "The Chemistry of Leaf-movement in *Mimosa pudica L.*" *Tetrahedron* 55 (36): 10937–48.

Vallemont, Abbé de. 1707. *Curiosities of Nature and Art.* London: D. Brown, A. Roper, and Francis Coggan.

Wakefield, Priscilla. 1805. *Domestic Recreation; or Dialogues Illustrative of Natural and Scientific Subjects.* London: Darton and Harvey.

———. 1806. *Domestic Recreation; or Dialogues Illustrative of Natural and Scientific Subjects.* 2nd ed. London: Darton and Harvey.

Webster, Charles. 1966. "The Recognition of Plant Sensitivity by English Botanists in the Seventeenth Century." *Isis* 57: 5–23.

[Wynne, J. H.] 1773. *Fables of Flowers, for the Female Sex.* London: George Riley.

CHAPTER SEVEN

Reading Natural History at the British Museum and the *Pictorial Museum*

AILEEN FYFE

In 1852, the journalist Blanchard Jerrold recommended that British Museum visitors go in the company of "a methodical and homely guide" in order to acquire "various bits and scraps of pertinent information" while passing from one object to another (Jerrold 1852, 229). Fifty years earlier, the guide would have been human, but personal guided tours had been abandoned in the British Museum. Rather, visitors who desired a companion could purchase one of several guidebooks—including the one written by Jerrold. Thus, printed books and museum objects came together to provide an enhanced visitor experience.

In the mid-nineteenth century, both museums and print were finding new audiences among the lower middle and working classes. Extending access to reading material and museums was widely promoted as essential to working-class self-improvement. These ideas gained government support with the passing of acts in 1845 and 1850 which permitted local councils to raise money from rates for the construction of public libraries and museums. Although the effects of these acts would not be fully apparent for some decades, the spread of the Saturday half-holiday and the cheap railway excursion ticket made it easier for new audiences to visit existing museums (Pimlott [1947] 1976, chaps. 5–7; Bailey 1978), and the range of travel handbooks and advice literature indicates a rising enthusiasm for domestic tourism among all classes of the population (Simmons 1970; Vaughan 1974; Buzard 1993, chap. 1). These changes in visiting patterns happened at the same time as print became the first of the mass media (Secord 2000, chap. 2; Fyfe 2004, chap. 1). The lower middle classes and some of the working classes had been the target of a few publishers since the 1820s (Topham this volume), but by the 1840s and 1850s, they were widely recognized as an important market for printed material. Improving literacy rates helped

(Schofield 1981; Vincent 1989, 2000), but so too did the lower costs of production made possible by new industrial printing technologies (Eliot 1994; Weedon 2003).

In contrast to the well-educated and affluent readers and visitors of earlier generations, the new readers and visitors included people with the most basic of educations. Once upon a time, typical readers or visitors could have been expected to possess a wide general knowledge, which helped them understand what they were reading or looking at, and which meant that writers and curators could take some things for granted (Ousby 1990; Mandler 1997, chaps. 1–2; Mandler 1999). When large numbers of the working classes began to purchase books and periodicals, let alone to visit museums, it was much more difficult for writers or curators to presume any background knowledge, whether of subject-specific content or the vocabulary needed to explain it. Later in the nineteenth century, improved educational provision and the ubiquity of print would mean that subsequent generations were better prepared, but the new readers and visitors of the 1840s and 1850s had few intellectual resources to draw upon and were thus more dependent than their predecessors or their successors on the information and interpretation offered as part of the specific reading or visiting experience.

Publishers had initially responded to the expanding reading audience by providing cheap reprints, but by the mid-nineteenth century, many had come to realize that such reprints often assumed too much. Those who specialized in cheap instructive works, including Charles Knight and W. and R. Chambers, recognized that the new audiences needed works written specially for them and adapted to their educational backgrounds. The end result could be high-quality introductory works of instructive nonfiction, often with illustrations, available at low prices. One of these works was *The Pictorial Museum of Animated Nature* (1848–49), written by William Martin and published by Charles Knight. It was a sophisticated product from a writer with considerable experience of introductory works and a publisher committed to the use of illustrations for instructive purposes.

The British Museum had been developed as a resource for the well-educated and gentlemanly public of the late eighteenth century, but by the mid-nineteenth century, it was receiving numerous visitors from all social classes. An 1835 government investigation had found that the museum's exhibits were not presented in such a manner as to be attractive or comprehensible to the general public, but the museum's move into new buildings during the 1840s provided an opportunity for redisplay. The keeper of zoology, John Edward Gray, was widely credited with a vast improvement in the

appearance of the natural history collections. Nonetheless, he was working amid spatial and architectural constraints, and the transformation was
not as dramatic as advocates of public education would have liked (Wilson
2002, chaps. 3–4; Stearn 1981, 22–26).

Mid-nineteenth-century publishers had already realized that it was
more effective not to try to satisfy different audiences with the same product and had developed new products for the new audiences. Museum curators, however, had barely started to consider the analogous problem, that of
balancing the needs of scholars with the ambition of public education for
the masses. Gray would propose a solution in the 1860s, but it was several
more decades before the idea of the "New Museum," with its separate collections and displays adapted to the needs of different museum users, became widely accepted (Stearn 1981, 37; Bennett 1998). This was why printed
products were so important for the new visitors at midcentury. Although
the museum displays were still intended for educated visitors from polite
society, guidebooks and advice books could add a supplementary narrative
of information and explanation for the benefit of other classes of visitor.
Alternatively, a work such as the *Pictorial Museum* could create an idealized museum experience, incorporating both specimens and narrative, for
readers who stayed at home.

Whatever sort of museum they were visiting, visitors drew upon their
own mental resources to create meaningful experiences from the exhibits,
images, and texts presented to them. These would be, of necessity, individual, personal experiences. The carefully managed combination of images
and text in the *Pictorial Museum* tried to close down the range of potential
readings to the primarily instructive (although there were ways to read
against that aim). At the British Museum, curators had far less control over
visitors, who were free to wander through the galleries as they pleased,
encountering objects in no particular sequence. Those visitors who desired
order and meaning in their visits could acquire a guidebook to identify
specimens or an advice book to supply a connecting narrative. The philoso-
phies underlying these books varied so substantially, however, that visitors
who chose different guides would be likely to see the British Museum in
quite different lights.

Reading books and visiting museum collections were both key ways of
encountering natural history in the nineteenth century. They tend, however, to be studied separately by historians of science (Topham 1998, 2000;
Secord 2000; Fyfe 2000; Forgan 1994, 1999; Yanni 1999; Alberti 2002). One of
my aims in this chapter is to undermine that separation by examining a mu-

seum constructed solely from paper and printer's ink alongside a collection
of actual objects. Although a site such as the British Museum might appear to have obvious advantages over a book, by offering access to original,
three-dimensional, colorful specimens, the *Pictorial Museum* had the significant advantage of a connecting narrative, written specifically for the
new reading (and visiting) audiences. The first section of this chapter examines the *Pictorial Museum* and its claims to mimic and improve upon
the museum experience. We then turn to the British Museum and consider
how visitors might have attempted to make sense of its collections, using
the labels and guidebooks available in (or near) the museum. Finally, we
consider the role of other publications in informing visitors' experiences
of the British Museum. Key among these were advice books issued by publishers committed to working-class instruction, which offered their readers the narrative and interpretation that the midcentury British Museum
lacked. Used in conjunction, the collections and advice books could offer
an instructive and entertaining experience that neither the British Museum nor the *Pictorial Museum* alone could rival.

The Pictorial Museum of Animated Nature

For just threepence each week, through 1848 and 1849, the eight-page installments of *The Pictorial Museum of Animated Nature* (figure 7.1) offered
natural history enthusiasts the opportunity to purchase their own shares
in a museum. Half the pages were packed with wood-engraved animals, beginning with lions and tigers and continuing through giraffes and llamas,
hummingbirds and chickens, goldfish and sharks, before ending with sea
urchins, corals, and jellyfish. The museum's organizers claimed that its
skillful combination of letterpress and illustrations could perform "most
of the purposes of instruction and delight which belong to the greatest National Museums" (Martin 1848–49, 1:[1]). The author was William Martin
(1798–1864), a former curator of the Zoological Society's museum, who subsequently made his living as a science writer (Fyfe 2005). Its publisher was
Charles Knight (1791–1873), formerly publisher for the Society for the Diffusion of Useful Knowledge, with whom Martin had worked before. Knight
was committed to working-class self-improvement and specialized in the
production of cheap, illustrated, instructive reading material (Anderson
1991, chap. 2). He had published the *Pictorial Bible*, 1836–38, and the *Pictorial
Shakespeare*, 1839–42, and he applied the formula to other subjects in the
1840s.[1] These illustrated works were issued in weekly parts and sold by

Figure 7.1. The title page of *The Pictorial Museum of Animated Nature* illustrates the profusion of animals to be found inside. The text claims that this book can serve "most of the purposes of instruction and delight which belong to the greatest National Museums." Reprinted from Martin 1848–49, 1:[1].

itinerant vendors to bring them within the reach of as wide an audience as possible, particularly to those "in the neighbourhood of the great manufacturing towns and other populous districts," who did not frequent booksellers' shops (Knight [1864–65] 1873, 3:18).

The *Pictorial Museum* was slightly more expensive than most instructive part works, but it did contain an unusually large number of wood engravings (around fifty images per part). Given that the onepenny works rarely sold more than about sixty thousand to seventy thousand copies a week (Fyfe 2004, 49), the circulation of the threepenny *Pictorial Museum* must have been substantially less than that. The *Pictorial Museum* would have been most successful among lower-middle-class families, but it was probably beyond the reach of any but the most well-paid artisans. Undoubtedly, fewer people would have bought the complete *Pictorial Museum* than would have entered the British Museum, but equally, the opportunity to experience the *Pictorial Museum* was not limited to those able to visit London.

William Martin had already written books on natural history for Knight and the Religious Tract Society, both of whom understood the importance of writing at an appropriate level for the newly literate reading audiences. Yet this particular work had a grander scope than anything Martin had written before. Textually, the *Pictorial Museum* was similar to Martin's previous works and to other introductory natural history works on the market. What marked it out from its competitors was the inclusion of so many images, and particularly the way in which the images were made integral to the instructive experience, not merely decorative.

The title *museum* was used by some publishers for anthologies of short pieces of writing, but for Knight, it made explicit his belief in the educational power of museums and reiterated his commitment to assisting wider access to such resources (Anderson 1991, 57–67). The overwhelming impression in the *Pictorial Museum* was of hundreds of specimens dominating the text. Although there were, in fact, as many pages of text as of images, each section of text was structured around its images—rather than the images illustrating something mentioned in the text. The preface reinforced the dominance of the images when it described the ideal "visitor" as someone who looks *first* at the exhibits, and *then* "under a corresponding number of the text he finds a description of the animal, its structure, its habits, its localities, its use" (Martin 1848–49, 1:[1]). The contrast with another illustrated part work, *The Gallery of Nature: a pictorial and descriptive tour through creation*, published by W. S. Orr, with letterpress by Thomas Milner (Milner 1846) brings out the point. Although there were many attrac-

supply these cascades flow from two small lakes in the Catskill Mountains, on the West Bank of the Hudson. The upper cascade falls one hundred and seventy-five feet, and a few rods below the second pours its waters over a precipice eighty feet high, passing into a picturesque ravine, the banks of which rise abruptly on each side to the height of a thousand to fifteen hundred feet.

In the grandeur of their cataracts, also, the American rivers far surpass those of other countries, though several falls on the ancient continent have a greater perpendicular height, and are magnificent objects. In Sweden, the Gotha falls about 130 feet at Trolhetta, the

greatest fall in Europe of the same body of water. The river is the only outlet of a lake, a hundred miles in length and fifty in breadth, which receives no fewer than twenty-four rivers; the water glides smoothly on, increasing in rapidity, but quite unruffled, until it reaches the verge of the precipice; it then darts over it in one broad sheet, which is broken by some jutting rocks, after a descent of about forty feet. Here begins a spectacle of great grandeur. The moving mass is tossed from rock to rock, now heaving itself up in yellow foam, now boiling and tossing in huge eddies, growing whiter and whiter in its descent, till, completely fretted into one beautiful sea of snowy froth, the spray, rising in dense clouds, hides the abyss into which the torrent dashes; but when momentarily cleared away by the wind, a dreadful gulf is revealed, which the eye cannot fathom. Upon the arrival of a visitor at Trolhetta, a log of wood is sent down the fall, by persons who expect a trifle for the exhibition. It displays the resistless power of the element. The log, which is of gigantic dimensions, is tossed like a feather upon the surface of the water, and is borne to the foot almost in an instant. In Scotland, the falls of its rivers are seldom of great size; but the rocky beds over which they roar and dash in foam and spray —the dark precipitous glens into which they rush—and the frequent wildness of the whole scenery around, are compensating features. The most remarkable instances are

Falls of Trolhetta.

Figure 7.2. Thomas Milner's *Gallery of Nature* (1846) was also lavishly illustrated, but the images are essentially decorative, following the text rather than structuring it. Reprinted from Milner 1846, 285.

tive illustrations in the *Gallery* (figure 7.2), it was the text which directed the reader. The illustrations were placed amid the text and performed a largely decorative function. The illustrations in the *Pictorial Museum*, on the other hand, were its raison d'être and provided its structure and direction. Martin was careful to match the amount of text he wrote for each species to the number of illustrations available, enabling the text and images to function as a connected package. By purchasing the installments of the *Pictorial Museum*, readers were presented with a state-of-the-art product which attempted both to mimic museums as sites for instruction and entertainment in natural history and to use the museum conceit to compete in the literary marketplace with other natural history books.

The *Pictorial Museum* was designed for a family audience from the lower middle and working classes, people with basic literacy skills but not necessarily any further formal education. Its aim was unashamedly instructive. Its images of "all the important QUADRUPEDS, BIRDS, FISHES, REPTILES and INSECTS, which fill the earth" would catch the attention of readers and, with the description, would provide information about the structure, habits, place of origin, and uses to humankind of all these creatures. Martin claimed that the work would gratify "a rational desire for knowledge," advance "the best objects of education," and furnish "just conceptions to all, but especially to the young, of the wonders and beauties of God's Creation" (Martin 1848–49, 1:[1]). Although the *Pictorial Museum* had a much higher image-to-text ratio than most natural history books of the mid-nineteenth century, it also had far more text than would be found in contemporary museums, despite the prefatory analogy between its text and museum labels. The *Pictorial Museum*'s wordiness was essential to its instructional role, and readers were provided with almost as much information about species as if they were reading a standard natural history text. An examination of one entry will illustrate the range of information provided.

The sixteenth issue focused on llamas and giraffes. The nine images of llamas on the front page (figure 7.3) were followed by five columns of text on the second and third pages describing llamas. The final column on page three started to describe giraffes, and it was followed by a double-page spread of illustrations of giraffes (and another llama). The next two pages of text finished the description of giraffes and started on that of musk deer, while the back page (and the front page of the next issue) was filled with images of deer. Martin began the llama description by relating the llama to the camel (the previous week's topic), describing their anatomical and behavioral similarities and pointing out their differences. He then gave a history of European knowledge of the llama, from the conquest of Spanish

Figure 7.3. The front page of the sixteenth issue of the *Pictorial Museum*. In the final, bound version, this page would have appeared opposite the back page of the preceding issue, which featured camels. Reprinted from Martin 1848–49, 1:121.

America onward. He explained that the accounts given by travelers disagreed about how many species of llama there were, and he explained his
own conviction that there were three related species. The description of the
"true" llama was therefore followed by discussions of the alpaca and the
vicuña. After this discussion of taxonomy, Martin described the llama's
habits, its foodstuffs, and, particularly, its uses to humans: its ability to
survive long periods without much water, its capabilities as a pack animal,
and the value of its wool. He ended with the suggestion that llamas and
alpacas should be acclimatized in the mountainous areas of Scotland and
Ireland and bred for their wool, which would be ideal for the manufacture
of such delicate items as Paisley shawls (Martin 1848–49, 1:122–3).

The images helped convey the information about the appearance and
anatomical peculiarities of the llama—such as the different appearance
of the domesticated llama and the wild vicuña or the adaptation of the
llama's hoof for mountainous living. In addition to these purely descriptive details, the text also discussed habits and behavior, particularly when
they were of use to people. Dynamic and transient animal attributes, such
as behavioral traits, were, of course, more difficult to convey effectively in
text or single pictures. One of the advice books for the British Museum,
which we shall meet again later, noted perceptively that museums had
the same problem, their specimens being "but a poor representation of the
great world of animated existence" (Masson [1848] 1850, 169). This particular book recommended that "to the unscientific lover of animals, indeed,
a visit to the living mammalia in the Zoological Gardens would in many
respects be far more interesting and instructive than a walk amid these
stuffed and shrivelled specimens" (Masson [1848] 1850, 232). Like wood engravings, mounted specimens were static (and possibly distorted) representations of a living, moving animal.

Yet, for Knight and Martin, to admit that a visit to the zoo was essential would be to admit that the *Pictorial Museum* could not offer as complete an education as they desired. Part of their solution was the inclusion
of multiple images of the same animal, in various poses, but also, where
possible, in action (figure 7.4). For instance, with the camel, the reader
was presented with illustrations of caravans at an oasis and camels being
watered and mounted. The images showed camels being useful to humans
but also demonstrated what a camel is and does much more effectively
than a single "type" illustration could do (Martin 1848–49, 1:120). Another
solution was the use of quotations from eyewitnesses. The reader was invited to see llamas through Charles Darwin's eyes, when he encountered
them in Patagonia. Darwin's account from his *Voyage of the Beagle*, pub-

Figure 7.4. Illustration of camels in action in the *Pictorial Museum*. This page appeared in the middle of the fifteenth issue. Notice the different artistic styles of the images. Reprinted from Martin 1848–49, 1:117, detail.

lished in 1839, included descriptions of llama behavior in a herd, their love of rolling in the dust, their prancing antics when approached, and how they behaved when domesticated, told with all the vividness of the traveler's narrative (Martin 1848–49, 1:122–23).

Many of these features were characteristic of mid-nineteenth-century natural history books. What made the *Pictorial Museum* unusual was the number and variety of images and their tight integration into the text. Nonetheless, common as many of these features might be in books, they were not frequently found in museums. Visitors to the British Museum would not find labels offering them descriptions of the appearance, habits, behavior, and usefulness of particular species, let alone eyewitness accounts, or images to illustrate the typical behavior of the living relatives of the exhibit. British Museum visitors, indeed, were sometimes hard-pressed even to identify the species in question. These problems arose primarily from the lack of text available in the museum, and, as we shall see, specialized books arose to fill the need. But even in the realm of specimens, where the British Museum might be assumed to be preeminent, the *Pictorial Museum* could offer a more complete and better-organized collection.

In the British Museum, the visitor could wander through the rooms "as chance might direct." The arrangement of the galleries did not "compel the visitor to follow a certain route," nor did "the authorised catalogues point out any order in which the rooms should be visited" (Masson [1848] 1850, 19–20; Wilson 2002, 101). We will see later how visitors could arrange their tours, but the point is that organization of the visiting experience was not intrinsic to the museum. In contrast, the "visitor" to the *Pictorial Museum* was constrained by the sequential issue of successive parts and by the flow of the single narrative. Although a first glance at issue 16 might suggest a confusion of llamas and giraffes, with a couple of deer toward the end, this was a restrained confusion. The description of the animal kingdom had reached hoofed mammals; the previous issue had been about camels, and the next would focus on deer. In the space of this issue, therefore, the "visitor" encountered a limited selection of closely related animals and would not—unlike the British Museum visitor—risk being distracted by moose antlers or snails while trying to examine an eagle or flamingo. By reading the *Pictorial Museum* as its parts appeared, the reader would make over a hundred separate "visits" to the museum, focusing attention on a different class of animals each week. Over the course of two years, the reader would have been gradually and carefully introduced to the entire animal kingdom. (Those who gave up partway were no worse off than all those who went to the British Museum for the giraffe and elephant and ignored the shells and corals.)

We should note, however, that it was also possible to read the *Pictorial Museum* in another way. Readers who had collected all of the issues—or, indeed, who purchased it complete as two thick quarto volumes of 825 pages—were not constrained by the weekly issue of parts. Whether bound or loose, the completed *Pictorial Museum* would have been an ideal "dipping in" book for its intended family audience, with the pictures providing an immediate cue for a curious reader opening the book at random and skimming through the pages till something caught his or her eye. These subsequent readings could be almost infinitely diverse, perhaps even chaotic. Each time the book was opened could be the occasion for a short fascinating visit to a different section of the animal kingdom, with no need to extend the visit beyond the boredom threshold, for it would always be possible to return the next day. Read in this way, the instructive intention to display the animal kingdom in its hierarchical classes and orders would have been lost. Visitors would still learn about specific animals, but they would lose sight of "their affinities to other portions of the great whole" and thus of the overall picture Martin wished to present (Martin 1848–49, 2:425).

The *Pictorial Museum* was also a more geographically accessible and enduring experience than the British Museum. Readers did not have to travel to London to see its animals. They might have to visit a bookseller's shop or a library, but the idea behind the low price and the itinerant vendors was to make it available to readers with the minimum effort on their part. The literate urban working-class populations of northern England were unlikely to be regular visitors to the British Museum and would not benefit from local civic museums for another few decades. If they were fortunate enough to make a trip to London—for instance, for the Great Exhibition of 1851—many sights would have to be crammed into a short space of time. The British Museum might have to be done in a single afternoon. A visit to the *Pictorial Museum*, on the other hand, need not be rushed and could be easily repeated.

As one of the contemporary guides to the British Museum explained, a real museum could not hope to contain "a collection representing the entire range of animated nature," and this fact gave paper museums yet another advantage (Masson [1848] 1850, 169). While common local creatures could be easily acquired, rare creatures from distant countries were expensive and in short supply. The British Museum, as a national collection regularly receiving donations from military and diplomatic travelers, was in a far better situation to attempt an exhaustive collection than were the nascent provincial museums. Yet even the British Museum's collections were uneven, and the quality of the specimens varied. One of its earliest llamas, for instance, had been donated by the Royal College of Surgeons and was in extremely poor condition and decapitated. Fortunately, by the 1840s, the museum had acquired nine llama specimens, representing adults and young of several varieties. Two recent donations had come from Captain Fitzroy of HMS *Beagle* and from Charles Darwin, Esquire ([Gray] 1843, 171).

A printed museum could in theory achieve a complete collection simply by sending its artists to the major museums and zoos and filling any gaps by copying images from published sources. The *Pictorial Museum* claimed on its title page (shown earlier in figure 7.1) that it was "by far the most extensive collection that has ever been produced of pictorial representations" (Martin 1848–49, 1:[1]). It is, however, probable that many, if not most, of the images came from works previously published by Knight and whose engraved blocks he still owned. Knight was, after all, aiming to produce a cheap product, and while dispatching artists was cheaper than buying the specimens for real, recycling existing images was cheaper yet. The practice was not uncommon, and it had the effect, in the case of the *Pictorial Museum*, of creating a collection of images executed in various styles and with varying skill, in which some species were better represented than others. This unevenness of representation unintentionally enhanced the imitation of a museum, with its different numbers of each species, in varying states of preservation and display.

Another problem physical museums had was that only durable, macroscopic specimens could be displayed. Tiny invertebrates, such as rotifers, were invisible to the naked eye. Only the shells of molluscs were "large enough and durable enough" to exhibit, but they gave little idea of what the animals themselves were like. Jellyfish were even more awkward, since they had no shells, bones, or skins which could be preserved (Masson [1848] 1850, 169). In a printed museum, however, artists could illustrate the entirety of a creature and do so at an increased (or reduced) scale. A printed museum, then, could display microscopic creatures as well as lions and could present the soft and fragile parts of molluscs and jellyfish (figure 7.5).

Using images which were scaled representations of the originals also made it easier to organize the exhibits both efficiently and taxonomically. Curators of real museums had to worry about the most efficient way to use their limited space to display specimens ranging enormously in size. In the British Museum, large giraffes and rhinoceroses stood on the floor of the gallery; medium-size birds and small mammals were in wall-mounted cases; and the smallest specimens, such as molluscs and insects, were

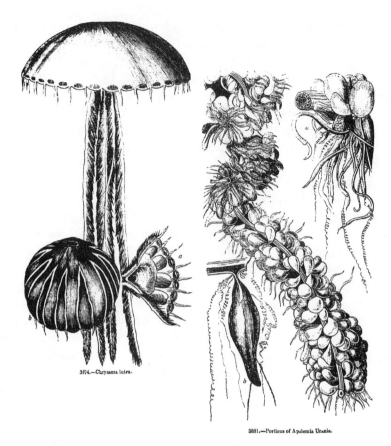

3374.—Chrysaora lutea.

3831.—Portions of Apolemia Urania.

Figure 7.5. Even jellyfish can be displayed in a printed museum. Illustration
reprinted from Martin 1848–49, 2:420, detail.

in table cases standing on the floor. This arrangement created a tension
between the curators' desire to organize their exhibits in a taxonomic sequence and the need to use the available space as efficiently as possible.

At the British Museum, the Eastern Zoological Gallery (figure 7.6) lay
between the mammal collections in the southern wing and the reptile and
fish collections in the northern wing; as we might expect, its key specimens
were the mounted birds, which filled no fewer than 166 wall cases. Yet, as a
guidebook explained, birds were not the only exhibits: "The smaller table
cases contain bird's eggs, and the larger collection of shells of Molluscous
animals. Over the cases containing the birds are a series of horns of hoofed
quadrupeds; and suspended over these is a collection of paintings" (Scott

and Kesson 1843, 22). The Northern Zoological Gallery was a similar mixture, with reptiles and fish on the walls and butterflies, beetles, shells, and sponges on the tables (Mason 1847, 5–7). In the *Pictorial Museum*, the variation in size between the illustrated specimens was relatively small, and there was only one sort of space to be filled. The illustrations all appeared on dedicated pages, and since the pages were quarto (somewhat larger than those of a typical book), it was easier to arrange the images both taxonomically and efficiently.

The *Pictorial Museum* offered its readers access to the entire animal kingdom, arranged in a logical order, undistorted by the demands of spatial efficiency. Unlike the British Museum, it presented exhibit, identification, and information all in the same place. Thus, by providing a complete instructional package blending exhibits and information, the *Pictorial Museum* could convincingly claim to surpass most natural history books and most physical museums.

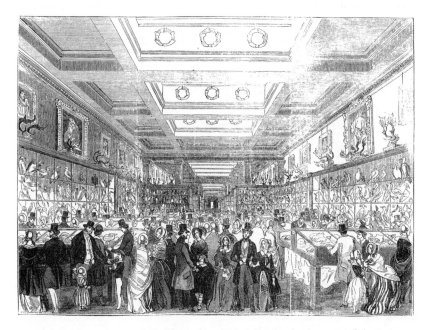

Figure 7.6. In 1845, more than fifteen thousand visitors took advantage of the Easter Monday opening to visit the British Museum, where "the birds, unquestionably, form the most popular attraction." In addition to the wall cases containing birds, notice the antlers and portraits high on the walls and the shells in the table cases. Illustration reprinted from *Illustrated London News*, March 29, 1845, illustration at 201, quotation at 202.

The British Museum Natural History Collections

At midcentury, "crowds of all ranks and conditions besiege[d] the doors of the British Museum, especially in holiday times" (Jerrold 1852, 5). The British Museum's fame at this time was based largely on its collections of classical and Egyptian antiquities and the newly arrived Assyrian remains (Wilson 2002; Crook 1972; Bohrer 2003, chap. 4). Nevertheless, almost half the gallery space was still devoted to natural history, where Hans Sloane's collections were displayed alongside more recent acquisitions. It was suggested that the natural history collections be moved to a separate museum (Rupke 1988), but they were still in Bloomsbury when the museum completed its move into Robert Smirke's new quadrangular building, where they occupied three of the upper galleries.

In the early nineteenth century, it had been difficult to gain admission to this "public" museum, but by midcentury the system of advance tickets and very limited opening hours had been removed. The museum was open to all comers, at no charge, on Mondays, Wednesdays, and Fridays from 10:00 a.m. until 4:00 p.m. in winter and from 10:00 a.m. until 7:00 p.m. in summer (Cunningham 1851, 149). Its special holiday openings, such as Easter Monday, attracted throngs of working-class visitors. Compared with annual visitor figures of around 20,000 at the start of the century, there were 750,000 visitors in 1846, 1 million in 1850, and an incredible 2.5 million in the exhibition year of 1851 (Knight 1847–49, 1:33, Wilson 2002, 99–100). This was, however, the peak of the museum's nineteenth-century popularity, and numbers declined despite the introduction of Saturday openings (in summer) in 1856. Visitors to London may have preferred the increasingly sophisticated and modern displays of the Crystal Palace, the Museum of Practical Geology, or the Royal Polytechnic Institution, to the traditional and scholarly British Museum (Wilson 2002, 100). The large numbers of visitors to the British Museum around midcentury probably had more to do with the expansion in domestic tourism, the availability of cheap excursion tickets, and the organization of trips to the Great Exhibition than with any competitive urges among the museum's staff to attract these new audiences.

Commentators generally assumed that the British Museum ought to be both a repository for the researches of "the studious and scientific" and an institution of public education enabling the rest of the population to "obtain a certain amount of information and improvement of taste" (Fergusson 1849, 3–4). The 1835 comments of the museum's director, Henry Ellis, about the undesirability of admitting the lower classes are often

taken to indicate that the museum's staff were more interested in its scholarly mission than in public education. Yet Ellis had changed his mind by
1840 (Wilson 2002, 86–7). Moreover, in 1835, the museum's natural history
staff had admitted that the collections were not sufficiently attractive or
explanatory for the public, and John Edward Gray (then an assistant) argued
that "to encourage a taste for science among the people generally" was
indeed part of the museum's role (quoted in Stearn 1981, 22). But even in
the new building, the natural history departments were always competing
for collection and display space with the antiquities and the printed books
and manuscripts. Gray was commended for vastly improving the display of
the zoology collections, but he was still constrained by lack of money and
space and by the architecture of a building designed primarily as a library
(Stearn 1981, 25; Fergusson 1849, 27–29; Yanni 1999).

The British Museum's natural history collections were abundant in
skeletons, stuffed skins, horns and antlers, shells and fossils. A visitor who
turned right at the head of the principal staircase would pass through several rooms displaying mammals before turning left into the ornithological
gallery and left again into the reptile and fish displays. The fossils (and
minerals) lay beyond the fish. Viewed in this order, the visitor encountered
the extant vertebrates of the world in roughly taxonomic sequence. Until
1810, all visitors had been led around the galleries by a member of staff,
thus ensuring that they followed the correct route and noticed the most
interesting specimens. The option of human guides was not reintroduced
until the end of the century, so midcentury visitors were free to turn left at
the head of the stairs, and they might find themselves following the giraffe
with statues of Chinese and Japanese divinities and the fossil ichthyosaur.

The potential confusion would not be aided by the museum's mini-
malist approach to labeling. Some cases had no labels. Most were adorned
with "numerous inscriptions, within and in front," but these hand-written
labels frequently gave little more than the country of origin in a "tolerably
clear" manner and the names of the animals in Latin, which was "rather
a stumbling-block" for some visitors (Knight 1847–49, 1:38–39). The mineralogical displays were unusually fortunate in their printed labels, which
allowed "a patient visitor . . . to pick up much interesting information"
(Wilson 2002, 101; Masson 1850, 114). The main purpose of the museum's
labels was to give the taxonomic name of the specimen, thus enabling
experts to use the collection to identify new specimens or to study the
similarities and differences between species and genera. They were not
intended for public education.

Another of Gray's achievements was to publish detailed catalogs to the collections, organized taxonomically without reference to the displays. These catalogs were sold in the museum, as was a *Synopsis of the Contents of the British Museum* (Wilson 2002, 67, 101; Grewal 1996, chap. 3). The *Synopsis* was updated each spring by the curatorial staff and was an abridged list of the museum's holdings. By 1852, the fifty-ninth edition ran to 260 closely printed pages and sold for one shilling. There was no official guidebook until 1859, but unofficial guidebooks were sold in nearby coffeehouses. Henry Green Clarke had begun his guidebook series with the London art galleries, and issued a sixpenny *The British Museum: a Hand-Book Guide for Visitors* in 1843.[2] It was reissued throughout the 1840s and 1850s and was also subdivided into shorter guides to specific divisions of the museum. Other guidebooks included R. T. Scott and J. Kesson's threepenny *Complete Guide to the British Museum*, published by G. Vickers (Scott and Kesson1843) and William Mason's *A Guide to the British Museum* (Mason 1844). These unofficial guides drew heavily upon the *Synopsis* but were far shorter. Judging by frequency of reprinting, Clarke's was by far the most successful guidebook, followed by Mason's.

Guidebooks as a genre are believed to have originated to enable countryhouse visitors to identify items in the painting and sculpture collections when the housekeeper was absent or inadequately informed (Harris 1968). The British Museum guides retained this emphasis on identification rather than description or explanation. There were section headings for each room of the museum, and under those headings followed a list of the items in that room, often numbered according to the numbering of the display cases. The succession of sections within the pamphlet would suggest a possible route through the museum: Clarke and Scott and Kesson began on the ground floor with antiquities, while Mason followed the *Synopsis* by starting upstairs with natural history.

The guidebooks mentioned far fewer items than did the *Synopsis*, and their editors ruthlessly pared the entries down to the bare minimum. Apart from an introduction to the history of the museum and its public hours, most of the pages simply listed objects (figure 7.7). The *Synopsis* itself was the most detailed, but even its entries were hardly highly informative. In the Southern Zoological Gallery the visitor was told:

> Cases 1 and 2 contain the different varieties of Llama; the wild ones are brown . . .
>   Cases 3 to 16 contain the different species of Oxen and Elephantidae . . . (Anon. 1852b, 2)

Figure 7.7. Although H. G. Clarke's guidebook described the mammalian collections with minimal detail, the editions from the mid-1850s onward had the advantage of a few woodcuts—though the print quality was noticeably low. Illustration reprinted from Clarke 1855, by permission of the Bodleian Library, University of Oxford [G.A.Lond.8°772, opening pages I-20/12–24].

It was routine for natural history specimens to be listed in groups, which meant that the guidebooks often provided even less help than the labels. All the guidebooks devoted far more pages to antiquities than to natural history, but even though antiquities were usually listed as individual items, the descriptions were still very short. In the Phigaleian Saloon, Greco-Roman antiquities were listed thus:

No. 8*. A group of Bacchus and Empelus. Found A.D. 1772, near La Storta, on the road to Florence, about eight miles from Rome.
A head of Minerva, helmeted . . . (Anon. 1852b, 98)

For cultured, well-educated visitors, such brief tags were, perhaps, enough to call up memories of readings in the history of Rome and recent archaeological finds. They provided very little information, but they might be enough to link the exhibit to something the visitor already knew. For the new audiences of the mid-nineteenth century, with fewer mental resources available, the labels would be of little help.

Occasionally the guidebooks drew attention to a particular exhibit or, through their choice of words, suggested an appropriate reaction from the visitor. Unusual specimens, such as the kangaroo, were "deserving of the visitor's attention," while the giraffe, like the elephant and the walrus, "commands the attention of every visitor" (Mason 1847, 5). The assumed interest in exotic creatures (Baratay and Hardouin-Fugier 2002, 150–51; Ritvo 1987, chap. 5) had also been apparent in the *Pictorial Museum*, whose very first sentence announced, "We enter within the doors of our Pictorial Museum, and our eye is at once arrested by a group of fierce yet beautiful animals, among which stands one prominent in stern grandeur and majestic bearing—we cannot mistake him,—the lion, the king of beasts" (Martin 1848–49, 1:2). Like the giraffe at the head of the British Museum staircase, the lions and tigers in the opening issue of the *Pictorial Museum* drew the reader in, encouraging him or her to dwell, "like the visitor to a Museum, upon their curious forms and their picturesque attitudes" (Martin 1848–49, 1:[1]). Very rarely, guidebooks offered a snippet of information about the history of an individual, such as the revelation that one of the elephant specimens was "the skeleton of Chunie, the stupendous Elephant shot about twenty years since at Exeter 'Change" (Mason 1847, 5).

The guidebooks routinely implied that visitors should express awe at the collection, as illustrative of both the wonders of creation and the skill of those who amassed the specimens. Thus, the bird collection was "most perfect, beautiful and wonderful," the reptiles were "extensive and splendid," and even the weasels were "very complete and highly interesting" (Mason 1847, 5, 7). An underlying assumption that the animals were designed by a creator surfaced only rarely, as when, for instance, Mason commented that "every mind" would be impressed with "the wonders of the Creation" when presented with the display of "Bears, Camelopards, Camels, Horses, Deer, Elephants, Rhinoceroses, Hippopotami, Porpoises, Seals, Armadilloes &c" (Mason 1847, 5). In the *Pictorial Museum*, the conclusion had reminded its readers to reflect upon "the cause of causes, the Almighty Creator, who in wisdom has formed our globe and all that live upon its surface," but such sentiments were usually taken for granted in the British Museum (Martin 1848–49, 2:425).

In the British Museum, taxonomic groups of animals were not separated as carefully as in the *Pictorial Museum*, and visitors were free to wander among the exhibits, creating their own experiences. Visitors hoping for an education in natural history may well have been disappointed.

The British Museum labels, even if supplemented with the guidebooks, did not offer much information beyond the names of specimens, and they did not relate one specimen to another to enable the visitor to deduce anything from the contrast. On the other hand, if the visitor's reason for being in the British Museum was for curiosity or entertainment, the visit might be highly successful. The range and variety of specimens was unequalled, and the experience of standing next to a real giraffe or to Chunie the elephant was altogether different from scrutinizing a black-and-white illustration (Benjamin 1970).

Creating a Multimedia Visiting Experience

It is already apparent that visitors to the British Museum did not view its exhibits in isolation: there were labels on the cases, and visitors might have acquired a guidebook. The guidebooks, however, were only part of the print culture that might have shaped a British Museum experience. Visitors would have had to consult travel handbooks and railway timetables to get to Bloomsbury at the right time on the right day to gain admittance. If they had recently read an account of African exploration, or an article on the wonders of marine life, or seen a reproduction of a painting of Cook in Australia, such memories could inform their encounter with the mounted lion or kangaroo or the display of molluscs and corals. After the visit, they might seek out further information on the exhibits they had seen, perhaps consulting an encyclopedia or even the *Pictorial Museum*. The British Museum experience was enmeshed in print, and individuals' experiences of the collections would be as unique as their personal reading histories.

While it would be difficult to track down all the potential influences on visitors, the books which were published specifically to advise them are an obvious place to start. These "advice books" differ from guidebooks in containing far more narrative text and making more effort to inform reactions to the collections. They might have been read before or after but were probably only skimmed during a visit. Moreover, while the guidebooks were issued by specialist publishers, the advice books were generally produced by philanthropic publishers committed to promoting self-improvement. Charles Knight's description of the experience of visiting the British Museum indicates why he believed visitors needed help: "Many persons feel, that when they leave the Museum after a visit of two or three hours, their thoughts are so filled with a chaos of minerals, stuffed monkeys, Greek statues, beautiful shells, Hindoo idols, vases, humming-birds, Egyptian mummies, monstrous fossil animals, and Polynesian trinkets, that it is difficult to retain a clear idea of any of them. This is a pity" (Knight 1847–49, 1:35). Knight included sixteen pages on the British Museum in *The Land We Live In* (1847–49), his illustrated guide for excursionists, originally published as a part work (Knight [1864–65] 1873, 3:69–70). Nor was he alone: at least three other publishers produced books on the British Museum in the years around 1850. A 64-page guide appeared in a series calling itself the "New Library of Useful Knowledge" (Anon. 1852a); Bradbury and Evans issued a somewhat longer one by Blanchard Jerrold (Jerrold 1852); and a 432-page volume titled *The British Museum: historical and descriptive* appeared in W. and R. Chambers's "Instructive and Entertaining Library" series (Masson [1848] 1850). As the author of this last work, David Masson, made clear, the aim of such advice books was to turn museum visiting from a "tedious and bewildering" experience, caused by the visitor's lack of "the necessary amount of preliminary knowledge," into a constructive and educational occasion (Masson [1848], 1850, v). These authors and publishers felt the guidebooks discussed in the previous section did not supply enough information, nor did they help visitors make sense of what they saw. Masson's book aimed not only to guide visitors but also to teach them basics of natural history. The very existence of such advice manuals around the mid-nineteenth century is yet another indication of the arrival of large numbers of new visitors to the British Museum.

Aberdeen-born David Masson (1822–1907) was a writer and journalist who had recently moved from Edinburgh to London and who would become professor of English literature at University College in 1852. Despite his move south, he maintained his connections with various Edinburgh publishers, including Chambers, while forging new links in London's literary circles. In contrast to Masson's Scottish Free Church inclinations and scholarly interests, Blanchard Jerrold (1826–1884), the son of the journalist and editor Douglas Jerrold, moved in the circle of young bohemians and was quintessentially a journalist. After gaining some public success in 1851 with *How to See the Exhibition, in four visits*, he wrote guidebooks for several other exhibitions as well as *How to See the British Museum, in four visits* (1852). Where Masson had written a manual of basic scientific instruction, Jerrold interpreted the British Museum through the lens of the 1844 *Vestiges of the Natural History of Creation*, presenting all its collections (historical as well as natural historical) as evidence for progressive development.

One of the key advantages of a museum was the access it offered to encounter genuine specimens. Masson claimed this was a "more decisive method of actually presenting to the eye the objects spoken about" than text or images could be (Masson [1848] 1850, 113), and Jerrold noted that "the mind is particularly impressible" while looking at objects (Jerrold 1852, 229). Although objects were often presumed to speak directly to the eye (Bennett 1998), as natural history writer J. G. Wood pointed out, "To one who has not learned to read, the Bible itself is but a series of senseless black marks; and similarly, the unwritten Word that lies around, below, and above us, is unmeaning to those who cannot read it." The skill of "reading" natural objects had to be learned, and many of the people who ignored natural objects did so not because of a lack of interest but "for want of proper teaching" (Wood 1858, [1]). Among the new visiting audiences of the mid-nineteenth century, literacy itself could barely be taken for granted, and the skills for reading natural objects were even rarer. With the British Museum doing little to help these visitors, the advice books were crucial in suggesting what to look at, and how to compare objects.

Like the guidebooks, the advice books drew attention to curious, exotic, and unusual species: the giraffes and elephants featured again, and Knight speculated about the last moments of the rhinoceros, noting the "numerous bullet-holes in his tough horny hide," which suggested "a fierce battle for his life" (Knight 1847–49, 1:38). Visitors might well admire the large exotic creatures without prompting, but there were other cases where advice books were needed to "direct the eye to objects, and at once to interest the visitor in them, by shortly explaining their points of interest" (Jerrold 1852, 30). For instance, Masson, Jerrold, and the New Library of Useful Knowledge all pointed out the "ornithorhyncus" or "mullingong" (platypus), an animal of "extraordinary aspect," which, being smaller than the giraffes, might go unnoticed (Masson [1848] 1850, 237). Masson's book included numerous small images to show visitors what they ought to be looking for (figure 7.8). The platypus deserved attention for possessing "the body of an otter" yet "the beak of a duck and the claws of a bird," and it was particularly important for Jerrold as evidence for the existence of gradations between classes, as the bird merged into the mammal (Jerrold 1852, 9, 61).

Jerrold wished his reader to obtain "clear impressions of the different departments or classes" of natural history "by guiding his eye consecutively to those objects which bear relation to each other" (Jerrold 1852, 229). Likewise, Masson suggested that visitors contrast the "strong hooked-bills, stout legs, and sharp, powerful talons" of the raptors with the "long stilt-like legs" of the wading birds and consider how the birds' appearance related to their way of living (Masson [1848] 1850, 221). Knight recommended that visitors to the saloon of hoofed mammals should pay particular attention to "the different degrees of vigour, of size, and of strength, in animals

Figure 7.8. Masson's advice book used illustrations to highlight particular specimens. While the auroch (Lithuanian bison) donated by the emperor of Russia might be striking enough to be noticed by the visitor without further prompting, the image of the platypus encouraged visitors to seek out this curious creature. Reprinted from Masson [1848] 1850, 236.

brought from different countries" (Knight 1847–49, 1:38). By suggesting informative comparisons, the advice books were helping the visitor make meaning from the exhibits. Later in the century, in sites such as the re-located British Museum (Natural History) at South Kensington, this role would be taken on by the museum itself (Stearn 1981; Forgan and Gooday 1994–95, 1996; Yanni 1999, chap. 5). But at midcentury, visitors to the British Museum were very much left to make such meanings as they could, with whatever assistance they could find.

The recommended way of viewing exhibits, with an emphasis on un-usual creatures and the drawing of comparisons, was common to most of the advice books. So too was an insistence on self-discipline, to avoid the confusion arising from random wanderings. Ideally, the museum should be divided into four or five carefully planned visits, one for each division of the collections (Knight 1847–49, 1:35; Jerrold 1852). When visiting the natu-ral history collections, the tourist should ignore all mummies, sculptures, and Roman oil lamps. From the entrance hall, the visitor was advised to climb the main staircase, and look "the giraffe boldly in the face" as he or she emerged into the Central Saloon with its collection of hoofed animals (Knight 1847–49, 1:38). The visitor should then proceed counterclockwise around the upper galleries, through more mammals and the birds, and then into the Northern Gallery, which contained reptiles and fish, followed by the mineral and fossil collections.

The curators' need to make efficient use of wall and floor space meant that the Eastern and Northern Galleries displayed a mixture of vertebrates on their walls and invertebrates in their table cases, undermining the idealized route from mammals down to invertebrates. The advice books, however, could restore order. When Masson's book was first published in 1848, the Northern Gallery (figure 7.9) had been a particular problem. The move from old Montagu House to Smirke's new building took place in installments, and the curators could not rearrange properly until the move was complete. In the meantime, the Northern Gallery was in "a very confused

Figure 7.9. Illustration of the coral room in the Northern Gallery of the British Museum. In addition to the corals (in the wall and table-cases), this room displays reptiles and fish high on the walls. Reprinted from *Illustrated London News*, April 3, 1847, 221.

state, being used as a kind of receptacle for superfluous radiata, articulata, mollusca, and vertebrata indiscriminately" and looking rather like "a stow-room for whatever could not be accommodated elsewhere" (Masson [1848] 1850, 423). Masson wanted his readers "to single out, first, the specimens of the lowest classes of animal life . . . and thence gradually to ascend to the higher organisms." His tour, therefore, began with the sea urchin, and moved upward through corals and annelids to the vertebrates. In 1848, this required a complex route through the gallery, starting in Room II with echinodermata and moving into Room III for corals, then back to Room I for annelids, then forward to Room IV for crustacea and insects (Masson [1848] 1850, 170–78). By 1850, rearrangements had made the Northern Gallery "decidedly more pleasant to visit than it used to be" (Masson [1848] 1850, 423). Thus, Jerrold could offer a relatively simple suggestion for imposing taxonomic order: the visitor should start by examining all the specimens in the mammal galleries but should walk through the Eastern and Northern galleries looking only at the walls and then return through the same galleries, looking at the tables on the second pass (Jerrold 1852, 16–57).

Masson and Jerrold both believed that it was crucial to view the specimens in taxonomic order. Masson claimed this would encourage visitors to pay more attention to the "peculiarities of the individual animals" and that it would uncover the "hierarchical arrangement of the whole" (Masson [1848] 1850, 169). Jerrold agreed that it would bring out their connections, but he believed that it was "the gradations of life" which would become evident to visitors (Jerrold 1852, 61). The contrast in philosophical lessons Masson and Jerrold wished readers to draw illustrates how two advisers could "read" the same collections quite differently, and it suggests that a visitor's choice of advance reading material could predispose him or her to quite different interpretations of the British Museum. A visitor who took Masson to heart would see the museum as an instructive scientific experience, consistent with a designed, created world, while a visitor influenced by Jerrold would see evidence for progressive development wherever he or she looked.

Masson assured his readers that viewing the strange and varied creatures of the British Museum would "be eminently instructive, calculated as it is to enlarge and diversify his notions of the world and its wonders" (Masson [1848] 1850, 169). He included introductory essays to all his sections so that his readers were equipped with basic knowledge in each science. There were, for instance, thirteen pages explaining mineralogical classification so that the label on the first case in that gallery ("electropositive native metals and their unoxidised combinations") would make sense (Masson [1848] 1850, 127). Masson was also committed to a vision of completeness, which led him to include a chapter on botany, even though the botanical collections of the British Museum could be viewed only by special request (Masson [1848] 1850, 156).

Masson typically devoted a few sentences or even a paragraph of discussion to each species, and he made use of the numbering on the display cases to connect his text to the objects, just as Martin did with the figure numbers in the *Pictorial Museum*. If, for instance, visitors were intrigued by the inhabitants of cases I and II in the Southern Gallery, Masson's book would provide them with the species name and the information that these were "the camels, as they have been called, of the new world," thus linking the unknown to the known (Masson [1848] 1850, 236). In this instance, Masson provided little more information about llamas than Jerrold did (according to whom, "the wild are generally brown, and the tame of mixed colours"; Jerrold 1852, 8), but Masson's desire to instruct usually made him more detailed. However, having warned visitors that they would be better off learning about mammals at the Zoological Gardens, Masson skipped through the Southern Gallery much more rapidly than the others. Jerrold, in contrast, placed less emphasis on his instructive content, referring to it as merely "various bits and scraps" and "little points," which would entertain visitors as they moved through the galleries. He was willing to acknowledge that many visitors were "chiefly bent upon enjoying a few hours amusement," and his book was only half the length of Masson's (Jerrold 1852, 229).

The philosophy underlying Masson's account of the natural world was only occasionally apparent in his text, but it was clearly a vision of a created world, which had undergone "ages of preparation" for the arrival of humankind, which had been "gifted with reason" and represented the pinnacle of a hierarchical creation. In Masson's view, man was the "climax" and "the end," and there was an "enormous interval between himself and all beneath him in the scale of animated nature" (Masson [1848] 1850, 266). Masson regarded the exhibits as demonstrating the hierarchies of the animal kingdom and denied that they suggested gradation between the classes. He was particularly critical of Gray's decision to arrange the mammals according to the quinary system, which, according to Masson, had as its "principal object . . . to exhibit the manner in which the orders seem to be connected together by a common principle of gradation running through them." To counter this method of display, Masson explained Cuvier's classification, enabling his readers to subvert the curator's arrangement (233). He was also quite explicit that humankind represented the final stage in the creation and that any discussion of man's future progress was a discussion of "his hopes of immortality" and not any future physical development (266).

In contrast, Jerrold encouraged visitors to read the British Museum as proof of the regular gradations between animal species, of the existence of ancient life-forms which no longer lived on the earth, and the gradual progress of human civilizations, from Egyptians through Greeks and Romans. It was nothing less than a demonstration of the progressive development hypothesis as laid out in *Vestiges of the Natural History of Creation*, a work which Jerrold quoted several times (Jerrold 1852, 47, 64; Secord 2000). The structure of Jerrold's book, with introductory and concluding remarks to each of its four "visits," gave him ample scope to expand on the connections between the exhibits. He assured his readers that philosophers had definitely "settled the scheme of the world" and that it was a scheme of progression, or development. "The gradations of life may be clearly apprehended by the visitor" in the transition between galleries, from mammals to birds, to reptiles and fish, and finally to invertebrates, including the polyp to which "philosophers" trace all animal life (Jerrold 1852, 60).

Following the author of *Vestiges*, Jerrold used the language of awe and wonder in a manner akin to a discourse of design. He wrote of "the grandeur of the scheme" and claimed that "all these great and complicated developments are the beautiful works of the Great Unseen." As in Masson's account, mankind can say "I am the King" (Jerrold 1852, 62). But Jerrold pointed to "the consequent littleness of individual manhood" and allowed man to be—not created in the image of God—but merely the "highest type of the Articulata, Radiata, Mammalia, or any order of vegetable or animal life" (61–62). For Masson, the primates were evidence of the great gap between humans and other animals, but for Jerrold they were "an inferior brotherhood"—lower down the scale, but kin nonetheless (62). Rather than by a Christian God, Jerrold argued, man should be humbled by thoughts of "the great laws that govern the universe . . . , and the great progresses of Nature" (122), which, rather than climaxing with man, hold out "promises of future developments" (227).

Looking at the same exhibits, Masson and Jerrold saw quite different visions of the natural world. Those who read their books would be encouraged to view the collections in the same way as their adviser. I would not wish to argue that all those who read Jerrold became devout believers in progressive development, for that would ignore readers' creativity and selectivity. But when visitors came face-to-face with the British Museum collections, their reactions, impressions, and memories would have been shaped by their prior reading.

Conclusions

Historians of tourism have started to talk about a "multimedia experience" in the nineteenth century, when the experience of viewing a historic site or picturesque view brought back (and was thus colored by) memories of novels and travel tales, or Royal Academy paintings, or West End dramas (Mandler 1999; Carroll this volume). The use of a vivid imagination, playing off previous reading or viewing experiences, often made sites and views far more impressive to some travelers than to others (Ousby 1990, chap. 4). At the British Museum, visitors mentally combined the immediate stimulus of a llama or a platypus with the information available on the spot and with any relevant past experiences. The uniquely personal nature of those past experiences necessarily meant that visitors would interpret the British Museum in a uniquely personal way. Nonetheless, some sources were more obvious potential influences upon a visit than others, and the advice books demonstrate some of the more likely frameworks for viewing the museum.

A visitor who had read the *Pictorial Museum* before visiting Bloomsbury would gain an overview of the animal kingdom even more extensive than that offered by the advice books. Indeed, any natural history book could provide a structure into which individual museum exhibits could later be placed, as well as providing information about the appearance, habits, and uses of specific creatures. The difference was that whereas the *Pictorial Museum* tried to be self-sufficient—providing text and images in one place to replace the visiting experience—the advice books reveled in their connection to the British Museum and encouraged readers to combine reading and visiting. Yet, just as the advice books might become souvenirs, so a reader who encountered the *Pictorial Museum* after the British Museum might find its black-and-white images recalling the museum's originals. Memory could color the images and give them scale; and if the trip to London had also included the Zoological Gardens, that memory might bring the *Pictorial Museum*'s animals to life.

Examining the *Pictorial Museum* and the British Museum together demonstrates how important print was to the natural history visiting experience in the mid-nineteenth century. The producers of the *Pictorial Museum* would even have hoped that print's superiority was such that it could replace the actual visiting, but more realistically, print supplemented the visiting experience. Guidebooks identified specimens; advice books recommended routes, pointed out selected specimens, and encouraged visitors to make revealing comparisons; and all other natural history books and articles provided background information which could fit into the visiting experience in myriad ways.

The complexity and careful organization of the *Pictorial Museum* resulted from its publisher's substantial experience of the needs of lower-middle-class and working-class readers and his freedom to innovate in the use of texts and images. The *Pictorial Museum* was a sophisticated print product combining the exhibit-dominated feel of a museum with the information-rich feel of a book. It demonstrates how Charles Knight's awareness of the importance of museums for public education, combined with his experience as a publisher, could produce a work more complete, better organized, and more widely accessible than the British Museum, and which provided more guidance for inexperienced visitors through its tight linkage between image and text.

In contrast to the relative freedom of publishers, curators had only one architectural space available to them, and they lacked the staff or money for extensive reorganization. The eventual move to South Kensington in 1881 provided an unusual opportunity for a complete reorganization of the British Museum natural history collections, but by then, print had become a more central component of the museum experience, providing narrative and interpretation as well as identification. The increasing presence of print in museums in the later nineteenth century illustrates another aspect of curators' efforts to assist and educate, but also to control, their visitors (see Alberti this volume).

Although some British Museum curators of the mid-nineteenth century did appreciate the growing importance of providing an experience adapted to the new generation of visitors, this was a relatively new development at a museum which was protective of its scholarly reputation. The British Museum displays of this period were still constructed on the assumption that visitors had the resources (from education, private reading, or acquaintances with scholars) to create their own narrative in which to situate the specimens. For all the hundreds of thousands of visitors from the working and lower middle classes, who had only limited opportunities for prior reading, the extra assistance offered in the unofficial guidebooks was not enough. These books indicated a simple route and basic identification of specimens but little more. The advice books were published precisely to address that gap, with their publishers demonstrating an entrepreneurial awareness of changing audiences and market conditions, which curators were slower to acknowledge. At midcentury, publishers had greater experience, as well as greater freedom, than curators did, which is what made their products so effective with the new reading and visiting audiences. It should be no surprise that it was the philanthropic publishers who, acknowledging the advantages and limitations of existing museums, took the initiative in offering readers and visitors the best of both worlds by producing advice books to supply the information, interpretation, and narrative direction that were the strong point of the *Pictorial Museum* but absent from the British Museum.

NOTES

I wish to thank Juliana Adelman, Sam Alberti, Sophie Forgan, Bernie Lightman, Elizabeth Neswald, and Jim Secord for their enlightening comments and suggestions. I also wish to thank the Irish Research Council for Humanities and Social Sciences for supporting the writing of this chapter by awarding me a Government of Ireland Fellowship.

1. Knight described *The Pictorial Museum of Animated Nature* as belonging to a four-part series which mimicked Comenius's *Orbis Pictus* of 1658. There were, however, no indications on the products themselves of this over-arching series. The others were *The Pictorial Sunday-Book* and *Old England*, both published in 1845, and *The Pictorial Gallery of Arts* in 1845–47. See Knight [1864–65] 1873, 3:18.

2. On Clarke, see advertisement appended to Clarke 1843. He issued guides to the National Gallery, the Dulwich Gallery, Greenwich Hospital, and Hampton Court.

REFERENCES

Alberti, Samuel J. M. M. 2002. "Placing Nature: Natural History Collections and Their Owners in Nineteenth-Century Provincial England." *British Journal for the History of Science* 35:291–311.

Anderson, Patricia. 1991. *The Printed Image and the Transformation of Popular Culture 1790–1860*. Oxford: Clarendon Press.

Anon. 1852a. *The British Museum, in Five Sections; or, How to View the Whole at Once*. London: Cradock.

Anon. 1852b. *Synopsis of the Contents of the British Museum*. 59th ed. London: Woodfall.

Bailey, Peter. 1978. *Leisure and Class in Victorian England: Rational Recreation and the Contest for Control 1830–1885*. London: Routledge.

Baratay, Eric, and Elisabeth Hardouin-Fugier. [1998] 2002. *Zoo: A History of Zoological Gardens in the West*. Trans. Oliver Welsh. London: Reaktion.

Benjamin, Walter. 1970. "The Work of Art in the Age of Mechanical Reproduction." In *Illuminations*, ed. Hannah Arendt, 219–53. London: Jonathan Cape.

Bennett, Tony. 1998. "Speaking to the Eyes: Museums, Legibility and the Social Order." In *The Politics of Display: Museums, Science, Culture*, ed. Sharon Macdonald, 25–35. London: Routledge.

Bohrer, Frederick. 2003. *Orientalism and Visual Culture: Imagining Mesopotamia in Nineteenth-Century Europe*. Cambridge: Cambridge University Press.

Buzard, James. 1993. *The Beaten Track: European Tourism, Literature and the Ways to "Culture," 1800–1918*. Oxford: Clarendon Press.

Clarke, Henry Green. 1843. *The British Museum Hand-Book Guide for Visiters [sic]*. London: H. G. Clarke.

———. 1855. *The British Museum: Its Antiquities and Natural History, a Hand-Book Guide for Visitors*. London: H. G. Clarke.

Crook, J. Mordaunt. 1972. *The British Museum*. London: Allen Lane.

Cunningham, Peter. 1851. *Murray's Handbook to Modern London; or, London as It Is*. 1st ed. London: John Murray.

Eliot, Simon. 1994. *Some Patterns and Trends in British Publishing 1800–1919*. London: Bibliographic Society.

Fergusson, James. 1849. *Observations on the British Museum, National Gallery, and National Record Office, with Suggestions for Their Improvement*. London: John Weale.

Forgan, Sophie. 1994. "The Architecture of Display: Museums, Universities and Objects in Nineteenth-Century Britain." *History of Science* 32:139–62.

———. 1999. "Bricks and Bones." In *The Architecture of Science*, ed. Peter Galison and Emily Thompson, 181–208. Boston: MIT Press.

Forgan, Sophie, and Graeme Gooday. 1994–95. "'A Fungoid Assemblage of Buildings': Diversity and Adversity in the Development of College Architecture and Scientific Education in Nineteenth-Century South Kensington." *History of Universities* 14:153–92.

———. 1996. "Constructing South Kensington: The Buildings and Politics of T. H. Huxley's Working Environments." *British Journal for the History of Science* 29:435–68.

Fyfe, Aileen. 2000. "Reading Children's Books in Eighteenth-Century Dissenting Families." *Historical Journal* 43:453–74.

———. 2004. *Science and Salvation: Evangelicals and Popular Science Publishing in Victorian Britain*. Chicago: University of Chicago Press.

———. 2005. "Conscientious Workmen or Booksellers' Hacks? The Professional Identities of Science Writers in the Mid-Nineteenth Century." *Isis* 96: 192–223.

[Gray, J. E.] 1843. *List of the Specimens of Mammalia in the Collection of the British Museum*. London: George Woodfall, by order of the Trustees.

Grewal, Inderpal. 1996. *Home and Harem: Nation, Gender, Empire and the Cultures of Travel.* London: Leicester University Press.

Harris, John. 1968. "English Country House Guides, 1740–1840." In *Concerning Architecture: Essays on Architectural Writers and Writing, Presented to Nikolaus Pevsner,* ed. John Summerson, 58–74. London: Allen Lane.

Jerrold, William Blanchard. 1852. *How to See the British Museum, in four visits.* London: Bradbury and Evans.

Knight, Charles, ed. 1847–49. *The Land We Live In: A Pictorial and Literary Sketch-Book of the British Empire.* 4 vols. London: Knight.

———. [1864–65] 1873. *Passages of a Working Life During Half a Century: With a Prelude of Early Reminiscences.* 3 vols. London: Knight.

Mandler, Peter. 1997. *The Fall and Rise of the Stately Home.* New Haven, CT: Yale University Press.

———. 1999. "'The Wand of Fancy': The Historical Imagination of the Victorian Tourist." In *Material Memories,* ed. Marius Kwint, Christopher Breward, and Jeremy Aynsley, 125–42. Oxford: Berg.

Martin, William. 1848–49. *The Pictorial Museum of Animated Nature.* 2 vols. London: Knight.

Mason, William. 1847. *A Guide to the British Museum; Fully Descriptive of All the Most Interesting Natural Curiosities, Wors [sic] of Art, Greek and Roman Sculptures, Egyptian Antiquities, and Other Objects, Worthy of the Attention of Visitors in General.* London: Mason.

Masson, David. [1848] 1850. *The British Museum: Historical and Descriptive.* Edinburgh: Chambers.

Milner, Thomas. 1846. *The Gallery of Nature, a Pictorial and Descriptive Tour through Creation, Illustrative of the Wonders of Astronomy, Physical Geography, and Geology.* London: Orr.

Ousby, Ian. 1990. *The Englishman's England: Taste, Travel and the Rise of Tourism.* Cambridge: Cambridge University Press.

Pimlott, J. A. R. [1947] 1976. *The Englishman's Holiday: A Social History.* Hassocks, Sussex: Harvester Press.

Ritvo, Harriet. 1987. *The Animal Estate: The English and Other Creatures in the Victorian Age.* Cambridge, MA: Harvard University Press.

Rupke, Nicolaas. 1988. "The Road to Albertopolis: Richard Owen and the Founding of the British Museum of Natural History." In *Science, Politics and the Public Good,* ed Nicolaas Rupke, 63–89. London: Macmillan.

Schofield, R. S. 1981. "Dimensions of Illiteracy in England, 1750–1850." In *Literacy and Social Development in the West: A Reader,* ed. Harvey J. Graff, 201–13. Cambridge: Cambridge University Press.

Scott, R. T., and J. Kesson. 1843. *A Complete Guide to the British Museum Forming a Correct Catalogue of the Statues; Marbles; Mummies; Vases; Greek, Roman and Egyptian Antiquities; Minerals, Birds, Beasts, Fishes, Works of Art, and Natural Curiosities.* London: G. Vickers.

Secord, James A. 2000. *Victorian Sensation: The Extraordinary Publication, Reception and Secret Authorship of "Vestiges of the Natural History of Creation."* Chicago: University of Chicago Press.

Simmons, Jack. 1970. Introduction to *Murray's Handbook for Travellers in Switzerland*, by John Murray and William Brockeden. Leicester: Leicester University Press.

Stearn, William T. 1981. *The Natural History Museum at South Kensington: A History of the British Museum (Natural History) 1753–1980.* London: Heinemann.

Topham, Jonathan R. 1998. "Beyond the 'Common Context': The Production and Reading of the *Bridgewater Treatises*." *Isis* 89:233–62.

———. 2000. "Scientific Publishing and the Reading of Science in Early Nineteenth-Century Britain: An Historiographical Survey and Guide to Sources." *Studies in History and Philosophy of Science* 31A:559–612.

Vaughan, John E. 1974. *The English Guide Book, C1780–1870: An Illustrated History.* Newton Abbot, UK: David and Charles.

Vincent, David. 1989. *Literacy and Popular Culture: England 1750–1914.* Cambridge: Cambridge University Press.

———. 2000. *The Rise of Mass Literacy: Reading and Writing in Modern Europe.* Oxford: Polity Press.

Weedon, Alexis. 2003. *Victorian Publishing: The Economics of Book Production for a Mass Market, 1836–1916.* Aldershot: Ashgate.

Wilson, David M. 2002. *The British Museum: A History.* London: British Museum Press.

Wood, John George. 1858. *The Common Objects of the Country.* London: Routledge.

Yanni, Carla. 1999. *Nature's Museums: Victorian Science and the Architecture of Display.* London: Athlone.

CHAPTER EIGHT

Illuminating the Expert-Consumer Relationship in Domestic Electricity

GRAEME GOODAY

During the last few years of the expiring century a change of vast, but as yet hardly realized, importance has been wrought in one of the chief distinctive appurtenances of town life in all civilized lands—the organized public and private artificial lighting. I do not propose to treat directly of the most notorious feature of this era, the successful introduction of electric lighting on a commercial scale, about which much has been written . . . —*Ex Fumo Lucem* 1900, 710[1]

Why was so much written in the late nineteenth century about the arrival of the electric light? What could have been so "notorious" about it—and who would have judged it so? "Ex Fumo Lucem" wrote in the *Contemporary Review* as a lobbyist for the coal gas industry to challenge the authority of the self-styled electrical expert then in competition for popular acceptance—and market sales—of the two rival illuminants. This pseudonymous advocate encouraged readers to see gas as the ideal medium for lighting and heating the home and to shun the parvenu force of electricity so vigorously promoted by its partisan promoters in periodical and handbook literature. From that perspective, the alleged notoriety of such literature arose from the purportedly self-interested, untrustworthy, and fallible nature of such electrical specialists as Thomas Alva Edison and William Crookes, as well as less-familiar names: Robert Hammond, St. George Lane-Fox, and the marital partnerships of James and Alice Gordon and Maud and Edward Lancaster. Whether readers of this piece actually agreed with Ex Fumo Lucem that such putative electrical experts had misled the public, or even trusted this writer's claims about gas, however, was quite another matter. Questions about the discretionary judgment of readers of "popular science" are, of course, at the heart of this volume.

My specific contention in this chapter is that the popularization of technical specialisms should not be understood (simply) as an encounter between self-evidently "expert" authorities and an inexpert or credulous laity. For the case of electricity, at least, I suggest that the terms of engagement were more unstable and contingent than that: those aspiring to public authority on electric matters were not unequivocally nor consensually granted recognition as "experts" by their audiences. In this context, writing persuasively for broad popular acceptance was arguably one of a range of strategies *necessary* for any individual attempting to attain "expert" status; such writing was not merely an auxiliary activity for public relations (Marvin 1988, 15–17, 39–46). Yet writing for the public and even being widely read were of themselves no guarantee of recognition as an expert: audiences could be skeptical, equivocal, or indifferent to attempts by individuals to present themselves as authorities. Moreover, as we shall see, the typically divergent judgments among "authorities"—even within the electrical community itself—positioned the laity as a tribunal of allegedly "expert" judgments, being obliged to take their own informed decisions about which "expert" to believe. Somewhat ironically, the "lay" consumer thus had to bear some of the responsibilities more commonly associated with "expert" status, such as making discretionary choices in the face of uncertainty and dissent.

Highlighting the volume's theme of consumer agency and discretion, I suggest that authoritative status was accomplished only to the extent that audiences accepted "expert" claims and/or acted in concordance with recommendations. None of the figures mentioned earlier achieved unqualified long-term success in electric advocacy—whether through undergoing bankruptcy, forced withdrawal from the field, or being outmaneuvred by gas publicists (Clendinning 2004). While these figures cautiously avoided categorical self-representation in order to cast a discreet veil over their financial interests, our descriptive categories need not be so coy. The second section will thus analyze the role of these aspirant experts as "popularizer-entrepreneurs" to signal how their popularizing activities aimed both to gain them credibility and to alter consumers' attitudes and behavior in favor of electricity.

That being said, these quasi experts did not unilaterally determine the agenda of popularization: the nature and scope of issues on which expert judgment was obliged to comment emerged in a dialogue with the laity. In the third section of this chapter, I show that the public demanded authoritative assurances that it would be safe to install electricity in the home. In the fourth section, I show how the public demanded to know the nature of the mysterious intangible "electricity," which people would be expected to pay for as consumers yet could never see, touch, or quantify for themselves (Gooday 2004b, 237–39). Even such a high-profile transatlantic icon as Thomas Alva Edison was unable to give definitive answers on issues that householders in the United Kingdom and the United States considered ought to be the domain of expert advice. Faced with such concerns, electricity popularizer-entrepreneurs encountered the limits of their knowledge—a point ruthlessly problematized by representatives of the gas interest, who *could* specify both the chemical constituents of coal gas and the signs of impending poisoning or explosion due to gas leaks. So instead of factual certainties, electrical "experts" offered prospective consumers diversionary narratives of technological conquest and utopia, bolstered by images personifying electricity as an obedient (female) servant. In order to market the new illuminant, the popularizer-entrepreneurs had to reconfigure their identity from being sources of technical knowledge to being imaginative storytellers, futurists, and iconographers. Moreover, I show that such entrepreneurs acknowledged the gendered expertise of engineers' spouses—notably Alice Gordon and Maud Lancaster—as an essential feature of bringing such popularizing messages effectively to the female manager of the home.

My conclusion will consider what the symbiosis of popular and technical discourses on electricity meant: popular concerns could set the agenda for technical writing just as much as vice versa. There was, indeed, neither simple hierarchy nor neat demarcation between popular and technical treatment.

Recovering the Audiences for Popularizing "Electricity"

Recovery of past audiences is now a key theme for any historical study of science popularization (Cooter and Pumfrey 1994; Topham 2000; Carroll this volume; Alberti this volume). The results of this research have shown that the historian cannot assume a monolithic or deferential public response to emerging forms of science. Late nineteenth-century consumers of electric technologies and texts were by no means all "inexpert," passive, or neutral audiences. In what follows, I suggest that two themes can usefully differentiate the character of audiences for popular electricity: socioeconomic class and gender. Although obviously not the sole parameters that are relevant, these are the issues most important for electricity, especially electric light, as a luxury commodity and one that was a contentious candidate for installation in the domain of gendered power relations that was the Victorian home.

Affluent dwellers in the British metropolis were already familiar with electricity from theatrical displays of its dazzling effects (Morus 1998; Morus this volume), and readers often encountered electricity in the up-market *Quarterly Review* and *Nineteenth Century* (Gooday 2004c). So when in September 1878 Edison first announced to American and British householders that he was on the verge of a "cheap and practical substitute" for gaslight, their response was not one of ignorant deference but informed optimism—and the price of gas shares plummeted accordingly (Bazermann 1999, 160, 180). And when they were still waiting two years later, Edison showed a discomfited acknowledgement of the American public's role and discretion in the *North American Review* in October 1880. While he acknowledged the people's right to criticize his "unaccountable tardiness" in bringing his working filament lighting system to the public, he trivialized their impatient questions as "unscientific"—as if they were unqualified to gainsay his pronouncements (Edison 1880, 295).

Such responses indicate something of the unstable power relations of "popular" electricity—it was not quite clear who, if anyone, was in charge of the process of constructing a popular discourse of electricity, either in the United States or in the United Kingdom. Edison and his team of assistants certainly worked hard to persuade the public in both countries to adopt his form of electricity in their homes (Bazermann 1999; Helrigel 1998). Indeed, as David Nye has noted, Edison's own secretary, Samuel Insull, referred to this writing as "propaganda" to promote the merits of electric over gas lighting—a clear indication that Edison did not treat popular embrace of electric culture as a foregone conclusion (Nye 1990, 3). The enormous effort Edison devoted to popular writing makes sense if we consider his insight that public skepticism and distrust were as potentially fatal to commercial success as were technical and bureaucratic pathologies (Hughes 1983).

From the testimony of one of Edison's British competitors, Robert Hammond, we get some hint of the comparably mixed response of the British public to the electric light—and indeed the ambivalent response of prospective electricity consumers. As an energetic popularizer-entrepreneur, Hammond had lectured around Britain during 1882 and 1883 trying to emulate the public interest—if not complete trust—that Edison was starting to cultivate in the United States. Hammond published his lectures in *The Electric Light in our Home* in 1884, observing that many of those who came to his lectures were willing "freely to admit the great advantages" of electric light. Yet he had also found many people afraid to admit electric current into their homes, expecting it to bring an "insecurity" that far outweighed any comfort in prospect. This fear had been cultivated, he suggested, by the shareholders in gas companies "plentifully scattered" across the country who had not forgotten the financial losses incurred by transatlantic reports of Edison's new lamp in 1878: there was "everywhere an active body of intelligent men ready to condemn the electric light and to magnify the dangers arising from it" (Hammond 1884, 57–58). Hammond's strategy to overcome such anxieties was to try to convince his audiences through strategically chosen display experiments and anecdotes that gas lighting was a greater threat to health, life, and furnishings than his electric system (see later for further discussion).

While there was widespread interest in electric light following the exciting spectacles of the electric telegraph and the telephone in preceding decades, very few members of Hammond's audience could have expected to install electric lighting in their own homes. The early electric light (just like the telephone) was enormously expensive to purchase and remained more costly than gas lighting in Britain at least until World War I. Thus, in contrast to the distinctly unglamorous commodity of coal gas, only upper-class households could have afforded to purchase what was explicitly regarded as a *luxury*—an electric installation in the 1880s costing much more than the average worker's annual income (Byatt 1979, 2–3). It is therefore important to differentiate between two kinds of audience for popular electricity. On the one hand, there was the very tiny if influential upper-class constituency that could afford to exercise discretion over whether to bring it into their homes. And on the other hand, there was a large majority that could only consume the popular literature and public displays, perhaps aspiring to indulge in this dazzling extravagance if budgets ever allowed, but not in the immediate future.

Certainly in 1880, only the very wealthy, powerful, and leisured—in other words, the aristocracy—could afford to try the experimental technology. One famous aristocratic location for electric light was Hatfield House, the grandiose hereditary home of Robert Gascoyne-Cecil, the third Marquess of Salisbury, a leading Conservative politician, and later prime minister (1885–92, 1895–1902). An Oxford mathematics graduate and a gentleman Fellow of the Royal Society (FRS) since 1869, he was a devotee of private researches in ways that cut across any nascent amateur-professional divide in the natural sciences (A. Roberts 1999, 111–13). Out of political office after his party's electoral defeat in 1880, Salisbury found time to add electric light to his extant pursuits in photography, chemistry, and telephony (Briggs 1988, 373–74, 389). While he had sufficient wealth to pursue such pursuits, even as former foreign secretary (1878–80), he could not get away with imposing the harsh brilliance of the pre-Edison Jablochkoff arc lamp on the influential dinner parties organized by the marchioness, Georgina. As their daughter Lady Gwendolyn Cecil records, "For a brief period [Salisbury's] family and guests were compelled to eat their dinners under the vibrating glare of one of these lamps in the centre of the dining hall ceiling. No exertion of goodwill or courtesy could silence the plaintive protests of his lady visitors, and he would gird with growing despondency at the obstructions which feminine vanity offered to the conquests of science" (Cecil 1931, 3–4). Salisbury was "saved from humiliating defeat" in this social faux pas in autumn 1880 by adopting Joseph Swan's electric filament lamp (the chief rival to Edison's lamp) on the advice of mutual Newcastle friend Sir William Armstrong—then also electrifying his northeastern mansion Cragside. Salisbury thus installed a domestic Swan system with the help of his son Jem (who became the fourth Marquess of Salisbury after Robert's death in 1903) and his friend and adviser Professor Herbert Macleod of Cooper's Hill Engineering College.[2] As we shall see later, the much-viewed installation at Hatfield House was the scene of a fatal accident in December 1881 that generated much of the public anxiety about electricity that Robert Hammond sought to quell in his ensuing lectures.

Those who could neither afford the extravagance of electric light nor expect to be on the dinner list at Hatfield House could soon enjoy the displays of electric arc light in London's public places. The House of Commons was (briefly) arc lit from winter 1881, with the British Museum, Savoy Theatre, and Royal Academy adopting filament lamps at around the same time (Hannah 1979, 4). But the popular image of electricity that Hammond and others wanted audiences to accept for their activities to yield any profit was that of a household commodity. So the key move for the popularizer-entrepreneurs to harness the aspirations of the upper middle class to the electric light was to find a domestic site in central London at which it could be displayed. In contrast to the defeasible power of popular lecturing and literature that were central to Hammond's strategy, his rival R. E. B. Crompton enticed his wealthier audience by offering free installations in politically expedient and conspicuously metropolitan locations. For example, in 1883, the Crompton company offered a system gratis to Lord and Lady Randolph Churchill at Marble Arch, calculating that his company would benefit greatly from the electrically lit dinner parties hosted by Lady Churchill (the American-born Jenny Jermaine). As Lady Churchill later recalled, exaggerating retrospectively the Churchills' claim to priority, "[Ours], by the way, was the first private house in London to have electric lights. We had a small dynamo in a cellar underneath the street, and the noise of it greatly excited all the horses as they approached our door. The light was such an innovation that much curiosity and interest were evinced to see it, and people used to ask for permission to come to the house" (Cornwallis-West 1908, 102). The glamorization of electric lighting by Lady Churchill's dinner parties certainly generated a demand for the electric light among the middle class: Crompton's customers soon afterward included William and Ellen Crookes and the dramatist William Schwenk Gilbert (Crompton 1928, 110). The visible performance of congenial electric lighting was thus undeniably a key factor to both the commercial viability of and the popular interest in electricity (Nye 1990; Morus 1998).

While the late Victorian encounter with popular cultures of electricity was not simply or even primarily through the published media of book, periodical, or pamphlet, a substantial literature emerged to address the kinds of *problems* encountered by upper-class audiences when dealing with the electric light. Given their dazzling experiences of the garish arc lamp and the wildly sparking lecture display machine, male and female householders alike could otherwise have found it very hard to envisage electricity incorporated into the cozy orderliness of the urban home. Concern about the allegedly glaring quality of incandescent lighting was partly an extended response to the experience of arc lighting that had so exasperated Lord and Lady Salisbury's dinner guests in 1879–80. Continued revulsion to arc lighting was revealed in the following "Lines to the Electric Light at the G. W. Railway Terminus," published in 1888 by the *St. James Gazette* with reference to J. E. H. Gordon's first public installation in London:

Twinkle, twinkle, little Arc,
Sickly, blue uncertain spark;
Up above my head you swing,
Ugly, strange, expensive thing.

Now the flaring gas is gone.
From the realms of Paddington,
You must show your quivering light,
Twinkle, blinkle, left and right.

Cold, unlovely, blinding star,
I've no notion what you are,
How your wondrous "system" works,
Who controls its jumps and jerks.

Though your light perchance surpass,
Homely oil or vulgar gas,
Still, (I close with this remark),
I detest you, little Arc.
(Parsons 1940, 42)

For those who could afford the electric light in the home, the *aesthetic* challenge was somewhat different: visual faculties accustomed to the soft flickering of gaslight or candle did not easily adjust to the unvarying beam of the filament lamp. An extensive popular literature was developed to help householders adapt their perceptual faculties and domestic decor to the new unflickering illuminant—which, as James Facey's 1886 *Practical House Decoration* put it, brought "cold tones" and rays of "uncompromisingly searching brilliancy" to furniture and art. Other sources recommended darker lampshades, notably the Tiffany glassware that borrowed kaleidoscopic designs from William Morris's Gothic revivalism, to stifle the glare (Schivelbusch 1988, 179–82).

Since middle- and upper-class women were conventionally responsible for household aesthetics, they played a major role in deciding whether to electrify the home. But as their decisions were not necessarily consonant with the wishes of menfolk who were more typically impressed by dazzling public illuminations, we can readily see gender as an important theme in analyzing audiences for the early electric light. As Mrs. J. E. H. Gordon was so much at pains to note in her *Decorative Electricity* of 1891, women had started to exercise a newfound discretionary power in vowing never to allow the "very glaring and disagreeable" electric light into their homes. The unflattering exposure of "every wrinkle and line" and the headaches of which women often complained when faced with the electric light at the dinner table could be prevented by application of suitable artistry to situating and shading the lamps. She thus promoted the art of "decorative electricity" primarily, although not exclusively, to an audience of wealthier middle-class women and thereby sought (with some degree of success) to reconfigure them as ideal critical consumers. This approach in turn encouraged the kind of purchasing behavior that would assist her husband's electrical engineering business. And as I have suggested elsewhere, it was only by winning over such women consumers that electric light came to have a long-term future at all in Britain (Gordon 1891b, 59–60, 146; Gooday 2004a).

This discretionary power of female audiences to accept or reject electricity was just as important in relation to the popularization of electric cookery from the 1890s. Introduced by St. George Lane-Fox[3] in 1874, this

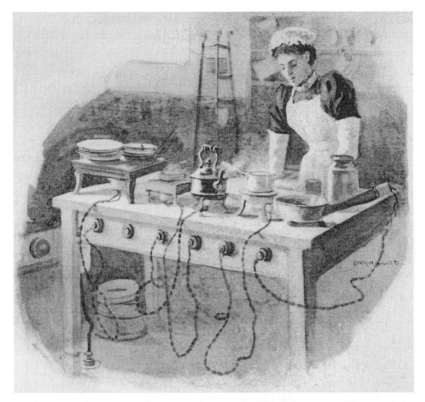

Figure 8.1. Illustration of a tabletop electric kitchen at Margaret Fairclough's School of Cookery in London about 1895: the spotlessly clean working environment and the ready mobility of the equipment employed are implicitly contrasted with the dirty and inflexible techniques of cooking on contemporary gas and coal stoves. Reprinted from A. W. 1895, by permission of Manchester Public Library.

new technology was first fully promoted by R. E. B. Crompton two decades later (Crompton 1895). Following the successful strategy of using the Churchills' home to showcase the electric light, his company promoted the art of electric cookery by supplying equipment to Margaret Fairclough's School of Cookery in London. The upmarket (aristocratic) fortnightly *Black and White* reported enthusiastically in January 1895 that students at this "School with Trained Lightning" produced food hygienically, safely, and efficiently (see figure 8.1)—if not obviously frugally (A. W. 1895). Yet in her later cookery textbooks, Fairclough dropped her advocacy of electric cookery (Fairclough 1911), and few contemporary cooks explicitly adopted it thereafter. Thus it was that Maud Lancaster—writing under the familiar authorial title of "Housewife"—sought to revive advocacy of this technology in collaboration with her engineer spouse Edward Lancaster in *Electric Cooking, Heating, Cleaning* in 1914. To a bourgeois society faced with the problem of servants becoming ever more difficult to recruit and yet more difficult still to discipline, they promoted electricity to a newly catego-rized audience of "housewives" as a novel and ultraobedient home help that never wanted a day off nor talked back (Davidson 1982, 39; Lancaster 1914, 8). As we shall see in further discussion, the Lancasters enjoyed only very limited success in attempting to popularize electric cookery as a standard feature of "modern" domestic life (Pursell 1993). Just as in the United States, albeit in somewhat different socioeconomic conditions, gas prevailed in the British kitchen long after World War I, notwithstanding the rival publicity of Edison and others to promote electricity (Helrigel 1998; Clendinning 2004). Once again the discretionary power of the female consumer was decisive—but this time largely to reject electricity in both material and cultural forms for decades.

Given the choices faced by householders between electricity and gas for domestic services, we can see that consumers both demanded "expert" knowledge and exercised discretionary judgment about whether to accept it. Now that we have seen that the agency and discretion of the consumer in encountering popularized forms of electricity was greater than perhaps anticipated, we can now return to consider what it was that supposedly constituted the "expertise" of those who sought to style themselves as pub-lic authorities on electric matters.

Visionary Futures? Problematizing the Electrical Expert

> Scientific experts have published their judgments, some of them pro-nouncing this [Edison] system to be the desiderated [*sic*] practical solu-tion of the problem of electrical lighting. . . . Still it must be confessed that hitherto the "weight of scientific opinion" has inclined decidedly toward declaring the system a failure, an impracticability, and based on fallacies. —*Edison* 1880, 295

A great deal has been written about the developing role and rising status of scientific "experts" in popular culture up to the early twentieth century. A common view is that such individuals ascended to a culturally hegemonic position even before the Great War had demonstrated their economic and strategic utility to the military-industrial state. Peter Broks, for example, has suggested that their "privileged status" by the early twentieth cen-tury afforded them a certain "paternalism" that borrowed from the earlier established model of the Victorian factory-owner. This view is that of a highly asymmetrical and sharply separated relationship: the public was left to trust that scientists did indeed possess "the heroic qualities of their magazine stereotypes" (Broks 1996, 39–40, 51–52). I suggest that there is scope for doubting whether this generalization extends to the electrical experts whose self-serving pronouncements the public did not so obviously choose to trust. So whereas Steve Shapin has proposed that Victorian "experts" conformed (or at least were thought to conform) to the norms of being disinterested and detached in interpreting the arcana of science to a presumptively subservient public (Shapin 1994, 412), we might instead consider them as cultural entrepreneurs seeking to capitalize in charismatic— yet fallible—ways on consumer concerns about electricity.

Historians of the public reception of electricity have tended to take at face value electrical writers' wishes to be authoritative "experts" and thus adopt uncomplicated unidirectional accounts, assuming these experts were generally writing to fill gaps or correct errors in popular knowledge of the subject (Marvin 1988; Nye 1990). And yet from Edison's testimony to the readers of the *North American Review* on the "Success of the electric light" in 1880, we can see that not even the technical specialist was so quick to presume omniscience. The embattled "wizard of Menlo Park" acknowledged criticism by rivals and public alike for failing to fulfill his promise in September 1878 that he could "divide" the dazzling brightness of Humphry Davy's electric arc light into units of illumination suitably modulated for the domestic environment. Still without a publicly demonstrable filament lamp in late 1880, Edison acknowledged that the weight of scientific opinion declared his system "a failure." His promotional *North American Review* piece on the "success" of the electric light aimed to gloss over this two-year delay and to win over both technical and nonspecialist audiences to his (self-serving) optimism that the new light would soon be available as promised (Edison 1880, 295).

To reclaim the ground of trustworthy expert in the face of skepticism from both the public and fellow electrical inventors, Edison adopted two kinds of extrapolative rhetoric. "Scientific men" had erroneously dismissed ocean steam navigation, submarine telegraphy, and duplex telegraphy as "impossibilities down to the day when they were demonstrated to be facts." Edison claimed that doubters made the same mistake in assessing the prospects of his filament lamp (Edison 1880, 295)—a generalization easily challenged by his better-informed readers. And responding to the public's repeated demand to know when it could see his lighting system demonstrated, Edison forecast on the basis of previous speedy successes with the phonograph and duplex telegraphy that it would be within two months (Edison 1880, 298–99). And indeed Edison's team finally mounted dazzling displays of electric filament lamps at Menlo Park on New Year's Eve 1880—carefully timed to highlight his prophetic powers and effectively staged to hush disbelievers and woo investors (Bazermann 1999, 183–84).

Edison's competitors in the United Kingdom found themselves under similar public pressure to deal with questions about the likely success of their plans for electric light and its relative safety in comparison to gas. They learned from Edison's example that popular forums were effective places to address these questions and also that narratives which framed solutions to lie in the future served as an important diversion from present-day failures. One such was St. George Lane-Fox, a member of the aristocratic Pitt-Rivers clan. In 1874, at the precociously early age of eighteen, Lane-Fox had submitted the earliest known patent for an electric cooking device.

From 1878 to 1883, Lane-Fox appeared to Edison to be the main British threat to attempts by him and Joseph Swan to patent and operate a filament lamp on a commercial scale. And indeed Lane-Fox did his utmost in that period to publicize his electric lighting and heating technologies to a range of UK audiences. On May 17, 1882, Lane-Fox lectured on the prospects of the electric light to the Royal United Services Institution in London; in so doing, he attracted the attention of the London *Times* ("Future of Electric Lighting" 1882). At this lecture Lane-Fox reported "one or two pertinent questions that the public are ever asking" with respect to his British patent for electric light: "They say, first of all, 'Shall we ever have the light in our houses?' Next they say, 'Shall we have to produce the electric current ourselves?' and thirdly, they ask, 'Will it be cheaper, or even as cheap as, gas?'" (Lane-Fox 1882, 1).

Adopting Edison's quasi-prophetic mode, Lane-Fox presented his answers to those three questions as an account of "The Future of Electric Lighting," a title he used also for copies subsequently circulated to private individuals. But with rather more rhetorical humility than Edison, Lane-Fox claimed not to be able to "withdraw the veil entirely"[4] from the future created by recent progress. The subject was "too vast and complicated" for him, so the members of his audience were (somewhat tellingly) invited to use their "imagination" to fill out his picture. Yet Lane-Fox informed them with remarkable self-assurance that from present knowledge, all could be answered "in favour of electricity." He had "not the slightest doubt" in his mind that homes would soon be electrically lit by a public supply, probably at one-twelfth the cost of gas—a much more optimistic forecast than that of other contemporaries. Lane-Fox also envisaged much scope for kitchen applications too, demonstrating to his audience a quart-size electric boiling pot (Lane-Fox 1882, 1–2, 14).

Shifting strategically between confident certainties and beguiling mysteries, Lane-Fox told the members of his audience that they could "hardly imagine" the changes that electricity would introduce and that it was only a matter of time before an exciting electric future came to pass. The meeting chair, William Spottiswoode, president of the Royal Society, was clearly convinced, suggesting that while generally speaking it was "dangerous" to prophesy about scientific progress, electricity was peculiarly different because it was "safer to prophesy what it can do than what it cannot do" (Lane-Fox 1882, 14, 17). Sadly for Lane-Fox, however, the future of his electric enterprise was short-lived. Acknowledging the practical superiority of Edison's rival system, he soon withdrew in gentlemanly defeat and took up spiritualism instead (Jehl 1937, 2:482). A photograph of Lane-Fox taken in his later years appears in figure 8.2.

In the 1880s, speculation about the many possibilities of an electric future became an important feature of "popular" discussions of electricity. As I have shown elsewhere, both the industry for electricity supply and "popular science" publishing developed a futuristic rhetoric symbiotically linked in a fantasy literature on electric utopias (Gooday 2004c). In 1883, the Scottish-American polymath John Macnie published a pseudonymous electric utopia of the ninety-sixth century entitled *The Diothas: or, A Far Look Ahead* in which electricity was the mobilizing force throughout industry, business, and the home—including electric lighting and labor-saving gadgets (Thiusen 1883). Soon afterward, its main premise was undermined by the financial collapse of schemes by Edison and fellow electrical engineers as too few householders took up electric lighting for their heavily mooted predictions of the future to become self-fulfilling.[5]

The situation turned in favor of Edison and fellow electrical engineers in 1888, however, when the American Christian Socialist writer Edward Bellamy published *Looking Backward 2000–1887* (Bellamy 1888). This best-selling book critiqued 1880s society from the standpoint of a millennial egalitarian utopia in which men and women enjoyed nationalized technology and electrified homes. The enormous popularity of *Looking Backward* inspired imitations among both technical and literary writers that constructed speculations about the future electric home and society produced—not least Macnie's 1883 *Diothas*, opportunistically republished as *Looking Forward* in 1889. And this burgeoning literature on electric futurism was repackaged by the socialist-Dissenter journalist William Stead in the new sixpenny *Review of Reviews* that, from its first issue in 1890,

Figure 8.2. St. George Lane-Fox (1856–1932) was one of Thomas Edison's major British rivals in the early development of electric lighting systems. Related to the minor aristocracy and from a strong military family, Lane-Fox was reputedly the first person to patent an electric cooking pan, which he did in 1874. Undated photograph reprinted by permission from Archives, Institution of Engineering and Technology, London.

took this electric literature to less-affluent audiences on both sides of the Atlantic (Gooday 2004c).

One of the upmarket shilling monthlies that published regularly and widely on the new literature of the future was the *Fortnightly Review*. In the early 1890s, two electrical specialists used the contemporary futurist idiom to present "popular" accounts of their expert knowledge, in which both had clear financial interests. One was Alice ("Mrs J. E. H.") Gordon's article on "The Development of Decorative Electricity" (Gordon 1891a) for February 1891, and the other was William Crookes's piece a year later on "Some Possibilities of Electricity" (Crookes 1892). Rather than offering technical engineering descriptions of electricity generation, each writer enticed sybaritic *Fortnightly* readers into a future in which electricity had eliminated most discomfort, fatigue, and inconvenience. As I have shown elsewhere, Alice Gordon's piece was written concurrently with her book on the same subject, *Decorative Electricity* (Gordon 1891a, 1891b; Gooday, 2004a). Both works presented wealthier middle-class men and women with the prospect of enjoying domestic electric light discreetly installed in fashionable (Japanese) style to modernize their homes—and of avoiding the glaring directness of filament lighting that had clearly exasperated many of her female readers. Thus the *Review of Reviews* presented Mrs. Gordon as contributing to the discourse of electric prophecy, summarizing her principal claim: "In the drawing-room the light of the future will be a reflected light" (Stead 1892, 165).

William Crookes's well-known futurist piece "Some Possibilities of Electricity" merits careful examination, and not only as an oft-cited early forecast of wireless telegraphy—five years before Marconi publicly demonstrated its practical possibility. Carolyn Marvin plausibly presents it as an important instance of how electrical experts forged their identity in relation to lay audiences by forms of social contract; specifically she sees Crookes as an "expert" asserting the engineer's capacity to conquer the "messy" disorder of nature for public benefit. Particularly she focuses on a claim that closes Crookes's paper—one that goes far beyond the utopian imaginings of either Macnie or Bellamy—that weather could be controlled electrically, reducing rainfall but concentrating it when needed by agriculture (Marvin 1988, 114). Importantly Crookes does not promise his readers that this outcome of taming nature will be accomplished with electricity. Rather, he appeals to their imaginations with *possibilities*—whether nearly within technical grasp, such as wireless telegraphy, or a remote fantasy, such as meteorological engineering.

Marvin's account treats Crookes's speculative forecasting in "Some Possibilities" as an unproblematic feature of electrical expertise, hinting that prophecy was a standard obligation of the expert (Marvin 1988, 156). By contrast, I suggest we cannot take for granted that all contemporaries accepted fantasy futurist narrative as a legitimate means for putative experts such as Edison, Gordon, and Crookes to engage with the public. Sensitive to this point in closing his paper, Crookes recognized that "I would perhaps, be styled a dreamer, or something worse, if I remotely hint at still further amending the ways of Nature." This cautionary note might have been Crookes's response to a critical comment from the *Spectator* to a public address he had given in November the previous year: "The scientific authorities of today have fallen into a rather provoking and tantalising habit of taking the public into their confidence, making known to it discoveries that are as yet only half known to themselves, and building upon them the basis of those discoveries a bewildering fabric of conjectural possibilities" ("Science and Conjecture" 1891, 723).[6] To curtail such thoughts, Crookes reminded readers of their nonexpert status by suggesting at the close of "Some Possibilities" that all matters concerning electricity could safely be left to the devices and "inspirations" of electrical engineers (Crookes 1892, 181). Yet William Stead and his evangelical staff at the *Review of Reviews* did not acquiesce in Crookes's framing of expert activity in this mode. In the "Reviews reviewed" section for February 1892, Stead's house writer (probably Stead himself) disparaged Crookes's *Fortnightly* piece as "sufficient to take away one's breath" for its indulgence in futurist popular speculation beyond what was warranted by contemporary electrical science. Beyond merely speculating that in the future, telegraphy and electric lighting could be wireless and electrical technique might rout agricultural fungi and parasites, "electricians, he thinks, should aim at nothing less than the control of the weather, and always make it wet at night time and sunshiny all the day; and when it was to rain, rain a downpour never a drizzle. Incidentally he would abolish London fogs and sterilise all germs in the water supply" (Stead 1892, 182; Gooday 2004c, 252).

For Stead, Crookes's undisciplined conjectures were symptomatic of the worldly hubris of electricians who tried to push the prerogatives of their expert status too far. Quite apart from any controversy over Crookes's continuing support for spiritualism, his journalism was distinctly problematic even within the terms of his more conventional activities. Revisiting Marvin's view, we can see that Stead considered Crookes's undisciplined conjecture as breaking the terms of the implied contract between expert and laity, distracting the latter from responsible and virtuous consideration of electric applications. And indeed the contestation of electric futurism as a legitimate expert activity might have been one of a number of factors in bringing about its demise in the United Kingdom—if not in the United States—early in the twentieth century (Broks 1996, 110–25).

The next section explores one of the reasons why popularizer-entrepreneurs found it so valuable to promote a utopian futurism about electricity in the 1880s and 1890s. This reason was the widespread fear noted by Robert Hammond in 1884 that electricity was too dangerous to be installed in the home—a fear more easily deflected by imaginative narratives of future safety and comfort than by contestable technical claims of risk-free technology.

Managed Lightning: Representing Danger and Electric Illumination

Whereas in the 1890s debates about the aesthetic qualities and cost of electric lighting became a key issue in the popularization of this new technology, during the preceding decade the literature on electricity had focused more specifically on widespread anxiety about the hazards posed by this barely tamed agency. This was indeed a more serious problem facing electrical popularizer-entrepreneurs in the early years of incandescent lighting. Although the popular press was not necessarily hostile, its habitual reporting of public accidents brought into general view the idea that electricity was constitutively dangerous—not merely dangerous if accidentally used inexpertly or in unauthorized ways. Journals and experts for the rival gas industry, as well as some newspapers, nurtured this cultural anxiety—notwithstanding the comparable weekly loss of life and property occasioned by gas explosions, which electrical enthusiasts accused reporters of passing over in culpable silence (Schivelbusch 1988, 33–37; Hammond 1884, 57–58; Gordon and Gordon 1891, 27).

The various hazards of electric lighting were rather dramatically apparent to those who experienced it firsthand in the early 1880s, the experimental technology clearly not being fully under "expert" control. Lady Gwendolyn Cecil recalled wryly that the "monotony" of domestic life at Hatfield House from 1881 to 1883 was often relieved by her father's haphazard experimenting with electric filament lighting. There were not a few evenings when the household had to "grope about in semi-darkness" illuminated only by a dim red glow, while on other occasions a "perilous brilliancy" culminated in "miniature storms of lightning" and then complete collapse. Both situations required the evening's entertainment to be concluded by candlelight (Cecil 1931, 6). In a similar vein, Lady Randolph Churchill later recalled the chaotic effects of electric lighting at her Marlborough Arch home: "I remember the fiasco of a dinner-party we gave to show it off, when the light went out in the middle of the feast, just as we were expatiating on its beauties, our guests having to remain in utter darkness until the [oil] lamps and candles, which had been relegated to the lower regions, were unearthed" (Cornwallis-West 1908, 102).

Even dignified meetings of the British Association for the Advancement of Science (BAAS) were not spared. When the BAAS met in York in 1881, many of its streets were lit by electric arc lamps. Some of these lamps were supplied by Robert Hammond, who complained that city officials had forced him to relocate his generator to an unhelpfully remote location. Hammond later noted that Sir John Lubbock would probably remember the delivery of his presidential address with a "very sad recollection of the unsteady working of those arc lights," given especially that Sir John was also infelicitously the chairman of the UK Edison Company (Hammond 1884, 201).

The most widely voiced concern was the threat to life and property from electrically induced fires and the equally lethal risks of electric shock from faulty (or even competently enacted) installations. Lady Gwendolyn Cecil recalled at least one occasion on which only the hunting ethos of the young male aristocrat saved Hatfield House from conflagration: "One evening a party of guests, on entering the Long Gallery after dinner, found the carved panelling near the ceiling bursting into flames under the contact of an overheated wire. It was happily a shooting party in which young men formed a substantial element. They rose joyfully to the occasion, and, with well-directed volleys of sofa cushions, rendered the summoning of fire-engines unnecessary" (Cecil 1931, 7).

While the upper-class huntsman might have thought little of such entertaining perils, such a response was not obviously typical for the public at large. One of the key questions put to electrical "experts" by the concerned household laity was: Could the electric light be considered safe enough for the home and personal health? Edison's 1880 paper appears to have set the pattern of expert reply to such questions forced upon them by doubtful householders: he simply denied the electric light was especially dangerous, drawing attention instead to the supposedly greater hazards of gaslight. Gas lamps notoriously consumed oxygen from living spaces and returned to them vapors injurious to health and corrosive to furnishings: construed thus, the comparison would always apparently be in favor of electric filament lighting. For this technology there would be "no poisonous or inflammable gases to escape," and thus the danger of fire was "reduced to *nil*," with a consequent reduction in the insurance premium (Edison 1880, 297–98).

Like Edison and many other less-than-disinterested proponents of electrification, Hammond in his 1884 popular treatise detailed how electricity fulfilled the conditions of perfect future lighting far better than gas: "We may begin by saying that *the perfect light for our homes should fulfil the following conditions,* 1—It should not rob the air of our rooms of oxygen. 2—Nor add noxious fumes to the air. 3—Nor be a source of danger in the house. 4—Nor be an unpleasant light. 5—Nor be difficult to control. 6—Nor be costly" (Hammond 1884, 14). While few denied that the first and second conditions could more easily met by electric than gas light, Hammond had to use all his powers of persuasion to show that the other criteria favored electricity.

Edison's claim about hazards from shock or fire was that the filament lamp could be manipulated even by the "most inexperienced domestic servant" and the "most careless person" could not cause injury to body or property through not understanding its mechanism (Edison 1880, 297–98). Yet in contrast to their regular attacks on the dangers of gas lighting, Edison's rivals were more cautious about positively asserting the safety of electric lighting, unless pressed to do so. In his 1882 address on the "future of electric lighting," St. George Lane-Fox omitted the safety issue in listing three concerns commonly raised by the public. But the question immediately put to him after his lecture by a certain Mr. Forster was whether there was "any danger connected with the use of electricity." Lane-Fox answered that he had heard "a great deal about" this topic but did not think there was anything to worry about since he anticipated that in the future, only low-voltage supplies would be used (higher "tension" supply being presented as the principal cause of risk). Thus, if the "proper" people took charge of the arrangements with proper precautions, there would be "no danger at all." Lane-Fox thus attributed all accidents to "the grossest carelessness and negligence" of users—acknowledging no responsibility lying with electrical manufacturers or advisers or in the basic nature of electricity itself (Lane-Fox 1882, 15–16). Unlike Edison, Lane-Fox was not prepared to tell his audience that there was no danger whatsoever, but he asserted that it was only the actions of nonexperts that created risks.

Lane-Fox mentioned no specific instances of harm caused by carelessness, but his audience was probably very familiar with the case of the worker William Dimmock, killed in an accident when installing electric cabling at Hatfield House on December 12 of the previous year (G. Roberts 1999, 111; A. Gay and Yeaman 1906, 467). A report in the London *Times* three days later reported that the inquest had determined an "accidental" death. It had been a rainy day and the worker had slipped on wet grass; he had then tried to break his fall by holding onto wires linking Hatfield House to the river-powered generator. Unfortunately, the particular section Dimmock touched happened not only to be bare but also "live," carrying alternating current at high tension. While Lady Gwendolyn Cecil's devotional biography of her father cast a discreet veil over this disaster, the British public picked upon it very quickly. Hence when Robert Hammond toured British cities in 1882–83 delivering lectures on the electric light, publishing them as *The Electric Light in Our Home* (Hammond 1884), he explained why his audience should not see this tragedy as any evidence of the inevitability of death on encountering electricity.

Hammond admitted that "serious accidents" had occurred in connection with very high-tension systems then under trial, but he demonstrated that death or serious injury did not follow directly from touching a live wire. He dramatically proved it in his lectures by grabbing the terminals of the battery used in his display experiment to show that its "low tension" current was safe even with uninsulated wires; it would thus be all the more safe with insulated layers of cotton, silk, india rubber, asbestos, bitumen, or gutta-percha. Such precautions, he said, could have prevented the fatal accident at Hatfield House (Hammond 1884, 59–61). Indeed, with proper insulated wiring, the homeowner need not worry about the "tension" of the currents involved at all; on this view, the future was thus a little more open-ended than Lane-Fox had suggested. Alas for Hammond, however, the British public was not sufficiently persuaded by his early efforts at popularizing and forecasting the future of electricity, and his company was declared bankrupt in early 1885. Only at the end of the decade did Hammond reemerge as a popularizer-entrepreneur ("Obituary Notices: Robert Hammond" 1916, 681; Parsons 1940, 111).

What kind of popularization did then eventually persuade the public that the Hatfield House death was atypical and thus that electricity posed no greater a hazard than the mundane household commodities? A key strategy was carefully managed *theatrical* display of the harmlessness of electricity, in ways like those documented by Morus for an earlier period. As Schivelbusch, Marvin, and Nye have each noted, the Edison company (and its rivals) often hired women dancers in the 1880s to perform onstage wearing outfits including headwear covered entirely with electric lights. In contrast to the notorious flammability of gas footlights on the stage, the representation of electric lights in direct proximity to the female body without harm resulting signaled clearly to audiences that the new illuminant was not only glamorous but also relatively hazard-free (Marvin 1988, 176; Nye 1990, 244; Schivelbusch 1988, 72). Similarly, Mrs. J. E. H. Gordon reported that she wore electric jewelery in her hair and on her body to promote her husband's electric consultancy business. Although the lights proved harmless, leaks in the lead-acid battery under her dress led to several discreet emergency changes of clothing—such were the hazards of demonstrating the safety of electricity (Gordon and Gordon 1891, 121–23).

Up to the time that Gordon's *Decorative Electricity* was published in 1891, there were indeed no reported deaths of householders from electric accidents in either the United States or the United Kingdom. But fatalities among artisans employed in electric projects were widely reported in the press. In New York alone—where overhead power lines dominated the skyscape as in no other city—twenty-two lives were lost to this cause between 1880 and 1890, compared to sixteen for all of Europe from 1880 to 1889 (A. Gay and Yeaman 1906, 3, 467–71). The New York deaths were heavily reported not because they were self-evident demonstrations of any unique danger arising from electricity (Marvin 1988, 119) but because of their relevance to the interexpert conflict in the "battle of the systems." Just like Londoners, New Yorkers had to choose between alternating current (ac) and direct current (dc) systems. As Mark Essig has recently explained, Edison cunningly declined to offer his dc system for the new U.S. punitive technology of the electric chair in 1889–90, promoting instead the rival Westinghouse company ac supply as ideally suited for the lethal shocks. Although Edison then opportunistically illustrated to the public how much more readily alternating currents led to death than direct current, such machinations lost him public credibility. By 1893, Edison's dc venture had collapsed; defeated by both commerce and public antipathy, he was forced to join the Thomson-Houston company in the new (ac) company General Electric (Hughes 1983, 107–9; Essig 2003).

In this period, William Crookes engaged in the popular promotion of direct current rather more successfully than Edison did. In his 1892 *Fortnightly* piece "Some Possibilities of Electricity," we see how Crookes similarly used a comparable dramatic appeal to the putatively deadly powers of alternating currents. On the one hand, Crookes played up the "natural" connection between electricity and life in the possibilities of interpersonal transmission of brain waves, cure of disease, and enhancement of agriculture—not to mention reports of an "electric slug" in some of the capital's private gardens. By implication, it was only a small step from the ubiquity of electricity in nature to the ubiquity of electricity in the home. Yet Crookes noted, on the other hand, that a major impediment to such an electrified future lay in the high comparative cost of electric lighting vis-à-vis gas, hence there had been widespread exploration of ac systems that offered greater economy of transmission. But, Crookes emphasized, they only did so with an increased risk of death: "Whilst we are seeking for cheaper sources of electricity, no endeavour must be spared to tame the fierceness of those powerful alternating currents now so largely used. Too many clever electricians have shared the fate of Tullus Hostilius, who, according to the Roman myth, incurred the wrath of Jove for practising magical arts, and was struck dead with a thunderbolt. In modern language, he was simply working with a high tension current, and, inadvertently touching a live wire, got a fatal shock" (Crookes 1892, 179).

His passing comment that alternating currents had "at best a somewhat doubtful reputation" signaled that Crookes was not writing for the *Fortnightly* as a disinterested natural philosopher. He was also a director of the recently launched and not yet profitable Notting Hill Electric Light Company founded the previous year to offer *direct* current to the plush district of London in which Crookes lived. As Crookes's biographer tells us, the company's dividends for 1891 had been nil, rising only to 1 percent within the next three years (Fournier 1923, 291–310). Crookes's aim in offering possible electric futures was thus arguably to divert *Fortnightly* readers from investing in rival ac supplies; indeed, the subsequent profitability of Crookes's dc lighting business well into the second decade of the twentieth century arguably owed something to his dextrous use of the futurist idiom (discussed earlier in the chapter) to persuade householders to electrify their homes.

Yet by the early 1890s, the risk of fire from electrical faults became of such palpable importance to consumers—and embarrassment to electrical experts—that attention to it was increasing in the popular literature. A whole chapter was devoted to the subject in *Decorative Electricity*, one of the most widely read and reviewed books on electricity of 1891 (and 1892 in second edition). Nominally authored by "Mrs. J. E. H. (Alice) Gordon," the editorial presence of engineer Mr. J. E. H. Gordon was signaled by the prominence of his name and qualifications on the frontispiece (Gooday 2004a). And James Gordon was named specifically as the author of the second chapter, "Fire Risks of Electric Lighting," with his professional credentials as a member of the Institution of Civil Engineers listed after his name (Gordon and Gordon 1891, 17–28). Readers would have had little doubt that his "masculine" engineering expertise gave him credibility to speak on issues of structural risk in ways that complemented Alice's expertise on domestic aesthetics and economics.

James Gordon conceded that there had indeed been occasional fires in electrically lighted houses but contended that such isolated incidents had caused "unnecessary alarm." Rather than following the strategy of Lane-Fox and Crookes in blaming careless users, Gordon attributed the provenance of fire risk to "unscrupulous contractors" who used too few fuses, insufficiently thick wire, and deficient insulation. By contrast, wiring installations that met professional standards would be approved by insurance companies, and consumers could thus rest easy in their homes. Any lingering fears among their servants could be put to down to lingering legends of Lord Salisbury's hapless employee in idle gossip and rumor: "I heard a tragic history the other day of 'a poor girl who had been killed by the current.' On asking particulars I was told that it was 'a housemaid at Arlington Street,' but the narrator was not quite sure when the accident had occurred. Further enquiry showed me that it was a developed account of the death of the gardener who was unfortunately killed at Hatfield some ten years ago" (Gordon and Gordon 1891, 27).

Faced with persistent skepticism about the comparative safety of electricity to gas, it is easy to see why popularizers in the 1890s borrowed from Bellamy the vision of electricity as harbinger of safety, fairness, and equality. Undiverted by such utopian rhetoric, however, and unconvinced of claims about the safety of electric technology, some ordinary citizens asked a further pressing question of the "experts": What exactly *was* this mysterious agency of electricity?

What Is Electricity? Popularizing Personifications of the Mysterious Agency

We know little as yet concerning the mighty agency we call electricity. "Substantialists" tell us it is a kind of matter. Others again view it, not as matter, but as a form of energy. Others, again, reject both these views. . . . High authorities cannot even agree whether we have one electricity or two opposite electricities. The only way to tackle the difficulty is to persevere in experiment and observation. If we never learn what electricity is; if like life or like matter, it should always remain an unknown quantity, we shall assuredly discover more about its attributes and functions. —*Crookes* 1892, 173

When members of the public asked, "What is electricity?" they were not engaged in metaphysical inquiry. They wanted to know what it was that they would be *paying for* if or when they became consumers of electricity: Edison's early customers were reluctant to accept that something so intangible could be trustworthily metered (Gooday 2004b). Moreover, how could "experts" even claim to be authorities on the subject if they could not give accounts of electricity to match the analyses of gas or water so readily offered by the laboratory chemist? Failing to meet such publicly set standards of authority and expertise, putative electrical experts adopted other tactics of literary diversion or of genial anthropomorphization to prevent this embarrassment from becoming a crisis.

William Crookes opened his 1892 "Some Possibilities" article for the *Fortnightly* with the foregoing epigraph, unambiguously deflecting any reader's expectation that his paper would answer the contested question. He offered instead as a rhetorical diversion a catalog of what electricity could be made to *do* in the hands of electrical engineers—a tactic very familiar to readers in the late Victorian period. This strategy of technological showcasing was adopted throughout much of *What is Electricity?* by the Harvard professor John Trowbridge (Trowbridge 1896, 1897). But unlike Crookes, Trowbridge had a more subtle approach to the question that hinged upon his attempts to become a *Maxwellian* expert in electricity. Although Trowbridge's book was explicitly a "popular" response to the question often put to him, he expounded instead the new Maxwellian orthodoxy. There was no such *thing* as electricity, just as there was no substance answering to the name of heat: both were modes of ether motion (Trowbridge 1896, v–vi, 308–9; 1897, v–vi, 308–9).

By the 1900s, however, a new discourse of subatomic matter, namely, the electron, seemed to some to offer an alternative answer to the question. In *The Autobiography of an Electron*, the prolific Scottish electricity popularizer Charles Gibson used this theory to anthropomorphize electricity as a cheerfully vigorous if nevertheless law-abiding swarming of negatively charged particles (Gibson 1911). Yet specialist electrical engineers were more cautious about following this reductionist approach. In surveying this electron theory for readers of the *Popular Science Monthly* in 1902, Ambrose Fleming of University College London pointed out that the question "What is electricity?"—often put to him by the "intelligent but nonscientific" inquirer—no more admitted of a "complete and final answer" today than did the question "What is Life?" (Fleming 1902, 8). Why might such expert popularizers have been reluctant to demystify electricity? The sense of electricity as magic spectacle and miraculous provider of novelties was key to maintaining public interest in the subject and to entrepreneur-popularizers' presentation of themselves as latter-day magicians with the power to divine the future.

This combined sense of electricity as mystery *and* destiny was a central theme in Henry Adams's famous encounters with the dynamo reported in his autobiographical *Education of Henry Adams*. Adams's first major encounter with electric technology was at the 1893 World's Columbian Exhibition in Chicago, and he surmised that millions of others like him drawn to this exhibit acquiesced as he did in incomprehension of how the dynamo drew its motive power from electricity: "They had grown up in the habit of thinking a steam-engine or a dynamo as natural as the sun, and expected to understand one as little as the other" (Adams 1918, 341). Two years later Adams toured France to see the spectacular medieval cathedrals of Chartres and Lourdes dedicated to the Virgin Mary. And it was to that particular mysterious icon of female power that Adams resorted when calibrating the potency of the comparably mysterious dynamo in his famous essay on the electrically powered 1900 Paris exhibition "The Dynamo and the Virgin" (Adams 1918, 381–89).

A number of artistic popularizers harnessed this discourse of mystery by personifying the agency of electricity in a predominantly female form. Perhaps the best known is "the fairy electricity" that appears in Continental electric literature (especially in France), the precursor to which can be seen in the mid-1880s frontispiece to La Lumière Electrique (Schivelbusch 1988, 77; see figure 8.3).

In various forms, this figure reified electricity as a cheerful, benign, and obliging female servant who mediated the agency of electricity first at the behest of engineers and latterly at the command of the (female) householder. By the early 1890s, various neoclassical forms of this figure holding electric lights aloft were adopted to promote filament lamps as safe, homely devices in marketing to male and female householders. For example, in the advertising for electric lighting in the front matter of the second edition of Mrs. Gordon's Decorative Electricity (Gordon and Gordon 1892) we see a figure bringing light symbolically to the entire globe—courtesy of the felicitously named decorative lighting company, Faraday and Sons (figure 8.4).

Some distinctly angelic forms were used in 1893 in a General Electric brochure to promote Edison's proprietary electric light. From 1900 onward, the U.S. and UK branches of General Electric adopted a more hedonistic manifestation of this form to promote the Edison "Mazda Drawn Wire Electric Lamps" as the best illumination for homes that had domesticated the electrical sprite (figure 8.5).[7]

During the early 1900s, we see the "electrical fairy" deployed to bring comfortable imagery to domains of electric domesticity other than the electric light. Now that a significant proportion of homes had adopted electric light, popularizing literature and displays did not just hint that householders could have a safe, clean, and economic future domestic utopia powered by electricity. Rather, this utopia was now available, with electric cookers, washing machines, irons, and vacuum cleaners bringing householders

Figure 8.3. New frontispiece to the popular French journal *La Lumière Electrique* in January 1884. While personifying electricity as an obedient female servant, her classical robes and star-shaped headpiece also evoke the heavenly muse Urania. Three years later, this figure was replaced by *la fée électricité*, a winged female bearing a torch in her right hand, the classical symbol of "truth" (see Cordulack 2005). Reprinted from *La Lumière Electrique* 1884, courtesy of University of Leeds Library.

modern industrial efficiency and hygiene into their home. The "electrical fairy" was harnessed to this task of promoting household electricity as a benign agency by "Housewife," the pseudonym of Maud Lancaster, in her book *Electric Cooking, Heating, Cleaning: A Manual of Electricity in the Service of the Home* (Lancaster 1914).

FARADAY & SON'S

ARTISTIC FITTINGS

3 BERNERS STREET

|OXFORD STREET W.

MANUFACTURERS OF—

The Dragon and Dolphin Brackets,

Cairene and Moresque Pendants,

Spanish and Venetian Church Lamps,

Embossed Shells and Adjustable Reflectors.

Figure 8.4. Illustrated advertisement for decorative lighting in the front material of Mrs. J. E. H. Gordon's *Decorative Electricity*. Although unwinged, the floating electrical fairy (perhaps also goddess) shown here bears classical garb and torchlight strongly reminiscent of the 1887 *fée électricité* of *La Lumière Electrique*. Reprinted from Gordon and Gordon 1892, in author's collection.

Figure 8.5. Illustrated British Thomson-Houston advertisement from about 1908 for the Edison Mazda electric lamp (Mazda being the ancient Persian god of light), a further elaboration of the "fairy of electricity" that uses both hands to hold up this exotic form of electric light as a luxuriant illuminant for all the earth's population. In implicit contrast with gas, paraffin, and candlelight, this model of electric lamp alone can rival the (benignly smiling) sun. Reprinted by permission from Archives, Institution of Engineering and Technology, London.

Edward Lancaster (1859–1952) started his career assisting Joseph Swan in demonstrating his electric lamps at the Paris Exhibition in 1881 and then bringing electrical light to railway stations in India. His first major encounter with domestic electricity beyond lighting seems to have come in a lavish "banquet cooked by electricity" organized by the City of London Electric Lighting Company in June 1894. The following year, as a widower, he married his cousin Maud Lucas, who took on not only his three young children but also his advocacy of electricity in the home. According to one of her son's biographers, Maud was fond of publicity, dividing her energies between philanthropic causes, Rosicrucianism, and spiritualism as well as cowriting a volume that apparently sold well as a "textbook for housewives" (Barker 1969, 15).[8]

The collaboration on *Electric Cooking, Heating, Cleaning*—the Lancasters' only such joint publication—is as mixed as its projected readership, and for important reasons. The authorship of British edition was identified quasi-pseudonymously as "Housewife (Maud Lancaster)," whereas the U.S. edition identifies her only as Maud Lancaster. In both editions, Edward Lancaster is listed as editor with his full professional membership credentials of the Institutions of Civil and Electrical Engineers. The most prominent authorial voice throughout the book is the female homemaker writing for women in a similar domestic situation about the delights and benefits of the many available domestic technologies for kitchen, dining room, and general household use. In self-deprecating mode, Maud opens by downplaying any claim to (the male world of) expertise, appealing to a specifically female audience: "The following pages are feeble efforts of mine to help my 'sisters in distress' and to convince them of the wonderful blessings provided for us by nature's gift of Electricity which, aided by scientific research and inventions, is capable of doing so much toward bettering the home life" (Lancaster 1914, 1). She soon offers a brief explanation of electricity "from a woman's point of view," noting jauntily that it is "impossible for me to tell *what Electricity is*" since not even the "greatest scientists" could tell her how to deal with this matter. All these experts could do was explain the generation of electricity and the many methods of utilizing it for "the benefit of mankind"—and indeed womankind. Significantly juxtaposed on the opposing page is a picture with the caption "Electricity—the Good Fairy," a terrestrially grounded young female figure engaging the viewer with a shy smile while pressing a light switch (figure 8.6). By implication, the indeterminate identity of electricity—represented metaphorically as the "Good Fairy"—need not be a worry to (female) readers. The important point was that this magical creature's intercessional power

Electricity—the Good Fairy.

Figure 8.6. Illustration of a distinctly domesticated and grounded "Good Fairy" of electricity shown pressing a light switch in Maud Lancaster's 1914 volume *Electric Cooking, Heating, Cleaning: A Manual of Electricity in the Service of the Home.* This was one of many half-tone images supplied to Maud and Edward Lancaster by the General Electric Company of America. Reprinted from Lancaster 1914, 6, by permission of Institution of Engineering and Technology, London.

epitomized how householders could trust the mysterious force of electricity to supply safe and convenient forms of household technology.

Yet it is not obviously Maud Lancaster who refers explicitly to the "good fairy of electricity" in the narrative of *Electric Cooking, Heating, Cleaning*. It seems rather to have been a result of the editorial intervention of Edward Lancaster, whose interpolated contributions spoke to electrical power station engineers (incidentally lending professional credentials to the rest of the work). Characteristically of the genre, fantasy anthropomorphization appears primarily for male consumption in the section entitled "The commercial aspect of electrical cooking" that concerned the duty of the power station engineer to share the task of promoting cooking with electricity, along with manufacturers and trade journals. Lancaster opined that it now only remained for the supply engineer to open up a campaign of "publicity" in favor of electricity for lighting, cooking, heating, and the many other uses it could be given in every home: "It is only a beginning to install lamps for lighting in a house, although an important step in the right direction, and what is wanted is an educational campaign to bring home to consumers who merely use their installation for lighting that the good fairy of electricity can do greater things than these for them" (Lancaster 1914, 26–27).

Contemporary reviews of the Lancasters' book acknowledged that it had both "popular" and technical dimensions—and was indeed the first serious work on the subject in English. A house writer for the *Electrician* observed that the "authoress" had written so as to ensure the subject could be "thoroughly appreciated by lay readers," while also supplying technical information for the specialist. While congratulating the "authoress" for the book as a whole, criticism for unexamined citation of manufacturers' data and the idiosyncratic calculations was directed at the editor—also referred to as "author"—who had palpably made "his influence felt" ("Electric Cooking, Heating, Cleaning &c. by Maud Lancaster" 1914).

Just as reviews of Mrs. Gordon's *Decorative Electricity* (with her husband's chapter on fire risks) revealed the complex variety of responses to the gendered expertise of a collaborative couple (Gooday 2004c), so more than twenty years later *Electric Cooking, Heating, Cleaning* evinced comparable journalistic response. The less sympathetic *Electrical Review* observed that while it would be useful to engineers, it was "not a book that would appeal to the average housewife" because little practical information seemed to be given from the housekeeper's point of view. Mischievously misreading this aspect of the book's authorship, this reviewer found it "somewhat amusing" to read a housewife appearing to offer a homily on power station work; he then immediately admitted such comments showed the "influence" of the mere male editor—it was clearly not difficult for readers to discern the gendered partitioning of expertise. More important, the house writer for the *Electrical Review* also agreed with what it took to be the male authorial voice of Edward Lancaster that the engineer really ought to electrically equip his own house. That way he could speak with the authority of "experience" on these points—experience to match the more evidently expert female voice on domestic technology ("Electric Cooking, Heating, Cleaning etc. by 'HOUSEWIFE' [Maud Lancaster]" 1915).

We can close by noting that *Electric Cooking, Heating, Cleaning* is littered with photographs and sketches of women at work using electricity, exercising technical facility with effortless fulfillment, with only the occasional supplementary image of indolent male luxury. Judging from this and other aspects of the book's physical production, its seems likely that publishers Constable and Company in London (and Van Nostrand in New York) saw women as the primary audience for this work—readers likely to respond most specifically to signals of female authorship. Consonant with this view, a carefully chosen and edited review of the book from the London journal *Truth* is extracted and cited opposite the title page of the UK edition: "Fancy cooking cutlets and frying pancakes with captured lightning! It really seems tremendous. . . . You would have fallen in love with the exquisite cleanliness of the process" (Lancaster 1914, ii).[9]

Here we see a redeployed variant of the "trained lightning" metaphor for electric current used in 1895 to describe Mrs. Fairclough's school of electric cookery for *Black and White* and a clear romanticization of the cleanliness of electric cookery. The carefully tailored gendered connotations of electricity in this work are not hard to see. And yet the women (and men) who read this book seemingly did not go out in large numbers to invest in electrical equipment for their homes. This lack of response might be explained by the outbreak of the Great War and consequent economies around the time the book was published. But even after the war, the campaign to popularize the "all-electric kitchen" was relaunched several times without comprehensive success (G. Roberts 1989; Pursell 1993). Clearly the readers of the Lancasters' *Electric Cooking, Heating, Cleaning* found it easy to resist the putatively expert recommendations of male and female authors and to stick with trusted gas or coal cooking technology instead.

Conclusion

In this chapter I have argued that it was the public demand for trustworthy knowledge of electricity that created the opportunity for certain individuals to carve a space for themselves as "experts" in the subject. Such figures did not appear unbidden from the ranks of the technical specialist as a natural by-product of professionalization, nor was their idiom of popular electricity inevitably brought into being in the train of an autonomously expanding "popular science." The very existence of a class of electrical "experts" was the product of both popular interest and the commercial expediency of cultivating consumer desire to purchase and own the products of electricity. Thus, I have examined the interaction of aspirant electrical "experts" with the audiences they wooed—not always or even often successfully—to illustrate the ambivalence and complexity of power relations in popularizing electricity. I have suggested that the form and content of the "expert" popular literature on electricity was not determined by "autonomous" authorities in the electrical professions appealing to well-established norms of expertise.

Edison, Lane-Fox, Hammond, the Gordons, and the Lancasters did not dogmatically dominate the process for popularizing electricity. The agenda of their literature was more typically set by the concerns of the "lay" reader—popular fears and doubts—that had to be acknowledged and addressed. To be recognized as some kind of expert, these writers had to respond in some way to the agenda of "lay" concerns about electricity and take on unprecedented forms of writing such as the forecasting of futures and the anthropomorphization of electricity as fairy magic that—at least in theory—diverted attention away from the absence of firm assurances about the safety of electricity. As I suggested in my introduction, this phenomenon somewhat subverts the common notions of "expertise" applied to late Victorian science and technology: the roles of the laity and the putative expert could be inverted in important ways when the latter sought to persuade the former to adopt a vision of electricity in the home that was neither familiar nor risk-free, and the laity had to judge whether to accept such "popular" representations.

As this study has focused entirely on the particular case of electricity in the home, it is important to consider whether there are any implications for other areas of "popular science" (or indeed "popular technology"). The model of "expert" status in the late nineteenth century that I have adopted here concerns a role created in the press or other popular arenas rather than one existing prior to engagements with the public; after all, it was the public that demanded the expertise and adjudicated upon its delivery. One might then ask questions of how, say, writers in popular botany, evolutionary theory, and astronomy came to be recognized (or not) as "experts" through the construction and reception of their work in the popular press—rather, that is, than writing such work from any presumptively established position of expertise. The historian can no longer suppose that the authoritative credentials of past popularizers were taken as granted by contemporaneous audiences.

Directly linked to this there are important questions about how to describe those who wrote on "popular science." If they were not publicly accredited experts before they began popularization, and might not have been accepted as experts even when they did, how can we describe their status in a non–"question begging" way? In this chapter, I have used the locution "popularizer-entrepreneurs" to emphasize the performative and aspirational nature of their role within the political economy of knowledge and material commodities. Equally inelegantly we can reframe the popular target of their commodity marketing—the laity—perhaps in such terms as "critical-discretionary consumers" of "popular science." Such a term would emphasize their nonpassive and nonignorant role (or at least to guard against traditional but nevertheless false assumptions about their quietude and inexpertness) and avoid overhomogenizing all consumers into some bland characterless "public" about which facile judgments are all too easily made.

This approach to studies of the popularization of science might help develop a better understanding about how it can be a fallible and fraught enterprise that by no means leads to expert-determined outcomes. Certainly the evidence I have presented in this chapter should make us think about how generally public demands and expectations steer the course of popularization activities and evaluate their success. Such is especially the case when consumers had to deal with competing and manifestly fallible public writers (e.g., Edison) on artificial lighting in arriving at their judgments about how to respond to—and indeed in formulating autonomous understandings of—popularized electricity. And it helps us understand in a deeper way why "Ex Fumo Lucem" described the controversies over power, uncertainty, and risk in the spread of electric illumination as he did in the *Contemporary Review* of 1900. There is indeed much that is "notorious" in "popular science" that is yet to be discovered by the historian.

&

NOTES

1. "Ex Fumo Lucem" is a reference to "Non fumum ex fulgore, sed ex fumo dare lucem / Cogitat, ut speciosa dehinc miracula promat." Horace *Ars Poetica* 5.143:

"Flame first and then smoke, but from smoke to bring light to illuminate / in order to enlighten and surprise us with dazzling miracles." The shorter version *Ex Fumo Dare Lucem*—to give light from smoke—was reputedly the motto of the Liverpool Gas Company.

2. For information on MacLeod, not specifically referring to Salisbury's electrical projects, see H. Gay 2003.

3. He was known as St. George Lane-Fox-Pitt after his father inherited the Pitt-Rivers title in 1880 ("Obituary: St. George Lane Fox Pitt," 1932).

4. Privately circulated copy available in Institution of Engineering and Technology (IET) Archives, London. My thanks to Anne Locker at the IET Archives for sharing this with me. While commonplace in nineteenth-century literature, this Platonic code for revelation of obscured truth was not customarily applied to forecasting the future.

5. For discussion of science and technology in American utopianism, see Segal 1985; Carey and Quirk 1970.

6. My thanks to Bill Brock for noting that Crookes's original lecture was reproduced in *Popular Science Monthly* 40 (February 1892): 497–500.

7. Ironically, Edison's 1912 piece on future womankind for *Good Housekeeping* personified electricity as a multitasking *male* imp—complete with lightbulb hat—as a miniature "Wizard of Menlo Park" giving the homemaker so much assistance that she could sit reading all day, just pressing buttons for domestic chores to be performed (Edison 1912).

8. My thanks to Anne Locker from the IET Archives and members of the Lancaster family for supplying me with additional information on Edward and Maud Lancaster.

9. In the U.S. edition, edited by Stephen Coles, this quotation appears on p. 329.

REFERENCES

Adams, H. 1918. *The Education of Henry Adams*. Ed. H. C. Lodge. Boston : Houghton Mifflin. (First 1906 edition privately circulated.)

A. W. 1895. "A School With Trained Lightning." *Black and White* (January): 47.

Barker, R. 1969. *Verdict on a Lost Flyer*. London: George Harrap.

Bazerman, C. 1999. *The Languages of Edison's Light*. Cambridge, MA: MIT Press.

Bellamy, E. 1888. *Looking Backward, 2000–1887 or, Life in the year 2000 A.D.* Boston: Ticknor; London: Ward, Lock.

Briggs, A. 1988. *Victorian Things*. London: Batsford.

Broks, P. 1996. *Media Science before the Great War*. London: Macmillan.

Byatt, I. C. R. 1979. *The British Electrical Industry, 1875–1914: The Economic Returns to a New Technology*. Oxford: Clarendon Press.

Carey, J., and J. J. Quirk. 1970. "The Mythos of the Electronic Revolution." *American Scholar* 39:219–41, 395–424.

Cecil, Lady G. 1931. *Life of Robert, Marquis of Salisbury.* Vol. 3 (1880–86). London: Hodder and Stoughton.

Clendinning, A. 2004. *Demons of Domesticity: Women and English Gas Industry, 1889–1939.* Aldershot: Ashgate.

Cooter, R., and S. Pumfrey. 1994. "Separate Spheres and Public Places: Reflections on the History of Science Popularization and Science in Popular Culture." *History of Science* 32: 237–67.

Cordulack, S. W. 2005. "A Franco-American Battle of Beams, Electricity, and Selling of Modernity." *Journal of Design History* 18:147–66.

Cornwallis-West, Mrs. G. 1908. *The Reminiscences of Lady Randolph Churchill.* London: Edward Arnold.

Crompton, R. E. B. 1895. "The Use of Electricity for Cooking and Heating." *Electrician* 34:793–95.

———. 1928. *Reminiscences.* London: Constable.

Crookes, W. 1892. "Some Possibilities of Electricity." *Fortnightly Review* 51: 173–81.

Davidson, C. 1982. *A Woman's Work Is Never Done: A History of Housework in the British Isles 1650–1950.* London: Chatto and Windus.

Edison, T. A. 1880. "The Success of the Electric Light." *North American Review* 131:295–300.

———. 1912. "The Woman of the Future: A Remarkable Prophecy by the Great Inventor." *Good Housekeeping* (October): 436–44.

"Electric Cooking, Heating, Cleaning &c. by Maud Lancaster, edited by E.W. Lancaster, A.M. Inst C.E., M.I.E.E." 1914. *Electrician* 74:399–400.

"Electric Cooking, Heating, Cleaning etc. by 'HOUSEWIFE' (Maud Lancaster), edited by E. W. Lancaster, M.I.E.E." 1915. *Electrical Review* 76:205–6.

Essig M. 2003. *Edison and The Electric Chair: A Story of Light and Death.* Stroud, Gloucestershire: Sutton.

Ex Fumo Lucem. 1900. "Gas Light." *Contemporary Review* 78:710–19.

Fairclough, M. A. 1911. *The Ideal Cookery Book.* London: Waverley Book.

Fleming, J. A. 1902. "The Electronic Theory of Electricity." *Popular Science Monthly* 61:6–23.

Fournier, D'Albe E. E. 1923. *Life of Sir William Crookes.* London: Unwin.

"Future of Electric Lighting." 1882. *Times* (London), May 18, p. 12, col. a.

Gay, A., and C. H. Yeaman. 1906. *Central Station Electricity Supply.* London: Whittaker.

Gay, H. 2003. "Science and Opportunity in London, 1871–85: The Diary of Herbert McLeod." *History of Science* 41:427–58.

Gibson, R. 1911. *The Autobiography of an Electron.* London: Seeley.

Gooday, G. 2004a. "'I never will have the Electric light in my house': Alice Gordon and the Gendered Periodical Representation of a Contentious New Technology." In *Culture and Science in the Nineteenth Century Media,* ed. Louise Henson, 173–85. Aldershot: Ashgate.

———. 2004b. *The Morals of Measurement: Accuracy, Irony and Trust in late Victorian Electrical Practice.* Cambridge: Cambridge University Press.

———. 2004c. "Profit and Prophecy: Electricity in the Late Victorian Periodical." In Cantor, Shuttleworth, Dawson, Gooday, Noakes, and Topham, eds. *Science in the Nineteeth-Century Periodical: Reading the Magazine of Nature,* Cambridge: Cambridge University Press, 238–47.

Gordon, A. M. 1891. "The Development of Decorative Electricity." *Fortnightly Review* 49: 278–84.

Gordon, A. M., and J. Gordon. 1891. *Decorative Electricity, with a Chapter on Fire Risks by J. E. H. Gordon.* London: Sampson and Low.

———. 1892. *Decorative Electricity, with a Chapter on Fire Risks by J. E. H. Gordon.* 2nd ed. London: Sampson and Low.

Hammond, R. 1884. *The Electric Light in our Homes.* London: Warne.

Hannah, L. 1979. *Electricity before Nationalization: A Study of the Electricity Supply Industry in Britain to 1948.* London: Macmillan.

Helrigel, M. A. 1998. "The Quest to be Modern: The Evolutionary Adoption of Electricity in the United States, 1880s to 1920s." In *Elektrizität in der Geistesgeschichte,* ed. Klaus Plitzner, 65–86. Bassum: GNT-Verlag.

Hughes, T. 1983. *Networks of Power: Electrification in Western Society, 1880–1930.* Baltimore: Johns Hopkins.

Jehl, F. 1937. *Menlo Park Reminiscences.* 2 vols. Dearborn, MI: Edison Institute.

La Lumière Electrique. 1884. January.

Lancaster, M. 1914. *Electric Cooking, Heating, Cleaning: A Manual of Electricity in the Service of the Home.* Ed. E. W. Lancaster. London: Constable.

Lane-Fox, St. G. 1882. "The Future of Electric Lighting: A lecture delivered at the Royal United Services Institution." Privately published; copy on file in Archives of the Institution of Electrical Engineers, London.

Marvin, C. 1988. *When Old Technologies Were New: Thinking about Communications in the Late Nineteenth Century.* Oxford: Oxford University Press.

Morus, I. R. 1998. *Frankenstein's Children: Electricity, Exhibition and Experiment in Early-Nineteenth Century London.* Princeton, NJ: Princeton University Press.

Nye, D. 1990. *Electrifying America: Social Meanings of a New Technology.* Cambridge, MA: MIT Press.

"Obituary Notices: Robert Hammond." 1916. *Journal of the Institution of Electrical Engineers* 54:679–82.

"Obituary: St. George Lane Fox Pitt." 1932. *Times* (London), April 7, p. 14, col. e.

Parsons, C. 1940. *The Early Days of the Power Station Industry.* Cambridge: Cambridge University Press.

Pursell, C. 1993. "Am I a Lady or Am I an Engineer?" *Technology and Culture* 34:78–97.

Roberts, A. 1999. *Salisbury: Victorian Titan.* London: Wiedenfield and Nicholson/Phoenix.

Roberts, G. 1989. "Electrification." In *Science, Technology and Everyday Life*, ed. Colin Chant, 68–112. London: Routledge.

Schivelbusch, W. 1988. *Disenchanted Night: The Industrialisation of Light in the Nineteenth Century*. Trans. Angela Davies. Oxford: Berg.

"Science and Conjecture." 1891. *Spectator* 67:723–24.

Segal, H. 1985. *Technological Utopianism in American Culture*. Chicago: University of Chicago Press.

Shapin, S. 1994. *A Social History of Truth: Civility and Science in Seventeenth Century England*. Chicago: University of Chicago Press.

Stead, W. T. 1892. "The Reviews Reviewed: The Fortnightly Review." *Review of Reviews* 3:165.

Thiusen, Ismar [John Macnie]. 1883. *The Diothas, or, A Far Look Ahead*. New York: Putnam. The second edition of 1889 was published as *Looking Forward or The Diothas*.

Topham, J. 2000. "Scientific Publishing and the Reading of Science in Nineteenth-Century Britain: A Historiographical Survey and Guide to Sources." *Studies in the History and Philosophy of Science* 31:559–612.

Trowbridge, J. 1896. *What is Electricity?* New York: Appleton.

———. 1897. *What is Electricity?* London: Kegan Paul Trench.

SECTION III

Display

CHAPTER NINE

Natural History on Display:
The Collection of Charles Waterton

VICTORIA CARROLL

The most common way in which people engaged with science in the nineteenth century was through viewing exhibitions. From world's fairs to private cabinets, from public museums and galleries to commercial demonstrations, panoramas, menageries, and freak shows, science was presented to nineteenth-century audiences as spectacle—as something knowable through visual experience (Altick 1978; Greenhalgh 1988). Millions of people, from all walks of life, visited exhibitions of science during the nineteenth century, yet very little is known about their experiences. Historians have continued to study exhibitions from the exhibitor's point of view, in terms of how displays were devised, promoted, and managed. This chapter aims to redress this balance. It argues for the possibility and desirability of studying exhibitions from the point of view of the visitor, in terms of how the exhibitions were variously viewed, interpreted, and appropriated by the people who went to see them.

It is often assumed that it is impossible to recover what visitors really made of exhibitions in the past—that all we can know is how exhibitors (collectors, showmen, curators) intended their displays to be received. This chapter aims to dispel this myth. The field of literary studies provides a useful point of comparison. Accounts of readers' responses to literary works have until recently tended to focus on responses implied by texts (Raven, Small, and Tadmor 1996, introd.; Suleiman 1980). However, a great deal can be learned about how people actually read and responded to books from reviews, reading diaries, marginalia, and other written traces of reading experiences (Jackson 2001; J. Secord 2000). The majority of nineteenth-century museum visitors came and went leaving barely a trace. A significant fraction, however, set down their experiences in writing, and some of their accounts circulated in the public sphere in magazines, newspapers, pamphlets, travel narratives, and memoirs. While it can be difficult to trace such accounts, it is not impossible. This study is based on twelve firsthand accounts of visits, made between 1834 and 1865, to Walton Hall, home of the notoriously eccentric, Roman Catholic traveler and naturalist Charles Waterton (1782–1865).[1] It argues that valuable information about the reception of nineteenth-century displays can be obtained from what the men and women who visited them wrote about their experiences.

Two methodological pitfalls I have tried to avoid are, first, that of assuming that everybody's experience was the same and, second, that of insisting that everybody's experience was unique. The first assumption I take to be obviously false, though it can sometimes be implied when, for example, prescriptive literature—in the form of catalogs, guides, and so on—is taken to be representative of real visitors' experiences. The second position is more insidious, for while it is probably true that no two visitors' experiences of a museum were identical, that does not entail that visitors were at liberty to project whatever meanings they liked onto the things they encountered. This chapter aims to steer a course between two extreme positions in order to offer a more nuanced picture of viewer response. Firsthand visiting accounts are useful to this end. On the one hand, they make possible audience differentiation right down to the level of the individual—they reveal diversity where homogeneity is so often assumed. On the other hand, they reveal commonalties among visitors' responses and provide clues as to how such commonalties might be explained. In particular, they illuminate shared generic frameworks and conventions of visiting which shaped how visitors interpreted their experiences. Interpretations are always shaped by preconceptions. Indeed, without preconceptions, interpretation would not be possible.[2]

Waterton was born at the family seat, Walton Hall, a few miles south of Wakefield in the West Riding of Yorkshire. He was educated at the Jesuit college of Stonyhurst in Lancashire and developed an interest in natural history as a youth. In 1804, having completed his studies, Waterton traveled to Demerara in British Guiana, where he administered estates belonging to his father and uncle until 1812, returning home at regular intervals. After 1812 he made four further voyages to the New World to collect natural history specimens. He published an account of these travels under the title *Wanderings in South America, the north-west of the United States and the Antilles, in the years 1812, 1816, 1820 and 1824* (Waterton 1825). He also contributed more than sixty essays to John Loudon's *Magazine of Natural History*; they were collected as *Essays on natural history, chiefly ornithology* (Waterton 1838, 1844, 1857).

Waterton displayed his natural history specimens at his home in York-shire.[3] While it was common for country houses to have a museum in this period, this museum was unusual in that Waterton had collected and pre-pared nearly all the specimens himself, rather than purchasing or exchang-ing them. Another unusual feature was that Waterton had transformed the grounds of the estate into what has since become known as "the world's first nature reserve" (Bell 1999, 7; Blackburn 1996). Having erected an eight-foot wall around the park to keep out foxes and poachers, he made numerous adaptations so that birds and wild animals could be observed in their natu-ral habitats. In 1829, in his late forties, Waterton married seventeen-year-old Anne Mary Edmonstone. Anne's father, Charles, was a close friend of Waterton's from British Guiana; her mother professed to be the daughter of an Indian princess. The marriage lasted under a year—Anne died shortly after the birth of their son, Edmund—but at Waterton's request, Anne's sisters, Eliza and Helen Edmonstone, came to live with him, keeping his house and overseeing the upbringing of his son. From the 1830s onward, these four constituted the family living at Walton Hall. Waterton and his family received many thousands of visitors. Some were close friends; many more were strangers who came to view the park and its inhabitants and to inspect the curious contents of the museum.

The central aim of this chapter is to explore, through visiting accounts, how people interpreted the natural history specimens displayed at Walton Hall. Working outward from this case-study, I aim to draw more general conclusions about nineteenth-century visitors' experiences of exhibitions. The chapter opens by examining how visiting conventions shaped Water-ton's visitors' interpretations of processes of travel, admission, reception, and hospitality. The second section investigates how they brought external information, narratives, and imagery—acquired from labels, guides, travel narratives, periodicals, shows, panoramas, prints, conversation, and so on—to bear on their interpretations of objects at Walton Hall. The third section considers Waterton's efforts to control his visitors' responses, in particular by promoting new conventions of taxidermic realism. The fourth section suggests that an encounter, or even a nonencounter, with the celebrated naturalist himself could profoundly affect how the visiting experience as a whole was understood.

Kinds of Visit

One of the key factors that shaped visitors' experiences of Walton Hall was the kind of visit they were undertaking. I shall begin, therefore, by identifying some of the kinds of country-house visit that were recognized as being possible in the nineteenth century.[4] The three kinds I shall discuss are the visit as an invited guest, the tour, and the excursion. This taxonomy of visiting is neither exhaustive nor exclusive, but it is helpful in bringing out some of the tacit conventions and more explicit regulations that governed how certain kinds of visit proceeded in this period.

Perhaps the oldest kind of country-house visit was that which took place in response to a personal invitation from the owner. In such cases, the visitor was likely to be of a similar social standing to the owner, and he or she would be known to the owner, either personally or through the testimony of a third person mediating the introduction. Special arrangements would be made in preparation for such a visit by the household—refreshments and sometimes lodging would be offered—and the visit would be conducted according to gentlemanly conventions of politeness.

The London physician George Harley (1829–1896), for instance, was invited in the spring of 1856 to visit Waterton in connection with the two men's shared interest in the possible medical benefits of "wourali" (also known as curare), a compound that South American Indians used to poison arrows; Waterton had acquired a quantity of the substance during one of his "wanderings" (Waterton 1825, 1–76). In his "Reminiscences of Charles Waterton," published in the *Selbourne Magazine* in 1889, Harley recorded his receipt of the invitation and his reactions to it. Initially he was worried at the prospect of visiting a gentleman of dubious reputation (more on that later), but, at the same time, he was desperate to replenish his stock of wourali, and so after much deliberation he decided to go ahead. Arriving on a bright and sunny morning, he was met by the gatekeeper. "In response to the coachman's summons," he recorded, "a middle-aged woman appeared, and after scanning both me and my portmanteau, asked if I were Dr. George Harley, and being assured of this, she 'dropped a curtsy,' and said that the Squire was expecting me" (Harley 1889, 35).

A second personal guest was the writer Julia Byrne (1819–94). Though they had never met before in person, Waterton invited Byrne via a mutual friend to retire to Walton Hall following the death of her husband. Byrne had heard about the premature death of Waterton's wife, and in her *Social hours with celebrities* of 1898 Byrne recalled how she was comforted by the prospect of spending time with "one who could so profoundly enter into my state of mind" (Byrne 1898, 34). Her reception at the hall was more familiar than Harley's: Waterton and his family came outside to greet her personally. "Anyone," she wrote, "who had witnessed the affectionate cordiality of my reception . . . would have thought we had been bosom-friends for years" (Byrne 1898, 33).

Processes of introduction, invitation, admission, reception, and hospitality were significant components of any country-house visit. If we are interested in understanding how private natural history museums were experienced by nineteenth-century visitors, such processes must be taken into account, for museum encounters were always understood in the context of a larger visiting event. While Harley and Byrne were treated differently on arrival because their visits were undertaken for different purposes, they were both invited guests who moved in high social circles; their visiting arrangements were thus relatively informal insofar as they were governed by etiquette rather than regulation. As Harley noted in his *Reminiscences*, however, such flexible arrangements would never do for the "many tourists and 'globe-trotters'" who tended to "treat the Hall and Park as a public show place," nor for the "large parties of holiday-makers from the surrounding districts" (Harley 1889, 35).

"Tourists" in this period were men and women from the leisured classes engaged in literally touring a defined region (Buzard 1993; Moir 1964; Ousby 1990). Since the 1790s the wars on the Continent had occasioned a boom in domestic tourism by closing off much of Europe and by stimulating a nationalistic enthusiasm for British culture. After 1815 Continental travel resumed, but tours of Britain remained popular. Within the class of domestic tours, there were several varieties. One was the picturesque tour, which, with its emphasis on viewing and sketching landscape, was popular from the late eighteenth century onward (Andrews 1989; Ousby 1990, chap. 4). Another was the antiquarian tour, which involved visiting buildings that were felt to capture the essence of key periods of British history: mansions dating from the "Olden Time"—roughly the period of the Tudors and Stuarts—were especially favored (Mandler 1997, Ch. 1). A third kind centered on sites of industrial production. Tourists such as the retired assistant commissary general George Head (1782–1875), who visited Walton Hall in 1835 in the course of his *Home tour through the manufacturing districts of England,* marveled at the country's rapidly developing agricultural, manufacturing, mining, and transport technologies (Head [1836] 1968). Such sites were rarely picturesque but could be appreciated aesthetically within the framework of the industrial sublime (Klingender [1947] 1968; Moir 1964, chap. 8).

Tourists were generally admitted to country houses on the basis of letters of recommendation signed by a person known to the owner. The vicar

and self-confessed "bibliomaniac" Thomas Dibdin (1776–1847), for example, visited Wakefield in 1837 with his daughter in the course of their *Bibliographical, antiquarian and picturesque tour in the northern counties of
England and in Scotland* (Dibdin 1838). The pair were alarmed to hear that
Waterton was abroad and "strict orders had been given that NO ONE should
be admitted upon any plea or pretence whatever." Armed with a letter of
recommendation from the Reverend Samuel Sharp, vicar of Wakefield,
they set off for Walton anyway, hoping to charm their way in, but when
they arrived, they found the gatekeeper to be a member of that "fierce and
peculiar race of human beings, of the feminine class, who are stern beyond
all softening." A debate ensued in which Dibdin and his daughter were
forced to plead and grovel; only after convincing the gatekeeper that they
were "both very peaceable and honest and would touch nothing" were they
reluctantly admitted (Dibdin 1838, 1:147–49). Admission procedures were
just as significant to nineteenth-century tourists as to invited guests: increasingly formalized yet occasionally flexible, entrance restrictions could
occasion anxiety, disappointment, or, as in this case, a triumphant sense
of achievement.

    The most notable change in visiting practices was the rise of the
daylong "holiday" or "excursion," which, emerging early in the nineteenth century but blossoming by the 1840s and 1850s, was the primary
medium through which middle- and working-class visitors gained access
to country-house museums. The rise of the excursion can be attributed
to changes in transport and to pressure from Whig reformers, who cam-
paigned for holidays for workers to allow them to pursue "improving"
leisure activities. Many country-house owners were willing to open up
their parks and gardens to excursionists for the sake of affording "rational recreation" to the working classes, and some even competed to provide the best facilities and attractions (Mandler 1997, chap. 2; Tinniswood
1989, chap. 7). At Walton Hall visitors were allowed to wander in the
grounds and could apply for permission to fish in the lake (see figure 9.1).
On festive occasions, bands would play in the grotto, and there would be
singing and dancing. Provisions were made for picnicking, as Waterton explained in a letter to his friend, the physician Richard Hobson: "If any
of your friends want a pic-nic, the building at the grotto will be at their
disposal the entire day, and they will have crockery, kettles, &c., from the
Hall. . . . The woman at the lodges, for the trifling gratuity of a shilling will
prepare their tea, and clean all up the following morning" (Hobson [1866]
1867, 268). Visiting parties could be sizable: more than a hundred patients
from the West Riding Mental Hospital visited Walton Hall each year for

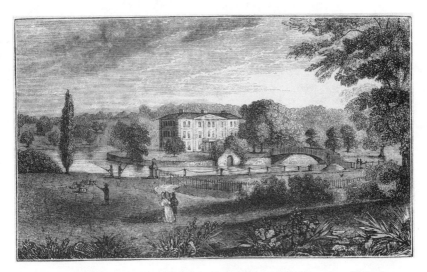

Figure 9.1. Walton Hall, by Edwin Jones. Frontispiece to Waterton 1844. Visitors can be seen strolling the grounds and fishing in the lake. Reprinted by permission of the Whipple Library, University of Cambridge.

their annual excursion, and wedding parties could be even larger (Edginton 1996, 192).

The reception of excursionists into the hall itself depended on the social status of the party: in accordance with long-standing customs of reception, while "gaudy equipages" were received at the main door, members of the lower classes were admitted "in a manner more congenial to their habits" by way of the servants' entrance (Head [1836] 1968, 156). In the early years, admission was granted freely to excursionists, but as numbers increased, stricter regulations were enforced. One visitor wrote in 1848 that "owing to the misconduct . . . of the many, that indiscriminate admission of all comers is now at an end" (Kinsey 1848, 36). In a move which reflected the growing formalization of country-house admission procedures across the country, visiting was restricted to certain days and times, and excursion parties were required to obtain tickets several days in advance. The ticket of admission to Walton Hall shown in figure 9.2, for example, entitled a party of six to "view the Museum and Grounds on Saturdays and Thursdays excepted from 10 till 12 & 2 till 5 in the afternoon."

Like all literary productions, nineteenth-century visiting accounts were shaped by generic constraints: in recording their experiences, visitors deliberately framed their visits according to the purposes they intended their written accounts to serve. For example, invited guests, tourists, and

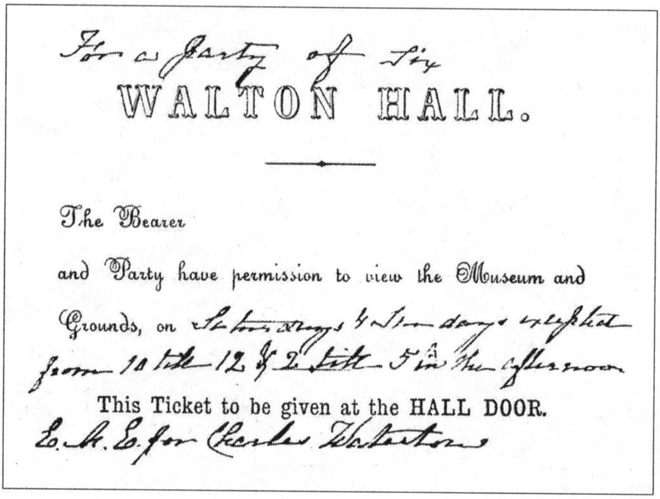
For a Party of Six
WALTON HALL.

The Bearer
and Party have permission to view the Museum and
Grounds, on Saturdays & Thursdays excepted
from 10 till 12 & 2 till 5 in the afternoon
This Ticket to be given at the HALL DOOR.
E. M. E. for Charles Waterton

Figure 9.2. Ticket of admission to Walton Hall, from a collection of thirteen letters to T. Allis (York Reference Library, accession number NRA9540). It entitles a party of six to "view the Museum and Grounds on Saturdays and Thursdays excepted from 10 till 12 & 2 till 5 in the afternoon." Reprinted by permission of City of York Libraries.

excursionists alike placed great emphasis on their journeys—the routes they took, the sights they saw along the way, and the company they traveled with were all dutifully recorded. However, different visitors stressed different aspects of their journeys. In an account published in the *Gentleman's Magazine* in 1848, the Reverend William Kinsey, who spent a day at Walton Hall as one of "an intellectual party" of excursionists, commented, "Passing over the paved streets of once 'merrie Wakefield,' now, alas! the abode of rampant and Republican Dissent, and distinguished by its tall chimneys, vomiting forth eternal smoke, we crossed the River Calder by a handsome bridge, on the right bank of which there is a beautiful chapel, happily rescued now from the hands of the Vandals, who had desecrated it to serve the purposes of commerce" (Kinsey 1848, 33). Kinsey's nostalgic rendering of historic Wakefield was calculated to appeal to the readership of the *Gentleman's Magazine,* who were accustomed to finding antiquarian and historical features within the magazine's pages (Darcy 1983). By contrast, a writer for the *Leisure Hour,* a cheap, weekly family journal of instruction and recreation produced by the evangelical Religious Tract Society, framed his journey as a lesson in self-improvement, taking every opportunity to exhibit appropriately informed responses to landscape, collections, wealth, and gentlefolk, and to demonstrate virtuous character traits and polite manners for the benefit of his readers.[5] "It was a lady," he wrote of a pompous traveling companion, "and we gallantly bore the infliction" (Anon. 1859a, 486); "We knew that he was a roman Catholic," he wrote of Waterton, "but that made no difference to us on the occasion of our visit" (Anon. 1859a, 487).

Visiting accounts were carefully tailored to meet the demands and expectations of editors, publishers, and readers. Moreover, because these readers were also potential visitors, visiting accounts were themselves important in establishing and maintaining the conventions that governed how certain kinds of visit were conducted and understood.

Reading Texts, Images, and Objects

Conventions were also important when it came to interpreting visual phenomena. For example, one very broad framework within which people viewed things in this period was the picturesque. Originally coined in eighteenth-century intellectual circles to describe views (primarily landscapes) which would look good in a picture, the term *picturesque* came, around the turn of the nineteenth century, to dominate discourses of travel, landscape painting, and garden design, and by the 1830s or so, the language of the picturesque had permeated practically all areas of visual culture (Bermingham 1987; Hussey [1927] 1967; Watkin 1982). Picturesque principles of viewing were propagated throughout the nineteenth century in aesthetic treatises, heavyweight journals, travel narratives, poems, penny magazines, and countless guidebooks; outside of the literary domain, the same principles were embodied in the designs of gardens, buildings, paintings, prints, theater scenery, and panoramas. It comes as little surprise, therefore, to discover that Waterton's visitors employed these principles in viewing the grounds and surrounding landscape at Walton Hall. George Harley, for instance, wrote, "Once within the park gates a pleasing view met my eyes: it was a richly verdured and well-wooded glade sloping down to a large lake, the placid waters of which were glistening with the rays of the sun," thus demonstrating his familiarity with the convention that landscape should be considered in terms of bounded views—of scenes which might form the basis of a picture. Similarly, the *Leisure Hour* journalist wrote, "You looked around you on all sides, and you found a most tasteful mixture of wood and water, forest and sward, wild and cultivated; here nature running riot, and there, art adorning nature: and the whole was circumscribed within this valley, hills, feathered to the tops, bounding your view. There was excellent taste in selecting and decorating such a spot" (Anon. 1859a, 486). Part of the purpose of this description is to record the writer's own picturesque viewing experiences. The description is also meant, however, to remind the wide-eyed readers of the *Leisure Hour* that, in viewing and commenting upon country-house estates, *they* must always remember to reflect upon the role played by art in improving upon the scenes provided by nature. Just as visiting accounts functioned both to record and to promote visiting conventions, so they both recorded and promoted viewing conventions too.

My primary concern in this chapter is not with very general interpretative frameworks such as the picturesque (which has, in any case, already received a great deal of scholarly attention); rather, my interest lies with the more specific frameworks that visitors to Waterton's museum brought to bear upon their interpretations of his natural history specimens. In what follows I will, however, keep in mind the thought, illustrated so clearly by the case of the picturesque, that visitors' responses were shaped not only by what there was to see but also by highly formalized discourses that contained within them elaborate protocols for looking at and responding to various kinds of objects, as well as generic rules governing how these responses should be recorded. Visiting accounts are important in this connection because they provide evidence of how these discourses were applied in practice and of how they were propagated through visual and textual media.

Text was becoming an increasingly important component of the museum visit during the first half of the nineteenth century, with the rise of the guidebook and the move toward fuller labeling of museum exhibits. Some account of the hermeneutic relationship between texts and museum objects is, therefore, essential to any understanding of museum visitors' experiences of natural history specimens. This relationship is peculiarly transparent in the case of Waterton's collection because of its explicit connection to the *Wanderings in South America*, first published in 1825 (figure 9.3). Visitors to the museum were handed a copy of the *Wanderings* on arrival and were encouraged to view the specimens in conjunction with the text in order to learn more about each of the species represented in the museum. Case labels gave page numbers of relevant passages; one visitor remarked favorably that the system provided a "ready means of identifying with the object present its habits in its native wilds" (Head [1836] 1968, 157). The experience of visitors was thus partly a learning experience, in which textually-based information was brought to bear on the objects on display. However, the *Wanderings* was not just a catalog of abstract information; it was also a collection of adventure stories about how Waterton had obtained the very specimens in the museum from the wild. In this respect,

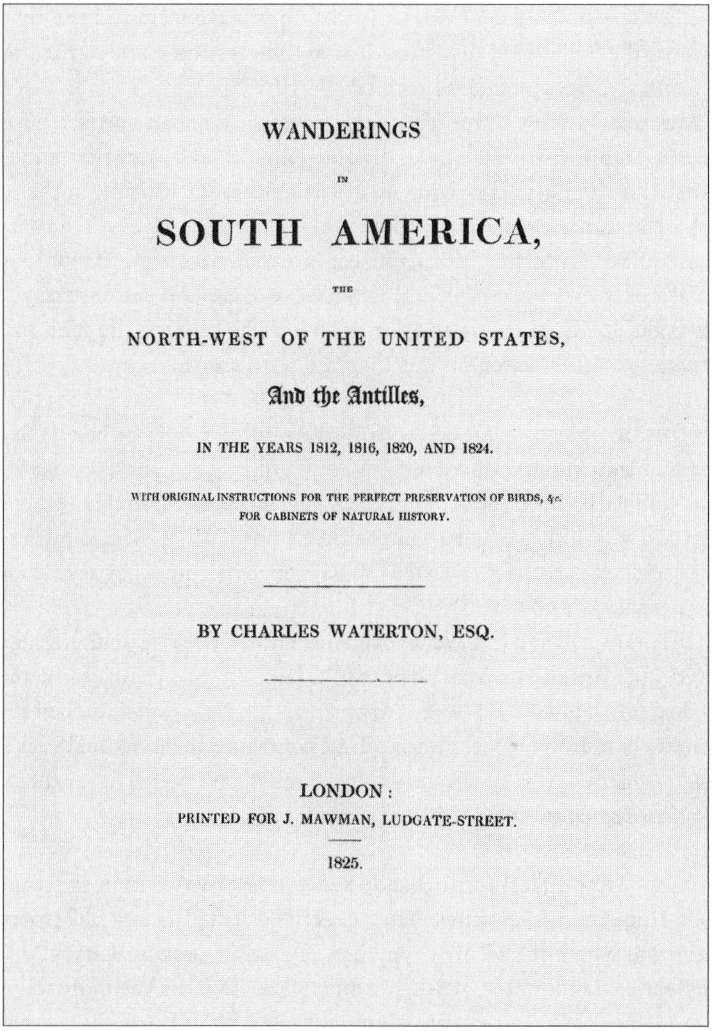

WANDERINGS

IN

SOUTH AMERICA,

THE

NORTH-WEST OF THE UNITED STATES,

And the Antilles,

IN THE YEARS 1812, 1816, 1820, AND 1824.

WITH ORIGINAL INSTRUCTIONS FOR THE PERFECT PRESERVATION OF BIRDS, &c.
FOR CABINETS OF NATURAL HISTORY.

BY CHARLES WATERTON, ESQ.

LONDON :

PRINTED FOR J. MAWMAN, LUDGATE-STREET.

1825.

Figure 9.3. Title page of *Wanderings* (Waterton 1825). The work was initially published as a costly octavo volume but was soon followed by cheaper editions. *Wanderings* remained in print until 1984 and is still regarded as a natural history "classic" today. Reprinted by permission of Nick Jardine.

the object-text relationship at Walton Hall was rather different from that experienced by visitors to other museums (cf. Fyfe, this volume). Visitors interpreted the specimens in light of these stories, thus attributing a further layer of meaning to the objects they encountered.

Visitors viewing Waterton's crocodile, or cayman, for example, which was exhibited at the top of the staircase, may have followed the instructions inscribed on a card displayed inside the glass case and turned to the *Wanderings* to read about its origins. During his travels in British Guiana, Waterton had set about obtaining a perfect cayman specimen for his museum. Trapping the beast was trouble enough, and Waterton and some Indians who were assisting him spent several nights waiting in the forest before a cayman eventually swallowed the baited hook they had cast into the Essequibo River. But how to kill it without damaging its skin? That was the really awkward problem. Eventually Waterton set on thrusting a canoe mast down the cayman's throat, but as he ordered the men to haul the beast out of the water, he was inspired with another idea:

> By this time the cayman was within two yards of me. I saw he was in a state of fear and perturbation; I instantly dropped the mast, sprang up, and jumped on his back, turning half round as I vaulted, so that I gained my seat with my face in a right position. I immediately seized his fore legs, and, by main force, twisted them on his back; thus they served me for a bridle.
>
> He now seemed to have recovered from his surprise, and probably fancying himself in hostile company, he began to plunge furiously, and lashed the sand with his long and powerful tail. I was out of reach of the strokes of it, by being near his head. He continued to plunge and strike, and made my seat very uncomfortable. It must have been a fine sight for an unoccupied spectator. (Waterton 1825, 231)

Visitors to Walton Hall immediately recognized the cayman specimen as the one from the *Wanderings*. They described it not as any old crocodile, but as "[t]he fierce ill-looking cayman or crocodile, on whose back Mr. Waterton fearlessly mounted, while his men were dragging the monster of the deep from his native element" (Menteath 1834, 31). One visitor described his encounter with the cayman like this: "In a commanding position, with a lowering countenance, and an eye as horridly frowning as I ever beheld, stands extended at full length the renowned crocodile, sufficient in his own person to recall to the mind of the spectator that gallant equestrian feat which brought before the notice of the world the latter part of his history" (Head [1836] 1968, 157). The ambivalent use of the word *recall* is telling, for what is it, exactly, that is being recalled? It could be the cayman *story*; but it could equally be the *event* as it happened in the flesh.

Imaginative reconstruction played an important role in nineteenth-century travel. Picturesque travelers were advised that, as perfectly picturesque scenes were rarely to be found in nature, they should employ their imaginations in erasing anomalies and combining elements from several views, thereby creating ideal, composite scenes, on paper or in the mind's eye (Gilpin 1792, 7–8). Antiquarian enthusiasts who visited historic mansions were similarly encouraged to engage their imaginations and be transported back in time to experience the thrill of the Olden Time for themselves (Howitt 1838, 1:322). Waterton's cayman specimen prompted visitors to imaginatively "recall" their experiences as "unoccupied spectators" on the banks of the Essequibo River. Just as crumbling architecture and picturesque landscapes functioned as gateways to an imaginatively reconstructed historic past, facilitating a kind of time travel, natural history objects could facilitate imaginative travel of a geographical nature, evoking the distant places from which they originated. This phenomenon is illustrated again in Thomas Dibdin's visceral response to the preserved remains of a large snake: "Yonder is a *Boa Constrictor,* coiled up to make his spring upon the unwary traveller. His scales glisten and he moves along in splendid lubricity. I tremble to approach him, and can hardly think I have passed him in safety" (Dibdin 1838, 1:150). The visitor first sees the snake coiled and motionless, as if preparing to spring—so far it is only the visitor who is moving—but then the snake begins to move too, its scales catching the light as it slithers along, terrorizing the "unwary traveller." In this word *traveler,* we see the crux of the visitor's fantasy; as seen by Dibdin, the visitor is both the traveler on the staircase as he motions toward the glass-cased specimen and the traveler in the wilds of South America as he happens upon a deadly serpent. The specimen reaches back to the forests of British Guiana to recall the place and the time of the original man-snake encounter and, for a moment, the visitor imagines himself in the picture with such vivacity that he trembles at the thought.

It is common enough to hear that collected objects can evoke an experience of the exotic; what is less common is to find any explanation of how this occult-evoking mechanism might work, yet it is clear that the power of collected objects to evoke certain responses does not simply emanate from the objects themselves (Pearce 1994). In this connection I have found it helpful to consider natural history specimens as analogous to souvenirs, as discussed by Susan Stewart in her influential book *On Longing: Narratives of the Miniature, the Gigantic, the Souvenir, the Collection* (1993). Stewart constructs an opposition between souvenirs and collected objects: while souvenirs reach back to their origins, she argues, collected objects are torn from their contexts of origin and derive their significance from their place within the taxonomy of the collection (Stewart 1993, 151–54).

Figure 9.4. Edwin Jones, *Charles Waterton Capturing a Cayman* c.1824–25. Oil on canvas. Wakefield Museum, Wakefield, UK. In this painting, the birds and other animals Waterton collected from South America look on, framing the scene.

Waterton's specimens, however, owed their significance as much to the circumstances of their acquisition as to their place in the collection as a whole. I mentioned that one visitor thought the cayman was "sufficient in its own person" to recall the event of its capture (Head [1836] 1968, 157). In fact this is not correct. As Stewart points out, a souvenir can only recall its context of origin when supplemented by narrative (Stewart 1993, 136); similarly with natural history specimens. Nineteenth-century museum visitors gave meaning to collected objects through forming complex associations between the objects themselves and a host of remembered facts and narratives that were seen to relate to them.

Of course, many visitors to Walton Hall, or indeed any other museum, would not have taken the time to read extensively during their visits. The cayman story may yet have been familiar to such visitors, however; indeed, it may have been familiar even to some illiterate visitors through its pictorial representations. The *Wanderings* was reviewed in a wide range of periodicals, which frequently excerpted the cayman passage as well as a number of hair-raising stories relating to snakes: the *Literary Gazette* excerpted the cayman story and three snake stories (Anon. 1826b, 4–8); Sydney Smith excerpted the cayman story and two of the snake-wrestling adventures into the *Edinburgh Review* ([Smith] 1826, 310–13); the *Quarterly Review* excerpted two "wondrous tough stories," one cayman-related and one snake-related (Anon. 1825b, 320–23); and by April 1826, the *British Critic* was declining to excerpt the passages in full, on the grounds that "we believe they have been transcribed into most of the newspapers" (Anon. 1826a, 103). At Walton Hall, the picture shown in figure 9.4, painted by a friend of Waterton's, was hung near the specimen. The image reached a wider audience through the engraving shown in figure 9.5 by the caricaturist and illustrator Robert Cruikshank (1789–1856), elder brother of the well-known caricaturist and illustrator George Cruikshank (1792–1878). This image could, according to George Head, "be seen in many shop-windows." "Everybody is acquainted with the story of the crocodile," he wrote confidently (Head [1836] 1968, 158).

Finally, of course, visiting accounts themselves circulated these same narratives. Most of the accounts I have drawn upon here were published during Waterton's lifetime. They appeared in books and magazines with wide and varied readerships, from the staid, affluent subscribers to the *Gentleman's Magazine*, to the natural history enthusiasts who pored over the pages the *Magazine of Natural History* and the *Selbourne Magazine*, to the middle-class families who gained their weekly doses of instruction and entertainment from the *Leisure Hour*. Several of the accounts were reprinted

Figure 9.5. "It was the first and last time I was ever on a cayman's back." Hand-colored engraving by Robert Cruikshank. Images of celebrated persons and events were circulated in the nineteenth century through print shops. Sometimes the same images were reproduced in the periodical press, reaching even wider audiences. Reprinted by permission of Jim Secord from his private collection.

and thus reached even wider audiences: Stuart Menteath's contribution to the *Magazine of Natural History* was reprinted in the *Youth's Instructor and Guardian* for 1835 (Menteath 1835); the *Leisure Hour* journalist's piece reached a new, more local audience when it appeared in the *Wakefield Express* (Anon. 1859b). Accounts written in response to Walton Hall enticed others to visit. They popularized narratives from the *Wanderings* and encouraged the next generation of visitors to view Waterton's specimens in connection with these narratives so that they too could experience the appalling cayman, the terrifying boa, and all the other animals, just as they appeared in the *Wanderings*.

Taxidermic Realism: Art and Ideology

Visiting accounts propagated and reinforced general conventions of visiting and broad frameworks for interpreting visual phenomena. They also propagated rather more specific viewing practices. For example, very many visitors remarked that the specimens at Walton Hall were "prepared like life itself" (Peverell 1865, 600). Figure 9.6 is a photograph of the staircase at Walton Hall, with specimens mounted in glass cases attached to the banisters. British naturalist the Reverend John Wood (1827–89) observed, "Numbers of parrots and parakeets are displayed in all the attitudes which those mercurial birds assume, spreading their beautiful wings for flight, climbing up the boughs with their hooked beaks, ruffling their feathers, and scolding each other lustily, and, in fact, wanting nothing but movement to seem gifted with life" (Wood 1863, 122). Richard Hobson agreed: "You can scarcely conceive it possible that even *living* nature could surpass, nay could equal the simple dead representations here displayed, to such a state of perfection has art attained" (Hobson [1866] 1867, 189). It might be assumed that visitors' remarks about the lifelikeness of Waterton's specimens were based upon "objective" judgements of verisimilitude between specimen and living animal. Here I shall argue to the contrary that the perceived realism of Waterton's taxidermic specimens was, like all forms of artistic realism, conventional.[6] I shall argue, furthermore, that Waterton actively manipulated visitors' responses to his specimens by setting new standards by which taxidermic preparations were to be judged.[7]

Notions of realism were central to visitors' experiences of all kinds of exhibitions in the nineteenth century. Panoramas were designed to reproduce the experience of viewing real landscapes rather than painted imitations; ethnographic displays were carefully stage-managed in order to present audiences with convincing scenes of battle, work, and family

Figure 9.6. Specimens on the staircase at Walton Hall. This photograph is one of a stereoscopic pair produced in 1866 by James Wigney of Huddersfield. Viewed through the right apparatus, they would have given a three-dimensional impression. Nineteenth-century viewers may have found stereoscopic images more "lifelike" than traditional, flat representations. From the library of Rose Busk (Natural History Museum, call number 85.o.W). © The Natural History Museum, London, reprinted by permission.

life (Qureshi 2005); and theatrical illusions, such as the famous Pepper's Ghost, hinged on the ability of showmen to make their audiences believe they were seeing the impossible (Morus, this volume; Lightman, this volume). While part of the appeal of such exhibitions was, at some level, the prospect of witnessing Constantinople, a Bushman's home, or a ghost for oneself, in many cases a still greater pleasure was derived from marveling at the workmanship behind the exhibition—at the skill and ingenuity with which the real was, after all, represented (Morus, this volume). While it is widely accepted that realism on canvas, for example, or on the stage, has always depended on the skillful application of historically specific representational conventions, taxidermic realism is still too often treated ahistorically: what counted as "realistic" or "lifelike" in any given period is assumed to be self-evident, and the equation of "more realistic" with "better" is taken for granted.[8]

It has been argued that "the collectors of the nineteenth century no longer regarded taxidermy as a problem but considered it a technique" (Farber 1982, 562). While this was largely true of the growing class of "professional" naturalists, it was not true of Waterton, who in an essay of 1825 appended to the *Wanderings in South America* complained, "Were you to pay as much attention to birds as the sculptor does to the human frame, you would immediately see, on entering a museum, that the specimens are not well done. This remark will not be thought severe when you reflect that that which once was a bird has probably been stretched, stuffed, stiffened, and wired by the hand of a common clown" (Waterton 1825, 307). Waterton was critical of the taxidermic techniques in common use, and his complaints can be understood as responses to the increasing professionalization and systematization of natural history, which, as he saw it, tended to devalue the expertise possessed by gentlemen field naturalists such as himself.

In the 1830s Waterton became embroiled in a bitter dispute with William Swainson (1789–1855), self-made professional zoologist and adherent to William MacLeay's quinary system of classification, according to which animals were arranged within a set of nested circles, each of which consisted of five subsidiaries (Ritvo 1997, 31–35). The dispute began in earnest in 1836 when, in the introduction to his *Natural history and classification of birds*, Swainson disparagingly labeled Waterton "an amateur," insinuating further that in rejecting systematic nomenclature, Waterton was guilty of neglecting "all which could be truly beneficial to science" (Swainson 1836, 211). Waterton responded, in an 1837 pamphlet entitled *An ornithological letter to William Swainson*, by ridiculing Swainson's "fond conceit of circles" and accusing him of obtaining his zoological information from the "closet" rather than from the boundless fields of nature (Waterton 1871, 512, 513). "Let me tell you," he warned his adversary, "that the admeasurement of ten thousand dried bird-skins, with a subsequent and vastly complicated theory on what you conceive you have drawn from the scientific operation of your compasses, will never raise your name to any permanent altitude" (Waterton 1871, 521).

When it came to the issue of preparing specimens, Waterton found Swainson's method "wrong at every point" (Waterton 1871, 533). He insisted that anyone adopting Swainson's techniques would find his preparations to be "out of all shape and proportion": birds would "come out of his hands a decided deformity," while quadrupeds would "represent hideous spectres, without any one feature remaining similar to those which they possessed in life" (Waterton 1871, 519–20). Swainson employed traditional methods: he used arsenical soap as a preservative and generally kept his specimens as unmounted skins for convenience; when a mounted specimen was required, he stretched the skin over wires and wooden supports. During the 1810s and 1820s, Waterton had developed a rival method of taxidermy, which was published as an appendix to the *Wanderings* and was also described in many of his later publications. Waterton used corrosive sublimate, a compound of mercury, as a preservative; dispensing with wires and wooden supports, he allowed his skins to dry and harden very slowly until they could stand their own weight, and he adjusted their proportions daily as they dried, over many weeks or even months.

In contrast to professional taxidermists, who tried to "effect by despatch what could only be done by a very slow process" (Waterton 1871, 534), Waterton indulged each of his specimens with painstaking care and attention, which, as a gentleman of leisure, he could afford. Furthermore, on the basis of his field experience he was able, he claimed, to re-create naturalistic forms, attitudes, and expressions in his skins. "You must possess Promethean boldness," he advised his students, "and bring down fire and animation, as it were, into your preserved specimen" (Waterton 1825, 308). While the motivation behind Swainson's choice of technique was that specimens should be easy to prepare, store, and handle, the intended outcome of Waterton's process was that visitors to his museum at Walton Hall should exclaim, "That animal is alive!" (Waterton 1871, 536).

In his polemical writings—which were circulated through a range of printed media, including travel narratives, pamphlets, and magazines—Waterton actively encouraged his readers to discriminate against a whole host of traditional taxidermic techniques and the visual effects of these techniques, insisting that the results were unrealistic representations of living animals. In doing so, he sought to establish new conventions for taxidermic realism. These conventions were ideological, for they placed value on field knowledge and on leisure, as possessed by the naturalist, and implicitly devalued systematization and professionalism. His taxidermic methods remained controversial; Swainson was not the only opponent. The *Monthly Review*, for example, dismissed Waterton's technique as a "curious *mélange* of wireless cotton, sublimate, and sentimentalism" (Anon. 1825a, 71). Nevertheless, Waterton succeeded in convincing many of his visitors that his own specimens were more lifelike than those displayed elsewhere, that "the natural and appropriate attitudes . . . , together with the exquisite beauty and preservation of the plumage, surpassed all that we had ever witnessed in any museum we have visited" (Anon. 1824, 245). Though these visitors had probably seen only a fraction of the species represented in the museum in a living state, they nevertheless delighted in seeing, in each parakeet, each snake, and each crocodile, the results of a superior knowledge of the field and of a method which had "attained a state of perfection" hitherto unknown (Hobson [1866] 1867, 189).

The Celebrity as Specimen

Walton Hall was home to a remarkable natural history collection, but it was also home to people, and, of course, the person everybody wanted to see most of all was "Charles Waterton, the famous traveller and naturalist" (Harley 1889, 36), who by the 1840s had become sufficiently celebrated to merit a portrait in the *Illustrated London News* (see figure 9.7). For many visitors, an encounter with Waterton was as sought after as an encounter with his home. In particular, visitors felt they could only truly understand the collection once they had become acquainted with Waterton's "character," something that could be reliably known only through a meeting in the flesh. Here I shall explore how visitors' preconceptions about Waterton's character affected how Waterton and his collection were experienced.

Waterton's public reputation was slightly dubious, especially in the years following the publication of the *Wanderings*, when the truthfulness of his more extraordinary adventures, snake- and crocodile-related ones in particular, was doubted. The *Quarterly Review* labeled these accounts "wondrous tough stories" tending to "somewhat stagger our faith" (Anon. 1825, 319, 323); the *Edinburgh Review* agreed that Waterton's "stories draw largely sometimes on our faith," conceding that the book was all the more entertaining for this reason ([Smith] 1826, 315). In later years the periodicals tended to defend Waterton on the grounds of his perceived eccentricity and his gentlemanly status. By 1841 *Chambers's Edinburgh Journal* could explain, "It was the fashion among the countrymen of Mr Waterton, when his "Wanderings in South America" were first published in 1825, to laugh at the statements of the traveller, as being somewhat Gulliverian in their cast. But a more thorough acquaintance with the character of Mr Waterton has convinced the world of his being a man at once of talent and veracity, though with some oddities in his composition" (Anon. 1841, 188). To witness these aspects of Waterton's character was, for visitors, a way of assuring themselves of the authenticity of the stories contained in the *Wander-*

Figure 9.7. "Charles Waterton, Esq." Illustration from *Illustrated London News,*
August 24, 1844, 124. Visitors frequently expressed surprise at Waterton's cropped hair
and disheveled costume: "He was attired in a grey coat, very ill made, or at least ill
fitting. Blue inexpressibles, with a vest of no particular cut, shoes thick and clumsy,
stockings of blue, with a coloured neckerchief put carelessly round the throat, and a
shocking bad hat, completed the habiliments. We thought at the time that the entire
garniture, if put up for sale at a public auction, might probably have fetched about
three and fourpence" (Anon. 1859a, 487).

ings and, subsequently, of the authenticity of their own adventures in the
museum.

Guests who were personally invited to visit a country-house museum
were usually guaranteed an introduction with its owner; part of the purpose of such a visit would generally be to effect a meeting between the

parties, often with some specific end in mind. Julia Byrne, for instance, was shown around Walton Hall by Waterton in person. Furthermore, as an invited guest, she was allowed privileged access to areas of the house that were ordinarily closed to visitors. She was especially excited about being numbered among the "few confidential friends" whom Waterton invited into his bedroom. "He assured me I was one of the few strangers ever admitted," she explained. "'But then,' he added, gracefully, 'you are not, and never have been, a stranger to me; a common sorrow led us to understand and know each other from the first'" (Byrne 1898, 43–44). Byrne's experience of the bedroom was "not only interesting, but a surprise—indeed, a succession of surprises" (Byrne 1998, 44). The bedroom contained no bed—since the death of his wife, Waterton had preferred to sleep on the floor—but it did contain tools and a menagerie of half-resurrected birds and animals, for it was here that Waterton prepared his taxidermic specimens. During her stay of several weeks, Byrne was allowed to assist in transforming the withered skin of a cock pheasant into a beautiful specimen, which she afterwards received as a gift. She derived great honor from being admitted to Waterton's workshop and being allowed to participate in a process which most people would only ever read about in the magazines and newspapers.

Harley was also admitted to the "inner sanctum," and he was equally perturbed by his experience. Crossing the threshold, he was "almost struck dumb with astonishment": there, right before his eyes, was a baboon, swinging in the air, suspended from the ceiling by strings (Harley 1889, 165). Once he had recovered his composure, he gratefully allowed Waterton to initiate him in the mysteries of his art (Harley 1889, 167). He was further grateful for the opportunity to assuage his fears about Waterton's reputation as "an unblushing and unblushable storyteller" (Harley 1889, 22): "No one could catch a sight of his beaming smile, or receive a glance from his speaking eye without feeling that no matter how bizarre might be the appearance of the outer man, the inner was lit up by a genial, highly cultivated, and sympathetic mind. The cordial clasp he gave to my hand, and the words of warm welcome with which he greeted me, associated as they were with a winsome expression of truthful sincerity, at once drew my heart towards him" (Harley 1889, 37). As a gentleman himself, Harley could perceive Waterton's virtue in every look, every gesture, every word: the clasp of his hand was cordiality itself, the glance of his eye spoke straight to Harley's heart. The experience of meeting Waterton in person was sufficient to convince Harley that any doubts he may have had about the truthfulness of the *Wanderings* were unfounded and that the specimens in the museum were genuine artifacts of the adventures narrated therein.

While personal guests such as Byrne and Harley stressed their thorough acquaintance with Waterton's personality in their visiting accounts, many tourists and excursionists would have been happy to simply set eyes on him. "It was a sad disappointment to us both that such a man should be from home on the occasion of our visit," wrote Thomas Dibdin of his nonencounter with Waterton. "The master-spirit was wanting, to give pungency to anecdote and truth to conjecture. Were he only present to receive our bow and curtesy, it had been something" (Dibdin 1838, 1:147, 151). With the rise of the culture of celebrity in this period, the practice of visiting the homes of famous men and women in the hope of being granted an audience with them, or even of stalking their local haunts just to catch a glimpse of them, was becoming increasingly popular (Gill 1998, chap. 1). Sometimes the results could be disappointing. The *Leisure Hour* writer reported, "The actual Waterton did not at all correspond to the imaginary one that I had built up in my own fancy." He advised his readers, "If you would not break the charm which favourite living authors have flung around you, it may be as well, perhaps, not to pay them a visit" (Anon. 1859a, 486–87). Once he had conversed with Waterton, however, he was reassured: "Full of anecdote, he was a graphic describer, and we felt that in conversation we had the same man that appears in his books" (Anon. 1859a, 487).

Visitors and excursionists wanted to see Waterton because they wanted to see the hero of the *Wanderings*. Just as they desired to view the "fierce ill-looking cayman . . . on whose back Mr. Waterton fearlessly mounted," so they desired to look upon the figure of Waterton himself as "the first and only man who ever bestrode and rode a *cayman*" (Menteath 1834, 31; Anon. 1841, 188). In her reminiscences of Waterton, Julia Byrne related an anecdote which epitomized this kind of approach to visiting. In the years following her first sojourn at Walton Hall, Byrne accompanied Waterton and his family on travels to the Continent. On one occasion in Aix-la-Chapelle the party was supposedly seated for dinner opposite a confused English tourist who, having identified Waterton's Yorkshire accent but being unaware to whom he was speaking, began to inquire about Walton Hall. After asking about the visiting restrictions, he turned to questions concerning the hall's owner, still unaware that he was talking to the man himself.

"I've heard he lives among a lot of wild beasts and birds and things. Is *that* true?"

"Yes, sir, it's quite true; but they're very well-mannered and quite harmless beasts, and so is he."

"I'm glad of that, for I want to go there; and is the story about the
crocodile true?"

"What story? What crocodile?"

"Why, that he rides all over the place on a crocodile; that's what
I should like to see." (Byrne 1898, 76–77)

Byrne's anecdotal tourist learned everything he knew about Waterton from
his books, and it was the Waterton from the *Wanderings*—the celebrated
traveler who rode a crocodile in preference to more conventional modes of
transport—that he hoped to encounter at Walton Hall. Visitors viewed the
collection in light of the *Wanderings*, and they did the same for Waterton
too, objectifying him as a specimen in his own collection.

Conclusion

Published right at the end of the Victorian period in a volume of *Gossip of
the Century*, Byrne's anecdote can be read as a commentary on changing
attitudes toward travel, tourism, and exhibitions in the nineteenth century. As the possibilities for visiting extended increasingly to members of
the middle and working classes, visitors in the older, polite traditions began actively to define themselves in opposition to these new kinds of visitors, stressing their intimate, personal connections with country-house
owners, for example, and smiling down on those who believed they could
really know celebrated men and women through reading about them, and
perhaps glimpsing them at the ends of hallways, or riding out to distant
parts of their estates.[9] Different kinds of visitors had very different kinds
of experiences of country-house museums, and other displays of scientific
knowledge, in the nineteenth century. I have shown that through analyzing firsthand accounts, we can learn a great deal about how the conventions that governed visiting in this period shaped people's experiences.

Histories of collections too often leave collected objects to speak for
themselves. Traditionally, studies of natural history collections have been
concerned with where specimens came from, how much they cost, how
they were classified and cataloged, and perhaps how they were displayed.
More recent studies have enriched this picture by making visible the communication networks connecting collectors in all reaches of the globe; nevertheless, such narratives still tend to tail off as each specimen finds its
home in the museum under investigation. While there has of late been a
move in museum studies toward recovering the meanings which collected
objects had for people in the past, this work has tended to concentrate on
the meanings collections had for their owners; where visitors' responses

have been considered at all, they have generally been framed in terms of how owners intended their collections to be perceived. Yet, as everybody knows, it can never be guaranteed that people will view a display in exact accordance with the intentions of those responsible for the acquisition, organization, and presentation of its constituent objects.

Proponents of the New Museology have sought to address this issue with respect to present-day museums: through questionnaires, interviews, and participant observation, they have gathered and analyzed information about what people today actually do in museums and what they actually think of the exhibits (Vergo 1989). Such techniques are, of course, inapplicable to the study of nineteenth-century displays as perceived in their time. I would suggest, however, that visiting accounts can yield equally valuable information about the experiences of real people who attended exhibitions in the past. That visiting accounts have been so underused by historians stems, I think, from a misapprehension that such sources are practically untraceable. In fact, buried away in memoirs, biographies, guidebooks, pamphlets, periodical articles, and other "ephemeral" sources, there are countless reports of visits to museums and exhibitions waiting to be examined. Locating them relies on a certain amount of detective work. In the case of Charles Waterton, asking questions such as, "Who were Waterton's friends?" "Who shared his interests?" and "Who was in the right place at the right time?" can help identify authors who might possibly have visited Walton Hall and whose autobiographies and travel narratives are worth consulting. More laboriously, trawling periodical indexes can, with luck, yield interesting finds. Fortunately, the continuing development of searchable electronic indexes, such as the SciPer Index, promises to make this process easier.

The act of looking is never free from prejudice. In the field of art history, it is generally recognized that in different historical periods people looked at things differently, and the problem of how to recover what works of art meant in the past has been addressed through attempts to reconstruct "the period eye" (Baxandall 1988). It is my belief that this kind of approach is much more widely applicable. Studying nineteenth-century viewing conventions—general conventions such as those endorsed by proponents of the picturesque and more specific ones such as those embodied in discourses of taxidermic realism—can contribute greatly to understanding visitors' responses to collected objects. Furthermore, studying the processes by which new and controversial conventions for viewing became established can reveal a great deal about how, and to what ends, dominant individuals or groups sought to control the interpretations of others.

An account of what a handful of people made of one slightly peculiar museum would, of itself, be of limited historical interest. In this chapter I have aimed to use responses to Walton Hall as a springboard to a more general understanding of interpretative practices in this period. My primary concern has been not with what different objects happened to say to different people but with how different people made those objects speak. A chemically treated arrangement of knobbly skin and sharp teeth does not mean a great deal on its own: it becomes meaningful only by association. Waterton's collection was atypical in that it was very closely associated with a well-known book of adventures. Nevertheless, while the responses of Waterton's visitors were unusually focused (and thus unusually amenable to detailed historical analysis), they were not unique. The nineteenth century witnessed an explosion of circulated information, narratives, and imagery relating to an increasingly broad range of subjects; in museums, exhibition halls, galleries, theaters, gardens, and private houses up and down the country, men and women from all walks of life drew upon these new resources in interpreting an ever-more-diverse array of displayed materials.

NOTES

1. Biographical writings on Waterton include Aldington 1949; Barber 1980; Blackburn 1996; Carroll 2004b, 2004a; Edginton 1996; Hobson [1866] 1867; Irwin 1955; Sitwell 1958; Wakefield Museum 1982. Waterton prefaced each volume of his *Essays on natural history* with an autobiography; these were edited by Norman Moore, a friend of Waterton's, in a collected edition of the *Essays*: Waterton 1871. The twelve firsthand visiting accounts are Anon. 1859a; Brooke 1877; Byrne 1898, chap. 13; Dibdin 1838, 146–51; Dixon 1870, 304–11; Fitzgerald 1894, 225–38; Harley 1889; Head [1836] 1968, 153–71; Hobson [1866] 1867; Kinsey 1848; Menteath 1834; Wood 1863.

2. In thinking about the role of preconceptions in shaping interpretations, I have been influenced by Gadamer 1989.

3. Today the collection is on display at Wakefield Museum.

4. On country-house visiting, see, for example, Mandler 1997; Tinniswood 1989; Wilson and Mackley 2000.

5. I have been unable to trace the authorship of this article. However, it appears that the writer was male, as a reference is made to "our boyish pictures of the region" (Anon. 1859a, 486).

6. On objectivity and realism in scientific representations, see, for example, Daston and Galison 1992.

7. On Waterton's taxidermy, see Bann 1994; Grasseni 1998; Lee [1821] 1843.

8. Notable exceptions include Haraway 1989; Star 1992.

9. On antitourism in this period, see Buzard 1993.

REFERENCES

Aldington, R. 1949. *The Strange Life of Charles Waterton.* London: Evans Brothers.

Altick, R. 1978. *The Shows of London.* Cambridge, MA: Harvard University Press, Belknap Press.

Andrews, M. 1989. *The Search for the Picturesque: Landscape, Aesthetics and Tourism in Britain, 1760–1800.* Stanford, CA: Stanford University Press.

Anon. 1824. "New Method of Preserving Specimens in Natural History." *Kaleidoscope* 4:245–46.

Anon. 1825a. "Wanderings in South America etc." *Monthly Review* 108:66–77.

Anon. 1825b. "Wanderings in South America etc." *Quarterly Review* 33:314–32.

Anon. 1826a. "Wanderings in South America etc." *British Critic* 2:93–105.

Anon. 1826b. "Wanderings in South America etc." *Literary Gazette and Journal of Belles Lettres,* no. 468, 4–8; no. 470, 39–40; no. 471, 56–57.

Anon. 1841. "Waterton, the Wanderer." *Chambers's Edinburgh Journal,* no. 492, 188–89.

Anon. 1859a. "The Home of Waterton, the Naturalist." *Leisure Hour* 8:486–87.

Anon. 1859b. "The Home of Waterton, the Naturalist." *Wakefield Express,* August 27.

Bann, S. 1994. Introduction to *Frankenstein, Creation and Monstrosity,* ed. S. Bann, 1–15. London: Reaktion.

Barber, L. 1980. *The Heyday of Natural History, 1820–1870.* London: Jonathan Cape.

Baxandall, M. 1988. *Painting and Experience in Fifteenth Century Italy: A Primer in the Social History of Pictorial Style.* Oxford: Oxford University Press.

Bell, R. 1999. *Waterton's Park.* Wakefield, UK: Willow Island Editions.

Bermingham, A. 1987. *Landscape and Ideology: The English Rustic Tradition 1740–1860.* London: Thames and Hudson.

Blackburn, J. 1996. *Charles Waterton 1782–1865: Traveller and Conservationist.* London: Vintage.

Brooke, G. 1877. *Reminiscences of Charles Waterton, the Naturalist: From a Visit in 1861.* Wigan: Thomas Birch.

Buzard, J. 1993. *The Beaten Track: European Tourism, Literature, and the Ways to Culture, 1800–1918.* Oxford: Clarendon Press.

Byrne, J. 1898. *Social Hours with Celebrities: Being the Third and Fourth Volumes of 'Gossip of the century.'* London: Ward and Downey.

Carroll, V. 2004a. "The Natural History of Visiting: Responses to Charles Waterton and Walton Hall." *Studies in History and Philosophy of Biology and Biomedical Sciences* 35:31–64.

———. 2004b. "Charles Waterton." In *The Dictionary of Nineteenth-Century British Scientists*, ed. B. Lightman. Bristol: Thoemmes Continuum.

Darcy, C. 1983. "The Gentleman's Magazine." In *British Literary Magazines: The Romantic Age 1789–1836*, 136–40. Westport, CT: Greenwood Press.

Daston, L., and P. Galison. 1992. "The Image of Objectivity." *Representations* 40:81–128.

Dibdin, T. 1838. *Bibliographical, Antiquarian and Picturesque Tour in the Northern Counties of England and in Scotland.* 3 vols. London: Printed for the author.

Dixon, H. 1870. *Saddle and Sirloin: Or, English Farm and Sporting Worthies.* London: Rogerson and Tuxford.

Edginton, B. 1996. *Charles Waterton: A Biography.* Cambridge: Lutterworth Press.

Farber, P. 1982. *The Emergence of Ornithology as a Scientific Discipline, 1760–1850.* Dordrecht, Neth.: Reidel.

Fitzgerald, P. 1894. *Memoirs of an Author.* London: Richard Bentley and Son.

Gadamer, H.-G. 1989. *Truth and Method.* London: Sheed and Ward.

Gill, S. 1998. *Wordsworth and the Victorians.* Oxford: Clarendon Press.

Gilpin, W. 1792. *Three Essays: On Picturesque Beauty; On Picturesque Travel; And on Sketching Landscape: To which is Added a poem, on Landscape Painting.* London: R. Bladmire.

Grasseni, C. 1998. "Taxidermy as Rhetoric of Self-Making: Charles Waterton (1782–1865), Wandering Naturalist." *Studies in History and Philosophy of Biology and Biomedical Sciences* 29:269–94.

Greenhalgh, P. 1988. *Ephemeral Vistas: The Expositions Universelles, Great Exhibitions and World's Fairs, 1851–1939.* Manchester: Manchester University Press.

Haraway, D. 1989. *Primate Visions: Gender, Race and Nature in the World of Modern Science.* New York: Routledge.

Harley, G. 1889. "Reminiscences of Charles Waterton." *Selbourne Magazine* 2:20–23, 35–37, 114–117, 165–168.

Head, G. [1836] 1968. *A Home Tour through the Manufacturing Districts of England, in the Summer of 1835.* London: Frank Cass.

Hobson, R. [1866] 1867. *Charles Waterton: His Home, Habits and Handiwork.* London: Whittaker.

Howitt, W. 1838. *The Rural Life of England.* 2 vols. London: Longman.

Hussey, C. [1927] 1967. *The Picturesque: Studies in a Point of View.* London: Frank Cass.

Irwin, R. 1955. *Letters of Charles Waterton of Walton Hall, Near Wakefield.* London: Rockliff.

Jackson, H. J. 2001. *Marginalia: Readers Writing in Books.* New Haven, CT: Yale University Press.

Kinsey, W. 1848. "Random Recollections of a Visit to Walton Hall." *Gentleman's Magazine* 29:33–39.

Klingender, F. [1947] 1968. *Art and the Industrial Revolution*. New York: Augustus M. Kelley.

Lee, S. [1821] 1843. *Taxidermy: Or, the Art of Collecting, Preparing, and Mounting Objects of Natural History. For the Use of Museums and Travellers*. London: Longman.

Mandler, P. 1997. *The Fall and Rise of the Stately Home*. New Haven, CT: Yale University Press.

Menteath, J. 1834. "Some Account of Walton Hall, the Seat of Charles Waterton, Esq." *Magazine of Natural History* 8:28–36.

———. 1835. "Walton Hall." *Youth's Instructor and Guardian* 19:92–97, 125–29.

Moir, E. 1964. *The Discovery of Britain: The English Tourists 1540–1840*. London: Routledge and Kegan Paul.

Ousby, I. 1990. *The Englishman's England: Taste, Travel and the Rise of Tourism*. Cambridge: Cambridge University Press.

Pearce, S. 1994. "Objects as Meaning; or Narrating the Past." In *Interpreting Objects and Collections*, ed. S. Pearce, 19–29. London: Routledge.

Peverell, R. 1865. "Charles Waterton and Walton Hall." *Once a Week* 13:48–53, 57–62.

Qureshi, S. 2005. "Living Curiosities: Human Ethnological Display, 1800–1855." PhD diss. University of Cambridge.

Raven, J., H. Small, and N. Tadmor, eds. 1996. *The Practice and Representation of Reading in England*. Cambridge: Cambridge University Press.

Ritvo, H. 1997. *The Platypus and the Mermaid and Other Figments of the Classifying Imagination*. Cambridge, MA: Harvard University Press.

Secord, J. A. 2000. *Victorian Sensation: The Extraordinary Publication, Reception, and Secret Authorship of "Vestiges of the Natural History of Creation."* Chicago: University of Chicago Press.

Sitwell, E. 1958. *English Eccentrics*. London: Dobson.

[Smith, S.] 1826. "Wanderings in South America etc." *Edinburgh Review* 43: 299–315.

Star, S. L. 1992. "Craft vs. Commodity, Mess vs. Transcendence: How the Right Tool Became the Wrong One in the Case of Taxidermy and Natural History." In *The Right Tools for the Job: At Work in Twentieth-Century Life Sciences*, ed. A. Clarke and J. Fujimura, 257–86. Princeton, NJ: Princeton University Press.

Stewart, S. 1993. *On Longing: Narratives of the Miniature, the Gigantic, the Souvenir, the Collection*. Durham, NC: Duke University Press.

Suleiman, S. 1980. "Introduction: Varieties of Audience-Oriented Criticism." In *The Reader in the Text: Essays on Audience and Interpretation*, ed. S. Suleiman and I. Crossman, 3–45. Princeton, NJ: Princeton University Press.

Swainson, W. 1836. *On the Natural History and Classification of Birds*. London: Longman.

Tinniswood, A. 1989. *A History of Country House Visiting: Five Centuries of Tourism and Taste*. Oxford: Basil Blackwell and the National Trust.

Vergo, P., ed. 1989. *The New Museology*. London: Reaktion Books.

Wakefield Museum. 1982. *Charles Waterton 1782–1865: Traveller and Naturalist.* Wakefield, UK: Wakefield Museum.

Waterton, C. 1825. *Wanderings in South America, the North-West of the United States and the Antilles, in the Years 1812, 1816, 1820, and 1824, with Original Instructions for the Perfect Preservation of Birds, etc. for Cabinets of Natural History.* London: Mawman.

———. 1838. *Essays on Natural History, Chiefly Ornithology.* London: Longman.

———. 1844. *Essays on Natural History, Chiefly Ornithology.* 2nd ser. London: Longman.

———. 1857. *Essays on Natural History, Chiefly Ornithology.* 3rd ser. London: Longman.

———. 1871. *Essays on Natural History, Chiefly Ornithology. Edited, with a Life of the Author by N. Moore.* London: Frederick Warne.

Watkin, D. 1982. *The English Vision: The Picturesque in Architecture, Landscape and Garden Design.* London: John Murray.

Wilson, R., and A. Mackley. 2000. *Creating Paradise: The Building of the English Country House 1660–1880.* London: Hambledon.

Wood, J. 1863. "Modern Taxidermy." *Cornhill Magazine*, 1st ser., 7:120–25.

CHAPTER TEN

Science at the Crystal Focus of the World

RICHARD BELLON

The unprecedented physical magnitude of the Great Exhibition of 1851 dazzled, and often bewildered, its millions of visitors. Henry Cole, the civil servant whose indefatigable drive as publicist and administrator proved crucial in launching the exhibition, celebrated the Crystal Palace, its iconic glass-and-iron home, for teaching "the world how to roof in great space" (Cole 1853, 438). The exhibition's vast size was never an end in itself, however, but a necessary consequence of its core aspiration to encapsulate the entire world's industry underneath one (awe-inspiring) roof. A guidebook published specifically for the exhibition season opened with several pages extolling the "wonder-exciting" enormity of the Crystal Palace in dense statistical detail, but it stressed that one should not forget "that the universality in regard to contributors, and completeness in regards to objects to be contributed, are [its] striking characteristics" (Clarke 1851b, 3–7). Charlotte Brontë marveled to her father that "its grandeur does not consist in *one* thing, but in the unique assemblage of *all* things. Whatever human industry has created you will find there. . . . It seems as if magic only could have gathered this mass of wealth from all the ends of the earth" (quoted in Gaskell 1900, 521–22).

At its most basic level, this glorification of comprehensiveness rested upon a deep faith in the power of unfettered sight. The Crystal Palace promoted both philosophical and practical contemplation by serving (theoretically, if not practically) as a complete and accurate microcosm of the world's industry. The exhibition experience was meant to translate the distant and abstract notion of the "world" into a real and immediate lived experience. Visitors could contemplate the world of industry as a whole because they could literally *see* all of it. They could then compare and collate their understanding with the millions of others who shared in the experience. Science was at the very heart of this project, from its inception to its aftermath. The exhibition's organizers believed that constructing such a microcosm was the first step toward bringing productive human labor within the circle of natural law, so that men could someday explain success and failure in industrial pursuits as reliably as Isaac Newton had explained the dynamics of motion.

These aspirations for the Great Exhibition of 1851 were dramatic manifestations of the broader Victorian obsession with liberating perception from the shackles of time and space. This (putative) liberation was frequently devoted to education and research in institutions whose mission was the sustained intellectual, cultural, social, and economic improvement of the British people and their empire. Museum collections had conditioned Europeans to accept that the display of carefully collected and classified objects could make nature edifyingly visible from all times and from all spaces (Outram 1996, 256; Bellon 2006, 5–9). David Brewster insisted that "the naturalist, who arranges the objects of his research in splendid halls, and displays at one view the wonders of the remotest ages, and the most distant kingdoms, is justly honoured with a high degree of contemporary fame" ([Brewster] 1836, 265). Cooperative data collection promised, said John Ruskin, to transform solitary observers into "a part of one mighty Mind,—a ray of light entering into one vast Eye" (Ruskin [1839] 1903–12, 210). In these cases, Victorians painstakingly accumulated both the tangible and the intangible (visual records of landscapes, natural objects, standardized scientific data) which, when systematized, allowed for the methodical oversight of the world. In the most elaborate and ambitious of these spaces—and in the exhibition above all—social unity and common purpose followed directly from the widening of perception. As one commentator noted approvingly, "the philosopher and the savage stood side by side" inside the Crystal Palace (Tallis 1851, 1:207). Still another reflected that "even the despised savage is to be called on for his mite on this occasion, to prove his community of origin with ours, and to support his claim to a common destiny" (Clarke 1851a, 111).

The fixation with breaking the boundaries of time and space also drew audiences to commercial enterprises that depended upon a constant stream of paying customers. Popular entertainments, such as panoramas, routinely allowed spectators to "see" far distant sites from the comfort of a gallery. The exhibition embraced the idea that comprehensive acts of perception and simulated travel were the first steps in the productive accumulation of inductive knowledge; but the panoramas permitted, even if they did not necessitate, perception and travel to be pleasurable ends in themselves, productive of nothing but transitory enjoyment.

This chapter explores how exhibition organizers attempted, and largely failed, to create an enduring communal experience of research and education in the Crystal Palace. It does so, first, by examining speeches given by Prince Albert, William Whewell, and Henry Cole. These presentations, one before the exhibition and two after, distilled what the event was meant to accomplish and clarified the role of science in the endeavor. The following section locates the Great Exhibition of 1851 within the cultural and physical contours of London. The Crystal Palace absorbed and acquired meaning from the multifarious institutions of the British metropolis; sometimes this worked toward the stated ends of the exhibition and sometimes it did not. Finally, this chapter examines the disjunction between the rhetoric and the reality of the exhibition, particularly in its intellectual and social aspirations for science.

"A Period of Most Wonderful Transition"

Before 1851, industrial exhibitions had been national affairs, such as those sponsored by the British Association for the Advancement of Science (Morrell and Thackray 1981, 264). The idea to hold an international Great Exhibition of the Works of Industry of All Nations in Britain was originated in 1849 by a group centered on the Society of Arts and its president, Prince Albert (see figure 10.1). He delivered a widely reported keynote address to a grand banquet for British and foreign dignitaries held in March 1850 at Mansion House, the lord mayor of London's residence. This banquet was intended to raise enthusiasm for the upcoming exhibition by defining its moral and intellectual bearing for a still skeptical public. "We are living at a period of most wonderful transition, which tends rapidly to accomplish that great end, to which, indeed, all history points—the realization of the unity of mankind," Albert announced, to great cheers. He very carefully emphasized, however, that this was "not a unity which breaks down the limits and levels of the peculiar characteristics of the different nations of the earth, but rather a unity, the result and product of those very national varieties and antagonistic qualities" ("Exhibition of All Nations" 1850).

Albert and most of the other organizers never intended integration to be egalitarian. Everyone joined the common cause by submitting to the social, economic, and political role assigned to them by Providence. The exhibition would, he confidently believed, move men to discover the

Figure 10.1. Illustration of Prince Albert delivering an address during the opening ceremony of the Great Exhibition of 1851 on May 1, 1851. Reprinted from "Inauguration of the Great Exhibition Building, by Her Majesty, May 1, 1851," *Illustrated London News* 18 (1851): 351.

providential order without question or partiality. Albert's rationale deftly merged the dictates of religion, the precepts of political economy, and the philosophical authority of science:

> The great principle of division of labour, which may be called the mov-ing power of civilisation, is being extended to all branches of science, industry, and art. . . . The products of all quarters of the globe are placed at our disposal, and we have only to choose which is best and the cheap-est for our purposes, and the powers of production are intrusted to the stimulus of competition and capital. So man is approaching a more complete fulfilment of that great and sacred mission which he has to perform in this world. His reason being created after the image of God, he has to use it to discover the laws by which the Almighty governs

His creation, and, by making these laws his standard of action, to conquer nature to his use; himself a divine instrument. ("Exhibition of All
Nations" 1850)

He predicted that the exhibition would "give us a true test and a living
picture of the point of development at which the whole of mankind has
arrived in this great task, and a new starting-point from which all nations
will be able to direct their further exertions" ("Exhibition of All Nations"
1850).

Such grandiose ideas acquired an air of plausibility, even inevitability,
in an age accustomed to the fabulous being made real. As the *Illustrated
London News* reported, "the splendid achievements of modern science . . .
accustom the popular mind to the reception of great ideas" ("American
'Notions'" 1851). The "great idea" of the exhibition might have been un-
paralleled in its scale and scope, but its goals and organization were immediately comprehensible in an age where politics, economics, science, technology, communication, empire, and entertainment were all a maelstrom
of "universal activity, . . . fiery energy, and . . . indomitable aspirations"
([Hunt and Wyatt] 1851, 1198).

The large number of eminent men of science who dedicated themselves
to planning, running, and promoting the exhibition illustrates how thoroughly the British scientific community embraced and validated Albert's
vision. Eleven of twenty-four royal commissioners, appointed in January
of 1850 to oversee the ambitious project, were Fellows of the Royal Society
(FRS), including its president, Lord Rosse, and the president of the Geological Society, Charles Lyell. Three more commissioners would be elected
subsequently to the Royal Society's fellowship. Two had sons who would
be. Most of the crucial day-to-day organizational and promotional work
fell to two special commissioners, the civil engineer Lt. Col. J. A. Lloyd and
the chemist Lyon Playfair, both FRS. The long list of eminent British men
of science who served the exhibition officially in some capacity includes
Thomas Anderson, David Thomas Ansted, Thomas Bell, David Brewster,
Henry De la Beche, Augustus De Morgan, Michael Faraday, Edward Forbes,
James Glaisher, John Edward Gray, Arthur Henfrey, John Herschel, Joseph
Hooker, William Hooker, Edwin Lankester, John Lindley, Richard Owen,
John Forbes Royle, Edward Solly, and Nathaniel Wallich.

The call to join a grand project of social and intellectual cooperation
corresponded well with existing scientific practice. Few, if any, leading
Victorian men of science could have made sufficient headway in their pursuits without collaboration from the thousands of men and women who

collected and donated specimens and data. In this vein, organizers of the Great Exhibition of 1851 aspired to do more than diffuse their elite knowledge and social values to a general, and generally untutored, audience; the driving ambition was not merely to educate and improve but to enlist all segments of society in a common and sustained scientific, artistic, and industrial enterprise. Men of science, in this vision, did not form a distinct class but rather occupied an elevated rank within a finely graded social and intellectual hierarchy that mirrored the larger social hierarchy of Victorian Britain (Cannadine 1999; Alborn 1996). The Crystal Palace was meant to promote social interaction not simply as a good in itself but as a prerequisite for its scientific aspirations.

When Albert organized a post-exhibition lecture series at the Society of Arts to take stock of the exhibition, sixteen of the twenty-four talks were delivered by well-known men of science. The prince consort clearly intended to stamp the exhibition indelibly with the intellectual authority of science. He chose William Whewell, the Cambridge philosopher-of-all-trades, to deliver the first address in November 1851. Unlike so many of his friends and colleagues, Whewell entered the Crystal Palace as "a mere spectator" (Whewell 1852, 3–4). But he did provide the most coherent and systematic elaboration of the exhibition's underlying philosophical rationale.

From his base at the University of Cambridge, where he served as professor of moral philosophy and master of Trinity College, Whewell aspired to provide a comprehensive and conclusive theory of the scientific method. Two works in particular among his enormous output embodied this ambition: *The History of the Inductive Sciences*, published in 1837, and *The Philosophy of the Inductive Sciences*, published in 1840. Together they provided an extended discussion of how, in Whewell's view, the *scientist* (a term he coined) uncovered the "true bond of unity by which the phenomena are held together" (Whewell 1840, 2:211). Few of Whewell's contemporaries could have expounded on the "General Bearings of the Great Exhibition" with anything matching his philosophical authority and scientific gravitas (Yeo 1993; Edwards 2001).

Whewell brought the production of artistic and industrial excellence directly within the scope of his *History* and *Philosophy*: "To discover the laws of operative nature in material productions, whether formed by man or brought into being by Nature herself, is the work of a science, and is indeed what we more especially term Science." He conspicuously emphasized acts of comprehensive visual perception. The exhibition's organizers were praised as men of "wide views" who took a "lofty, and comprehensive, and hopeful view." The value of the exhibition itself, Whewell asserted,

was obvious "at a glance." The scientific heart of the exhibition experience was its facility in allowing the intelligent spectator to place the entire
world of human industry in the mind's eye by examining it with the bodily
eye inside the Crystal Palace (Whewell 1852, 3, 7–10).

Whewell believed that "travel" within the Crystal Palace was scientifically preferable to literally traversing the globe. An elaborate parable explained this position. He conjured up "an intelligent spectator" who travels
the entire world to study the totality of human industry. As he wanders
the globe, this spectator experiences a fantastic diversity of human works,
from the handicraft of the Chinese artisan to the enormous output of the
European workshop. He descends into mines as far removed as Mexico and
Norway. He marks the progress of the Western surge of art and industry
across the North American continent (Whewell 1852, 10–11).

Yet conceiving such a voyage required imagining away innumerable
practical difficulties and endowing the mythical scientific traveler with
the highest gifts of discernment, judgment, and vigor. But even if such a
heroic traveler and such a fanciful voyage were plausible, Whewell insisted
that the resulting survey would still be inherently flawed: "How far must
it be from a *simultaneous* view of the condition of the whole globe as to
material arts!" There is, Whewell insisted, a truer way of knowing:

> And now let us, in the license of epical imagination, suppose such an
> Ulysses—much-seeing, much-wandering, much-enduring—to come to
> some island of Calypso, some well-inhabited city, under the rule of pow-
> erful and benignant, but plainly, he must believe, superhuman influ-
> ences, and there to find that image of the world and its arts, which he had
> vainly tried to build up in his mind, exhibited before his bodily eye in
> a vast crystal frame. . . . And yet, in making such a supposition, have we
> not been exactly describing that which we have seen within these few
> months? (Whewell 1852, 12–13)

The "magic spectacle" of the Crystal Palace allowed for the creation of
truly scientific knowledge because it allowed the observer to transcend the
mere accumulation of episodic observations (Whewell 1852, 11–13).

Scientific knowledge thus demanded the creation of a site that did nothing less than annihilate the temporal and spatial distances that limited
human perception. The Crystal Palace functioned as a utopia outside of
space and time. According to Whewell, "by annihilating the space which
separates different nations, we produce a spectacle in which is also an-
nihilated the time which separates one stage of a nation's progress from

another" (1852, 14). As a result, Whewell concluded, a man of scientific spirit, with his "scrutinizing eye and judicial mind" unshackled from the limitations of particularized space, could analyze the material products of the entire world philosophically and arrive at scientific insight into the principles governing industrial pursuits (Whewell 1852, 13–15).

The Crystal Palace was intended not only to collapse physical and temporal distances but also to heal disruptive social rivalry. In a world of limited perception, even observers with the most honest intentions could act on little beyond self-love and self-interest because they could not see reliably beyond themselves. The utopian vistas inside the Crystal Palace, by contrast, allowed visitors to act in the interest of a world they could now perceive in totality. Whewell looked forward confidently to a future where goodwill and mutual respect would be cemented worldwide by people being able to say, "we were students together at the Great University in 1851" (Whewell 1852, 34). The exhibition would not simply create fellowship, however, but also, crucially, allow it to operate effectively. In this way, division of labor could foster human unity rather than discord and dislocation. As Steve Edwards observes, "Whewell's eye . . . was one guarantor of the supposed objectivity and social neutrality of [the exhibition's] structure of representation" (Edwards 2001, 46).

This integration was not only morally beneficial but also essential for the functioning of the exhibition as a site of scientific investigation. A single spectator would have had to walk between twenty and thirty miles to pass by all exhibits ([Askrill] 1851, 5). The "magic spectacle" of the Crystal Palace, even in its most idealized form, could never truly collapse time and space without cooperative social effort. The exhibition experience depended not merely on bringing objects together to be seen but also on bringing people together to see them. There was nothing pathbreaking about this aspiration. Richard Yeo and Timothy L. Alborn show that strategies for a division of labor that included an observant public had long been fundamental to scientific reform programs pushed by the likes of Whewell and Herschel in the decades before the exhibition (Yeo 1993; Alborn 1996). In a similar vein, Helen Small demonstrates that a desire to create a forum in which all classes united behind a common sensibility also pervaded mid-Victorian literature, particularly the writing of Charles Dickens (Small 1996).

Henry Cole's reflections on the international consequences of the exhibition, delivered in December 1852, wrapped up Prince Albert's lecture series. The talk provided crucial insight into the place of science in this unprecedented endeavor. He claimed that it had been the first ever "Parliament of Art, Science, and Commerce" in a world hitherto dominated by politicians, lawyers, and soldiers. He warmly endorsed his friend Playfair's much-admired, if admittedly rudimentary, classification scheme for exhibits in the Crystal Palace. Cole believed that this scheme would develop into a permanent scientific classification for all materials, instruments, and productions of human art and industry. He enjoyed amateur botany and maintained friendly ties with William Hooker, the eminent botanist and director of Kew Gardens. Cole obviously hoped that this nascent industrial systematics would provide the foundation for a new science of industry in exactly the same way that botanical systematics underpinned the science of plants. But, in Cole's view, "science appears to me only one of many ingredients which are necessary for the prosecution of successful industry. The idea of an International Exhibition was one of abstract science; . . . it was in itself of no use without other favourable circumstances enabling it to be realised." When the inspiration for an international exhibition originated in France, it languished as mere "philosophical theory," said Cole, smothered by the protectionist fears and jealousies of French industry. England, on the other hand, was prepared "by the cosmopolitan character of its people" and its embrace of free trade to transform the idea into tangible reality (Cole 1853, 421–28, 441–42, 445). Science, then, was an indispensable component to the exhibition, but it acquired its practical significance only when its abstract ideas became allied to the tangible practicalities of art and commerce.

The broad enthusiasm in the press for the Great Exhibition of 1851 in the nearly three years between Albert's and Cole's addresses illustrated the wide, if very far from universal, resonance of the event's modernizing agenda of temperate religious piety, liberal political economics, philosophical science, and social integration. Millions of people attended the exhibition, which supporters of the exhibition understandably interpreted as an endorsement of their ideals. The story is more complicated, of course. Not every attendee sympathized with, or even understood, the event's official motivations, and as Jeffrey Auerbach points out, many possible meanings emerged from the bottom up rather than emanating from the top down (Auerbach 1999, 2, 88). Not surprisingly, the reality of the exhibition did not live up to the rhetoric—particularly the more utopian flights of fancy which Albert and Whewell were far from alone in indulging. To understand both the provenance and the popularity of the exhibition's guiding ideals, and why they failed to achieve the desired practical results, the best place to start is not inside the Crystal Palace. Instead, we need first to reach the exhibition as the Victorians did—through the noisy, bustling,

tangled streets of London. The ideals of rational education and exact observation, so dear to Albert, Cole, and Whewell, lost their pristine quality in
this complex and diverse cultural landscape.

"A World of Its Own"

"The Crystal Palace is the centre of a system. Though being confessedly
the most wondrous 'sight' in the world, it is, nevertheless, surrounded on
every side with objects of interest to the stranger hardly inferior to itself.
London is a marvel among marvels" ("Guidebooks, etc." 1851, 654). This
conclusion from the *Athenaeum* captured the prevalent belief that the
Great Exhibition of 1851 was tied inexorably to the physical and cultural
geography of London (see figure 10.2). The exhibition site in Hyde Park had
been carefully chosen to integrate it into the cultural and economic space
of the British capital: it was convenient to "attractive parts of the metropolis" but still sufficiently removed to avoid interrupting its daily business
(Tallis 1851, 1:6–7).

As the German botanist Matthias Schleiden asked before the idea of
the exhibition originated, "could we choose a better starting-point [than
London], if we would, for any purpose whatever, make a survey of the
earth?" (Schleiden 1848, 228). By asking this rhetorical question, Schleiden
participated in the broader cultural construction of London as a place of
infinite variety and vast reach. Ironically, perhaps, these mid-nineteenth-
century constructions located a coherent meaning for London precisely in
the seemingly boundless commotion of people, capital, and goods which
characterized the imperial capital. The metropolis transcended local limitations to become, according to the Scottish writer John Fisher Murray, "a
great world—a world of its own—a 'great globe itself'" (Murray 1843, 1:20).
In a similar vein, a writer in *Fraser's* insisted that "London is not London
alone, it is the central point of the civilised universe, towards which rays
converge from every zone and meridian. London is a part of England, of
Europe, of America, of Africa, and of Asia. Beneath our feet is the focus
within which are concentrated the hopes, fears, rivalries, and jealousies of
all other nations of the globe" ("London from the Crow's Nest" 1849, 58).
Peter Cunningham, in an entry for John Murray's famous and influential
guidebook series, insisted that in London "everything is brought to a focus" (Cunningham 1851, xii). London's glory and authority rested heavily
on its unique assemblage of far-reaching cultural, social, and economic
activities. Innumerable printed works, from articles, to histories, to guidebooks, to novels, to architectural handbooks, reconceptualized London's

Figure 10.2. Detail of a map of London in 1851. From Cunningham 1851, © The British Library Board, all rights reserved [10028.c.21], reprinted by permission.

disorganized physical space by transforming it into a "universal empire of bricks and mortar," "a common centre—a reservoir—a pivot," and "the metropolis of the civilised world" (Murray 1843, 1:67; "Anecdotes of London" 1850, 361; Clarke 1851a, 5).

Publishers churned out a phenomenal number of guidebooks to London in 1851, "presumed to be suited," as a reviewer in the *Athenaeum* noted, to all the "tastes, intelligences, and pockets" of the unprecedented number of tourists visiting the city ("Guidebooks, etc." 1851, 655). These guidebooks and advice books became important tools for visitors to organize and interpret their experiences in London's museums and galleries, and their popularity illustrates how print culture could mediate an individual's response to the city as a whole (Fyfe, this volume). Henry Green Clarke alone published four London guides in 1851. The one-shilling, two-hundred-page *London: What to See, and How to See It* was already part of a series, which included a highly successfully sixpenny guide to the British Museum (Clarke 1851c). It was rereleased with an extra chapter on the Great Exhibition of 1851 under the title, *London In All Its Glory: Or, How to Enjoy London During the Great Exhibition* (Clarke 1851b). For four shillings and sixpence, the clothbound, octavo-size *London as It is To-Day: Where to Go, and What to See, During the Great Exhibition* offered more than twice as many pages, dozens of additional woodcut illustrations, and numerous colored engravings (Clarke 1851a). Finally, *Londres et ce qu'on y voit* (*London and What to See There*), priced at a shilling and a half, was one of many guides the publisher marketed to French visitors (Clarke 1851d).

When a writer in *Fraser's* concluded that the Crystal Palace and London were exactly alike in their labyrinthine magnitude, he drew upon the familiar understanding of the capital to locate the meaning and significance of the exhibition ("Memorabilia of the Exhibition Season" 1851, 123–24). In order to explore how the exhibition experience was embedded in the physical and conceptual experience of London, we will follow a route through London similar to one suggested in *London as It is To-Day* (Clarke 1851a, 7), starting at the Custom House on the bank of the Thames near the Tower of London and continuing to Piccadilly, which reaches west to Hyde Park Corner, just a short distance from the Crystal Palace.

At the Custom House, a small army of civil servants administered the foreign trade passing through the world's busiest port. Trade occupied a commanding position in London's dynamic system of commerce, finance, and finishing-trade manufacturing (Inwood 1998). When Thomas Howell examined the customs records from a single day (September 17, 1849), he found activity of staggering magnitude. In twenty-four hours, London accommodated the arrival of no fewer than 121 ships with a total registered tonnage of 29,699. His resulting lecture to the Clapham Athenaeum in 1850 celebrated not only the volume but also the "boundless variety" of the inbound foreign goods. He emphasized that science was integral to this economic enterprise at every level, from astronomy for navigation to chemistry and physics for manufacturing (Howell 1850, 7–9, 16). Science fed the ever-more-voracious appetite of industry by expanding dramatically the production and discovery of useful plant, animal, and mineral products. British industrialization could not have existed without a global network of mines, plantations, and farms, which, as British naturalists never tired of trumpeting, relied in turn on the technical expertise of geologists, zoologists, and botanists.

A quarter mile from the Custom House, the tourist passed the terminus of the London and Blackwall Railway, located on Fenchurch Street. The station was not simply an important hub in Britain's burgeoning system of rail transport but also a node in the revolutionary communications network created by the electrical telegraph. By 1851, the telegraph linked two hundred principal British towns. Britain and France were connected by the world's first successful undersea cable that year (Hunt 1997). The flow of goods monitored at the Custom House was connected intimately and inextricably with the flow of capital managed a short distance westward at two of the world's most powerful financial institutions, the Bank of England and the Royal Exchange.

If the Custom House and the railway symbolized economic and technological progress, and the Bank of England and the Royal Exchange embodied unrivaled financial power, then Christopher Wren's magnificent St. Paul's Cathedral represented an anchor of cultural and spiritual strength. Exhibition supporters repeatedly invoked this strength, from Albert's invocations of the Almighty to Richard Owen's assurance that humans are the divinely ordained masters of inferior nature ("Exhibition of All Nations" 1850; Owen 1852, 98). The cathedral also offered a uniquely commanding perspective of the metropolis for any visitor willing to climb the 616 steps to its ball and pay one shilling and sixpence for the privilege. *London as It is To-Day* explained that the arduous ascent would reward the visitor with a view of "the wide horizon, crowded as it is with men and their dwellings, [which] forms a panorama of industry and life more astonishing than could be gazed upon from any other point in the universe" (Clarke 1851a, 5). It is no coincidence that similar language was used repeatedly to describe the vistas inside the Crystal Palace.

Three hundred yards west of St. Paul's was Fleet Street, one of the capital's most ancient, important, and congested thoroughfares. It had long been a hub of intellectual activity, thanks to its large number of printers, stationers, booksellers, taverns, and coffeehouses. The offices of the *Economist*, one of the most emblematic periodicals of 1851, were located here. The exhibition's guiding faith in the power of the unfettered global exchange of goods and ideas to bring peace and prosperity was the *Economist*'s editorial policy made manifest. James Wilson founded the journal in 1843 to champion the broad principles of internationalism and free trade behind the contemporary movement to repeal the protectionist Corn Laws. In a remarkable essay published two weeks after the opening of the exhibition, Wilson's periodical embraced another defining characteristic of the exhibition, the faith in the transforming power of vision. "Speaking to the Eye" celebrated the recent advent of relatively inexpensive illustrated periodicals as a key expression of the "dawning and important" intellectual and technological changes revolutionizing society. Words, whether spoken or written, conveyed only what already existed in the mind. "But pictorial representation may at once convey totally different and totally new ideas to the mind. The artist speaks a universal language. . . . Pictures, then, . . . may make every one participate in the gathered knowledge of all. . . . The Great Exhibition itself, which is a representation to the eye, is a part of

the same progress. It is performing the office of a large illustrated newspaper. It is the history of modern art and invention taught by their actual
products. . . . It speaks all tongues" ("Speaking to the Eye" 1851). The essay
heaped praise particularly on the defining illustrated newspaper of the age,
the *Illustrated London News*. Its offices were nearby at 198 Strand.

An ambitious young printer and newsagent named Herbert Ingram
founded the *Illustrated London News* in 1842. In a long preface to the
first volume, Frederick Bayley, the editor, insisted with unabashed self-
congratulation that the new publication pursued "justice and the good of
Society, and above all, clasping Literature and Art together in the firm embrace of Mind" ([Bayley] 1842a). With such bravado ambitions, it was no
wonder that the newspaper's advent "*must* mark an epoch—give wealth to
Literature and stories to History, and put, as it were mile-stones upon the
travelled road of time" ([Bayley] 1842a). The lead article in the first issue
vowed grandiosely to follow its "great experiment . . . to keep continually
before the eye of the world a living and moving panorama of all its actions
and influences" ([Bayley] 1842b). The venture succeeded spectacularly. In
1851 circulation increased dramatically, thanks to extensive and enthusiastic coverage of the exhibition. A typical preexhibition print run was
seventy thousand to one hundred thousand copies; nearly two hundred
thousand copies were published for the issue covering the opening of the
Crystal Palace. The publishers accommodated increased demand with an
impressively large and fast steam-powered printing machine, put to work
on public display inside the Crystal Palace (Bailey 1996).

This printing machine was one of myriad symbols of power (political,
economic, cultural, scientific, and technological) displayed and celebrated
in the Crystal Palace. A visitor would have recognized immediately that
these symbols reflected—or, more accurately, idealized—life in the imperial capital. Exhibition organizers succeeded in appropriating these symbols but were much less successful in dictating how they were received.
The reason becomes clear in Leicester Square, the heart of London's entertainment scene, less than a half mile from the offices of the *Illustrated
London News*. Prince Albert had promoted the square as a site for the exhibition until it became apparent that the scale of the event required a
more spacious environment (Altick 1978, 229). Here Henry Mayhew found
that "new amusements were daily springing into existence, or old ones being revived. . . . All was bustle, life, confusion, and amazement" (Mayhew
1851, 132–33).

In Leicester Square, high-minded institutions of education and research
existed beside establishments marketing high-spirited frivolity. Others

promised an enticing mixture of self-improvement and entertainment, places where philosophy became spectacular and novelty reigned. Balancing commercial appeal with didactic value proved complicated, however. The Adelaide Gallery of Practical Science, located just off the West Strand, aspired at its founding in 1832 to educate the public in scientific and technological principles. As Iwan Rhys Morus points out, its proprietors did not consider edification and entertainment to be competing values and so strove to blend them into a popular and profitable whole (Morus, this volume). High-minded contemporaries did not always approve. One guidebook noted ruefully that "science alone, however, it was soon discovered, had not of itself attraction sufficient for the multitude" (Mogg 1846, 117). The Adelaide Gallery, such critics sniffed, pandered increasingly to vulgarity to stay afloat financially (Mogg 1846, 117; see also Altick 1978, 377–82).

The exhibition shared its promise of comprehensive vision and vicarious travel with the panorama, which had occupied a central place in the London entertainment scene for more than fifty years. In the late eighteenth century, Robert Barker perfected and patented a technique for representing a 360-degree scene on the cylindrical interior surface of a rotunda. This technique allowed a spectator to stand on a central circular gallery and partake in the illusion of actually seeing the locality represented by the painting. Barker, at the suggestion of a friend, christened this spectacle a "panorama." The neologism quickly acquired its more general meaning of any commanding, unbroken and comprehensive view or survey.[1] William Thomas Brande's authoritative *Dictionary of Science, Literature, and Art* noted that "when a painting of this kind is well executed, its truth is such as to produce a complete illusion. No other method of representing objects is so well calculated to give an exact idea of the general aspect and appearance of a country as seen all round from a given point" (Brande 1842, 877; see also Altick 1978; Hyde 1988; Comment 1999; Aguirre 2002).

In 1794, Barker opened a thriving business just off of Leicester Square. The enterprise continued to flourish after his onetime assistant Robert Burford became the sole proprietor in 1826. During the exhibition season, Burford enticed customers with dramatic new panoramas of Jerusalem and Niagara Falls. Barker's original idea spawned numerous imitators and competitors after his patent expired in 1801. In 1851, London audiences could marvel at panoramas of California, Ireland, Australia, Switzerland, Constantinople, Calcutta, and New Zealand. The *Illustrated London News* treated subscribers regularly to foldout engraved panoramas, starting with one of London in 1843 (Hyde 1988, 38, 184). The moving panorama promoted a different type of illusion. An enormous canvas (in some cases, thirty feet high and hundreds of yards long) was gradually rolled past a stationary audience, creating the semblance of motion. Spectators in 1851 could "navigate" the Mississippi, the Nile, the Dardanelles, and the Ganges, and "travel" the overland route to India.

London as it is To-Day was particularly enthusiastic about London's many panoramas because they allowed people to experience the wider world with "little fatigue or inconvenience, and what is equally important in these times of economy, at little expense." The New Zealand panorama was judged especially valuable for allowing emigrants "to catch a distant, faint anticipation of their land of promise." The author even suggested that foreigners "may visit London to obtain acquaintance with the peculiarities and impress of far distant countries" (Clarke 1851a, 267–78, 278, 281). The Illustrated London News further blurred the boundaries between the original and its facsimile by offering readers engravings of exotic and picturesque locales—such as the Taj Mahal and the Pool of Siloam in Jerusalem—drawn not from the actual sites but from their depictions in London dioramas ("Diorama" 1851).

James Wyld produced one of the exhibition season's most iconic variations on Barker's original idea with his Great Globe in Leicester Square (see figure 10.3). Wyld projected the earth's surface on the sixty-foot globe's concave interior. The attraction's guidebook explained that this allowed the spectator to "obtain a more enlarged or comprehensive survey of the surface of the globe" than possible with traditional forms of geographic representation. For a shilling, Wyld offered spectators nothing less than the chance "to survey the multifarious aspects of the globe with the eye of philosophy" (Wyld 1851, xix–xx, 1, 6). The London correspondent for Chambers' Edinburgh Journal enthused that the Great Globe was "geography-made-easy on a large scale," a place where "concrete ideas . . . may . . . be obtained in place of those abstract notions concerning the earth and its surface which mere reading often creates" ("Things Talked of In London" 1851, 14; "Great Globe Itself" 1851, 119). The Crystal Palace: A Little Book for Little Boys promised its young readers that a visit to Wyld's attraction would be "a year or two of the study of Geography mastered in an hour" (Crystal Palace 1851, 40–42). W. H. Smyth, the president of the Royal Geographical Society, heartily endorsed the Great Globe as "well worthy of . . . the spirit of the age" (quoted in "Mr. Wyld's Model" 1851, 512).

Alexander von Humboldt proposed in the second volume of Cosmos that all large cities should feature panoramas depicting typical landscapes from various latitudes and elevations. "The conception of the natural unity and the feeling of the harmonious accord pervading the universe can not

Figure 10.3. Illustration of Wyld's Great Globe in Leicester Square. Reprinted
from "Mr. Wyld's Model of the Earth" 1851, 511.

fail to increase in vividness among men, in proportion as the means are
multiplied by which the phenomena of nature may be more characteristically and visibly manifested," the illustrious German naturalist insisted
(Humboldt [1849] 1997, 98). This faith in panoramas accorded well with a
famously cosmopolitan naturalist who, in Schleiden's words, transformed
his voracious appetite for data into foundational science by "comprehending the whole earth in one intelligent glance" (Schleiden 1848, 239).
Humboldt's educational vision demanded an inclusive series of panoramic
scenes devoted to representative landscapes, from the picturesque to the
monotonous. This series would not concentrate exclusively, or even primarily, on spectacular landscapes, any more than a scientifically authoritative collection of minerals would contain only beautiful specimens or a
research herbarium would contain only horticulturally appealing plants.
In Humboldt's scheme, a panoramic scene would be valuable only if the
viewer could generalize the experience to other similar landscapes. Re-
vealingly, this educational vision existed in tension with commercial im-
peratives. Showmen boasted about their panoramas' educational value, but
they also realized how they made their money. London panoramas served
a steady diet of exclusive, exotic scenes and never bothered with the more

mundane "characteristic delineation of nature" that Humboldt proposed. Men of science might seek to observe the physiognomy of nature; most customers of panoramas wanted to marvel at the evocative, the newsworthy, and the extraordinary—or, to put it another way, at the uncharacteristic.

Humboldt's ideal called for the creation of orderly, durable, and cumulative educational experiences, but proprietors who depended upon the commercial marketplace could not afford to take such a long and methodical view. They needed a steady stream of revenue to survive, which meant that they had to have a much different relationship to science than those government-supported institutions whose success was defined by long-term service to the nation. We can perhaps best understand the difference by looking at two establishments a short walk from Wyld's Globe: the Egyptian Hall and the Museum of Practical Geology, which were neighbors on Piccadilly.

The Egyptian Hall had been one of the capital's most popular attractions since its opening in 1812. The name derived from its eclectic facade, nominally inspired by the great temple of Hat-hor at Dendra. It thrived for decades by offering novelty-hungry audiences a kaleidoscopic procession of displays and shows. The Egyptian Hall's seemingly endless diversity of curiosities led John Timbs to dub it an "Ark of Exhibitions" (Timbs 1867, 320). It accommodated everything from live orangutans to exotic cheese. Featured among the countless shows in the decade before the Great Exhibition were a speaking automaton, a machine for composing hexameter Latin verse, models of ancient and modern Jerusalem, and nine performing Ojibwa Indians in native costume. The Egyptian Hall diversified into enormous moving panoramas at the end of the 1840s. A compression of a 1,720-mile voyage up the Nile onto 800 feet of canvas was a popular attraction in 1851 (Altick 1978, 235–52; Noon 2003, 15–16, 71–73, 90–91; Timbs 1867, 319–20).

Prince Albert opened the Museum of Practical Geology in May 1851 in a ceremony attended by top government officials, members of the nobility, the leaders of Britain's scientific institutions, and other prominent literary and scientific gentlemen. This museum replaced the old Museum of Economic Geology, which had occupied obscure and increasingly cramped quarters in Craig's Court, Charing Cross. The newly christened museum "has now come forth into our most public thoroughfare," a correspondent for *Fraser's* reported, "and sits enthroned in its palace, to receive the admiration and abide the criticism of the public" ("Visit" 1851, 618). The head of the museum was no impresario but Henry De la Beche, the director of the Geological Survey of Great Britain. He personally chose the stone

Figure 10.4. Illustration of the Hall of Marbles at the Museum of Practical Geology in London. Reprinted from "Opening of the Museum of Practical Geology— The Great Hall," *Illustrated London News* 18 (1851): 446.

for the museum's resolutely respectable Renaissance facade. The museum literally turned its back on the hodge-podge commercial extravaganzas of the neighboring Egyptian Hall. The most singular feature of its Piccadilly facade was the absence of any doorway; the entrance was located on the building's more reputable Jermyn Street frontage (Yanni 1999, 51–61; Forgan 1999).

De la Beche's museum showcased the exactingly technical science exemplified by the Geological Survey. The visitor entered through the Hall of Marbles (see figure 10.4) and climbed a grand staircase to the principal exhibition room. Inside this massive chamber (ninety-five feet long by fifty-five feet wide), tightly packed specimens from all reaches of the globe ringed the visitor on all sides, rising steeply from the open expanse of the main floor through two mezzanine galleries to the thirty-two-foot-high ceiling. Most of the specimens were donated, illustrating the reliance of the museum's gentlemanly specialists on collaborators from all segments of society. At ground level, viewers could track the progress of a mineral from extraction to manufactured product. Climbing the stairs to the lower gallery, they could traverse geological history from the Cambrian to the Permian periods of the Paleozoic era. The upper gallery continued the historical progression through to the Oligocene epoch. By following a long spiral, spectators could trace the entire development of the earth and its improvements by human hands (Yanni 1999, 51–61; Forgan 1999; Clarke 1851a, 236–37). Unsurprisingly, given the significant overlap between exhibition officials and museum staff, the Crystal Palace aspired to provide its visitors with a similarly universal voyage through time and space.

Superficially, the Museum of Practical Geology had little in common with the Egyptian Hall. De la Beche strove to bring the entire range of geological phenomena before the attention of the public. While his museum did not entirely eschew the exotic or the extraordinary, the displays focused on the systematic arrangement of a complete series of typical British minerals, ores, and fossils. As Peter Cunningham explained, the museum gave "a proper and comprehensive view of the general subject" ([Cunningham] [1860] 2003, 179). This thoughtfully organized arrangement contrasted with the notoriously haphazard and ever-changing roster of exhibits at the Egyptian Hall.

The difference resulted from the fact that each institution submitted to very different forms of market discipline. The Museum of Practical Geology was funded by the government and housed the Geological Survey, the School of Mines, and the Mining Record Office. Playfair, one of the Great Exhibition's special commissioners, served the museum as its chemist. The establishment conformed to the standards of usefulness, precision, thoroughness, and rigor expected of Britain's increasingly technocratic science. De la Beche's establishment operated in the marketplace, but here the market it answered to was Britain's vast imperial economy, with its heavy reliance on the extraction, industrial application, and transportation of minerals and ores. The museum's existence depended on its ability to convince government paymasters that it provided useful long-term service to the imperial state. It could not afford to remain empty of visitors, but neither was it driven to maximize attendance. In the 1870s, Joseph Hooker, the director of Kew Gardens, a government-funded scientific institution with a very similar mission, resolutely opposed a vocal and well-organized campaign for earlier opening hours (Desmond 1995, 235–38).

The Egyptian Hall, on the other hand, had to make its way by coaxing shillings from consumers spoiled for choice in a crowded and competitive entertainment market. Not everything here achieved popular success. Tom Thumb, the famous American dwarf, routinely netted a gate of 125 pounds a day in 1844, while the simultaneous exhibition of Benjamin Robert Haydon's critically acclaimed paintings scarcely attracted a dozen viewers a week (Timbs 1867, 320). The enduring merit of Haydon's paintings mattered little in a context where success depended upon generating income on a day-to-day basis.

The significant differences between the Museum of Practical Geology and the Egyptian Hall should not obscure an equally profound similarity. Both attempted to transport spectators from their literal physical location in London's West End to places distant in both space and time. The hall's moving panorama of the Nile created the illusion of moving through space; the museum's winding display of geological history created the illusion of moving through time. In both cases, the representational strategy was largely the same, even if the applications and relationship to the marketplace were different. Just as Wyld's Globe allowed the spectator to stand at the physical center of the earth, the Museum of Practical Geology allowed its visitors to stand at the temporal center of the earth's history.

This illusion was absolutely fundamental to the Great Exhibition of 1851. Audiences inside the Crystal Palace had a choice, however. They could travel the world in the disciplined scientific spirit of the Museum of Practical Geology and unite sight with reason. Or they could seek the outlandish, unique, gargantuan, and bizarre as paying customers had been doing for nearly forty years at the Egyptian Hall.

London embraced and touched all parts of the globe with its economic activity, technological innovation, spiritual strength, intellectual attainments, educational opportunities, and entertainment spectaculars; at least that was the widespread notion. The Great Exhibition was in essence an organized intensification of the metropolis's boisterous cosmopolitanism. If London's "vastness is a wonder," then "one of the distinguishing characteristics of the Great Exhibition was its vast comprehensiveness" (Clarke 1851a, 7; Tallis 1851, 1:207). This vast comprehensiveness made immediate sense to any visitor even vaguely familiar with the city that thought of itself, and not entirely without reason, as the central point of the civilized universe. Thriving London institutions such as the Egyptian Hall and the Museum of Practical Geology provided templates not only for the organization of the exhibition but also for its public reception.

"The Crystal Focus of the World"

"There was yesterday witnessed a sight the like of which has never happened before, and which, in the nature of things, can never be repeated," the *Times* declared of the Great Exhibition's inaugural ceremony on May 1, 1851. "There were many there who were familiar with magnificent

Figure 10.5. Illustration of the exterior of the Crystal Palace. Reprinted from Tallis 1851.

spectacles; who had seen coronations, fêtes, and solemnities; but they had not seen anything to compare to this" ("London, Friday" 1851). Such euphoric reactions were rife. Mayhew's novel, *1851*, exuberantly described people in their bright Sunday attire "streaming along the road, like so many living rays, converging towards the Crystal focus of the world" (Mayhew 1851, 127). This pageant celebrated the (seemingly) spontaneous bonds of duty and affection which extended unbroken from the royal family to Mayhew's bright and cheerful working people. An illustration of the throng appears in John Tallis's three-volume *History and Description of the Crystal Palace* (Tallis 1851; see figure 10.5).

This "sunshine of popular favour" never waned ("Close of the Exhibition" 1851, 441). The largest single-day attendance, an astonishing 109,915, was on October 7, four days before the exhibition closed (Auerbach 1999, 148). The Crystal Palace's ostensible replication of the world was key to this popularity. Timbs reported that the Crystal Palace's transept "was declared *the equator of the world in Hyde-park*" and noted that the exhibition served exactly the same function for the study of art and industry as Wyld's Globe served for the pursuit of geography (Timbs 1851b, 100; 1851a, 380–82). The *Times* compared the exhibition to the Foucault pendulum on display at the Royal Polytechnic Institution: "It is the movement of a planet, not a mere mechanical rotation, but an intelligent and moral movement, which we shall see in the Exhibition" ("London, Thursday" 1851).

Tangibly witnessing the world on display did more than uplift. It made visitors active participants in a shared experience. Mayhew insisted that the exhibition, "looked at in its true light, is . . . a huge academy for teaching men the laws of the material universe, by demonstrating the various triumphs of the useful arts over external nature" (Mayhew 1851, 158). In a similar vein, a guidebook for Yorkshire visitors exhorted "all parties, and classes, and callings" to make every reasonable sacrifice to profit from the exhibition, "to look, and wonder, and question, and *learn*—aye, here they will discover the marvellous rapidity with which instruction and knowledge is conveyed by the eye and the finger—by the organs of sight and feeling." Inside the Crystal Palace "many of the advantages of extensive travelling are brought into a focus; for a vast amount of all that is worth seeing in most parts of the world is here spread out for examination and ready comparison, highly illustrative of the science, arts, character, customs, and habits of different nations" ([Askrill] 1851, 5).

Mayhew reported that "you might wander where you pleased—to 'France'—and see the exquisite tapestry; you might step across to 'Austria' —and wonder at the carvings of the furniture; . . . beneath the crystal roof were ranged all the choicest works of the whole world" (Mayhew 1851, 137). An exhibit replicating Tunisia is illustrated in figure 10.6. Travel inside the Crystal Palace was not just spatial but also temporal. "The history of the arts of life, and the progress of mankind will be traceable there," one guidebook promised, "from the lonely cave still inhabited by the African bushman, to the crowded city, where these multitudinous objects are collected" (Clarke 1851a, 111). These ideas had become conventional wisdom long before Whewell systematized them philosophically in his November lecture.

Every visitor had a responsibility inside the Crystal Palace. A review of the exhibition's *Official Descriptive and Illustrated Catalogue* in the *Athenaeum* explained the idealized nature of this collaboration. The review's authors were Robert Hunt, a protégé of De la Beche and professor of mechanical science and keeper of mining records at the Government School of Mines, and Matthew Digby Wyatt, secretary to the exhibition's executive committee and an architect whose designs for fabrics won a Prize Medal. "Philosophers"—men like themselves, in other words—had the task of systematizing the exhibition's results to kindle future moral and material improvement. Exhibitors submitted themselves to the collec-

Figure 10.6. Illustration from an exhibit replicating Tunisia inside the Crystal Palace.
Reprinted from "Tunis Court," *Illustrated London News* 18 (1851): 494.

tive verdict of their strengths and weaknesses so that they could learn to
direct their future exertions into the most appropriate and efficient channels. Artisans and artists learned to apply the laws of grace and symmetry
to their handicrafts. Consumers learned to differentiate truly tasteful design from frivolity and affected graces, and good workmanship from poor.
True improvement required the coordination of all classes. Work inside the
"Great Meeting-House of all Denominations should not but tend to a consolidation of all interests, and consequentially to the reciprocal benefit of
all" ([Hunt and Wyatt] 1851; see also Mayhew 1851, 158). Not surprisingly,
the official "verdicts" of the exhibition invariably reinforced dominant
views about class, region, race, gender, and occupation. For example, Hunt
and Wyatt quoted at length from the *Official Catalogue*'s judgment that
Indian arts had long stagnated, but the subcontinent remained a reliable
source of raw materials.

Leading men of science devoted countless and often grueling hours to
exhibition committees and juries. The botanist Joseph Hooker's jury responsibilities in Class III (Substances Used as Food) required him to work
"like a dragon," typically twenty hours a week.[2] His fellow jurors each
dedicated approximately two hundred unpaid hours to their task; he committed even more as author of the jury report (Hooker 1852, 51). The lavish
and expensive three-volume *Official Catalogue* drew upon the expertise

of such scientific luminaries as Isambard Kingdom Brunel, Augustus De Morgan, Edward Forbes, James Glaisher, John Edward Gray, Justus Liebig, John Lindley, and Richard Owen (Ellis 1851). It aspired to provide a permanent record of the exhibition's results after it closed and the exhibits were scattered.

The confident hopes and the stirring rhetoric often obscured a more ambiguous reality. Whewell's "simultaneous view" of the world's industry required not simply diverse exhibits, which the Crystal Palace certainly had, but also a truly systematic global cross section. This universality, so rhetorically central to the exhibition experience, was never achieved. The practical reality of filling the Crystal Palace sharply constrained what could be accumulated and how. Idealized scientific collecting practice demanded the creation and accumulation of type specimens that exemplified a much larger class of related individuals. Idiosyncratic specimens were worse than irrelevant because they could mislead the observer into mistaking an individual peculiarity for a common characteristic. The considerable contingencies of amassing displays for the Great Exhibition meant that the Crystal Palace was packed with odd and atypical items that, from a strictly philosophical point of view, had no business being there. Their presence lent the Crystal Palace the sensational air of the Egyptian Hall.

The work of soliciting British exhibits fell largely to the 297 local committees, with the liaison effort overseen principally by Lloyd, Playfair, and Wyatt. The quality and breadth of British displays depended largely on uneven local conditions, since enthusiasm for the exhibition varied considerably between regions. Foreign displays were even less subject to the event's organizational principles. The pattern of exhibits from the Zollverein, a customs union among certain German states, revealed much more about the tensions and rivalries in the participating German states than it did about their level of economic development (Auerbach 1999, 70–86; Davis 1999, 108–14). The problem with the "Chinese" section was even starker. It was organized by a British Board of Trade official after the Chinese government rebuffed the invitation to participate. The large majority of Indian displays were chosen by European merchants rather than the indigenous inhabitants. Many gaping holes existed in the coverage of certain classes of exhibits. The world's mining was imperfectly represented: Saxony and Hartz, cradles of mining and home to Europe's oldest mining schools, sent none of their products (Graham 1852, 1–2). The same was true in the exhibits of vegetable products (Hooker 1852, 51; Solly 1852, 246–47).

Critical evaluations of the Great Exhibition of 1851 surfaced in the press, even in such generally enthusiastic publications as the *Illustrated*

London News. One analyst for that paper concluded a week after the close of the exhibition that it had stumbled badly in trying to take the first steps toward a universal knowledge of science and industry. The organizers of the exhibition (faced admittedly with the daunting tasks of winning over a skeptical public and encouraging contributors to participate) threw displays together promiscuously "in such a manner as to be utterly useless for scientific research." In such a state of affairs, "everything was sacrificed to show and sound" ("Exhibition and Its Management" 1851). A correspondent for the *Athenaeum* acknowledged some of these problems but took a brighter view. He confessed that many well-heeled season-ticket holders spurned detailed investigation and instead treated the Crystal Palace "first and foremost [as] a lounge and a panorama." Yet even "the most listless lounger" gained an education at the exhibition: "Consciously or unconsciously, he will receive . . . lessons which cannot be altogether without effect in after life. The apparent idler may undervalue neither the edifice nor its contents:—he may wish only to enjoy them both in his own way" ("Great Industrial Exhibition" 1851, 605).

The most telling disjunction between hype and reality involved the *Official Catalogue.* This publication was meant to be the ultimate demonstration of the exhibition's practical utility and philosophical legitimacy. Seeing, touching, hearing, tasting, and smelling were the necessary first steps to true inductive knowledge, but they were not sufficient. Philosophical observers had to reduce, codify, and publish their sense perceptions in a uniform and scientifically rigorous record as the next step to the acquisition of truly panoramic understanding of phenomena. The *Official Catalogue* was designed to do so. Hunt and Wyatt explained its purpose as a *"memoria technical."* Merchants, manufacturers, philosophers, consumers, artists, and artisans would all use it to continue the studies of physical and material harmony begun inside the Crystal Palace ([Hunt and Wyatt] 1851). The *Official Catalogue* was dedicated explicitly to bringing a naturalist's insight to the study of industry (Brain 1993, 74). It was intended to serve exactly the same purpose in its sphere that such works as Whewell's *Researches on the Tides,* Hooker's *Flora Antarctica,* or De la Beche's *Report on the Geology of Cornwall, Devon and West Somerset* played in theirs.

Judging only by the euphoric reviews that greeted the *Official Catalogue,* it succeeded magnificently. A long advertising section at the end of the *Reports of the Juries* contained (often lengthy) excerpts from two dozen newspaper notices. The praise was breathless: "a work of immense magnitude"; "the *Annus Mirabilis* 1851 will live in its pages"; "nothing scarcely could be more clear than the whole arrangement of the contents"; "it may be truly said that this is the most extraordinary work that has been ever brought forth in this or any other country"; "it is not the work of a day, a month, or a year:—it is for all time" ("Advertisement" 1852).

Yet it seems unlikely that the *Official Catalogue* or any of the other official exhibition publications sparked even minor innovation or theoretical insights. The jurors, in hindsight, were almost exclusively backward-looking; they did a poor job in identifying the innovations that would drive future economic development (Auerbach 1999, 122–23). Herschel, at least, was skeptical that the exhibition's results justified the labor. He lamented to Humboldt that his service as juror in the class of Philosophical Instruments had been "a thankless and extremely laborious duty, which occupied time which ought to have been otherwise disposed of and interfered with every other object. Thank heaven it is over!"[3] A reviewer for *Fraser's Magazine* hit closest to the truth when he dismissed the *Official Catalogue*'s "heterogeneous collection of notices" as little more than "'illustrated' rubbish." The jury reports did not impress him, either. He thought that Hooker's report on articles used as food could have served as a "useful and interesting guide" to the actual exhibits, for example, but that it was impossible to understand as "an independent treatise" ("Exhibition Jury Reports" 1852, 491–92, 497). A skeptic in the *Illustrated London News* took an even harsher line, dismissing the *Official Catalogue* as "a heavy humbug, from which no information could be obtained" ("Exhibition and Its Management" 1851). The failure of the jury reports and the *Official Catalogue* represented, in essence, the failure of Prince Albert's grandiose goal of making the exhibition "a true test and a living picture" of human industrial development.

The grand aspiration to draw together everyone from the savage to the philosopher in a common enterprise fared little better. Jury awards infrequently went to popular exhibits. Spectators flocked to see the spectacular and bizarre, not the mundane and useful that were beloved of the scientifically minded juries and the moralizing commentators of the press. A writer for the *Illustrated London News* admitted that he had to work hard to appreciate the decidedly nonornamental Canadian timber trophy ("Canadian Timber Trophy" 1851; see figure 10.7). It is unlikely that visitors less dedicated to the exhibition's organizing ideals would have expended much effort to appreciate what looked like a giant, overbuilt sawhorse, even though it was a good type specimen of a commodity with tremendous commercial and military value. The mysterious Koh-i-noor diamond did attract large and enthusiastic crowds, but the very uniqueness that made

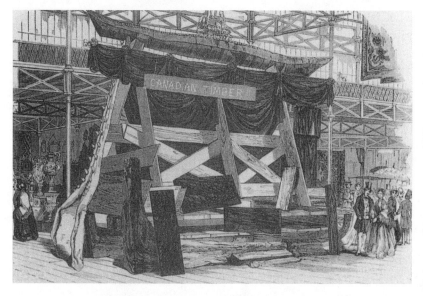

Figure 10.7. Illustration of Canadian timber exhibit in the Crystal Palace.
Reprinted from "Canadian Timber Trophy" 1851.

it popular also made it a poor scientific specimen, and so it was ignored in official publications. On the other hand, Newcastle's H. L. Patterson's display of a novel method for separating silver from lead was awarded a highly coveted Council Medal from the jury but attracted little popular acclaim (Graham 1852, 4). Indeed, when the geologist Gideon Mantell felt overwhelmed by the crush of the crowd, he ducked into an exhibit of minerals for respite (Mantell 1940, 274). Even if a visitor did earnestly approach the exhibition in the intended spirit, he or she faced a significant, perhaps even insuperable, problem. The placement of exhibits reflected practical necessity rather than the type of logical sequence central to scientific institutions like the Museum of Practical Geology (Auerbach 1999, 94–95; "Exhibition and its Management" 1851, 504).

The jury for Class I (Mining, Quarrying, Metallurgical Operations, and Mineral Products), led by De la Beche, was hindered by more than indifferent participation from the world's mining regions. Its report, written by Thomas Graham, contradicted Whewell's conviction that a collection could adequately substitute for visiting actual sites of industrial production. The value of any method of exploitation could be judged, said Graham, only by seeing firsthand the richness of the raw materials and the facility of extraction. "It is only in the works themselves, or in the mine, that this double condition of its value can with any certainty be appreciated." Displays of mining and metallurgy suffered from another inherent difficulty, Graham noted: "Its products possess immense utility, and yet offer but little brilliancy of appearance. Deprived of all that gratifies the eye by elegance of form and varieties of colour, they can scarcely neutralize these disadvantages and arrest attention." Strange forms or colossal size might capture public attention, but such attributes "are essentially fugitive, and generally barren of result." Despite the omissions and inadequacies in Class I, Graham still claimed that the Great Exhibition of 1851 formed "an epoch in the history of the industry and civilization of the world"; faith in the exhibition ideal coexisted, if somewhat awkwardly, with the realization that its results fell significantly short of its aspirations (Graham 1852, 1–2).

Mantell was one of the few who explicitly rejected the notion of a shared enterprise. He found the exhibition's effect "overpowering"—a word used repeatedly in his personal journal—and declared it "the most marvellous display the world ever beheld!" But his journal also scorned the idea that it "is or can be appreciated by the ignorant mobs who frequent it" (Mantell 1940, 267–74). Despite Mantell's dyspeptic snobbery, his private reflections captured more truth than the endless public homage to shared enterprise. Auerbach shows that the crowds tended to congregate around the spectacular exhibits rather than the more didactic and educational ones more consistent with the exhibition's stated mission (Auerbach 1999, 104–8).

In many cases, visitors no doubt sought only awe and sensual pleasure and gave little thought to their putative role in the inductive search for general laws of industrial production. In many other cases, the complexity and scale of the exhibits would have defeated even the best intentions to learn and participate. Charlotte Brontë failed to maintain the enthusiasm for the "magic" of the exhibition she had expressed so vividly to her father after her second visit. Subsequent visits were driven increasingly by a disenchanting feeling of duty rather than by either edification or enjoyment. "It is a marvellous, stirring, bewildering sight—a mixture of genii palace, and a mighty bazaar, but it is not much in my way," she admitted to a friend, just a few days after her rapturous report to her father. For Brontë, "its wonders appeal too exclusively to the eye, and rarely touch the heart or head." She believed that only those with a profound range of scientific knowledge, like David Brewster, who guided her through the Crystal Palace one morning, could rise above bewilderment and look "on objects with other eyes than mine" (quoted in Gaskell 1900, 525–26). De la Beche conceded publicly that "it was not to be expected that the mass of the thousands daily visiting the Exhibition should be capable of forming a correct judgment on [mining and metallurgical processes]." Worse, the educated classes, other than those with a direct professional interest, succeeded little better in grasping the lessons of the mineralogical exhibits (De la Beche 1852, 71–72). The Crystal Palace brought people together, but it did not integrate them in the way that its originators had expected, or at least dreamed.

Conclusion

The Great Exhibition of 1851 in London achieved popularity beyond even the most sanguine predictions. Six million people, or up to a fifth of the British population, attended (Auerbach 1999, 137). This triumph obscured the fact that the exhibition's grand scientific aspirations largely failed. The exhibition never gave birth to a science of human industrial dynamics. Exhibition organizers had to bow to practical and political contingencies in order to fill, in remarkably short order, the world's largest enclosed space with items from thousands of exhibitors worldwide. The resulting hetero- geneous and haphazard collection of industrial objects proved ill suited for scientific research. Organizers also had to accommodate popular tastes and expectations. The Crystal Palace in essence hybridized the rational education of the Museum of Practical Geology and the mass spectacle of the Egyptian Hall. Most visitors chose to enjoy the show rather than pitch into the hard work of inductive investigation; this is understandable, given the magnitude and haphazard arrangement of the displays.

The warm glow of triumph at first almost entirely obscured this truth. Organizers hoped to use the Great Exhibition as a springboard for a per- manent institution which, in the words of Richard Cobden, a leader of the free-trade movement and one of the exhibition's royal commissioners, combined "scientific accuracy with popular effect" (quoted in Auerbach 1999, 201). The Crystal Palace was dismantled and reconstructed on an even grander scale in Sydenham, a London suburb. Investors also hoped that it would generate the same robust profit at Sydenham as it had in Hyde Park. It did not. As James Secord notes, the conjunction between commercial capitalism and rational education proved short-lived. By 1855, the Crystal Palace's finances were in disarray. Justifiably anxious share- holders pushed for far-reaching changes. The Crystal Palace rededicated itself to the type of profitable mass spectacle found at the Egyptian Hall and moved away from the rational education of the Museum of Practical Geology (Secord 2004).

The exhibition itself was grounded in the ideal of science occupying an integrated, rather than an autonomous, place in British culture. In this vision, it was neither odd nor, at first glance, quixotic that the exhibition's champions wanted to create a grand experience that would align the activities of the broad public to the needs of scientific research. The practice and cultural position of science had changed dramatically in Britain during the previous half century, however, and this change in position cut fatally against the exhibition's hope for a grand cooperative enterprise. The increasing specialization and technical sophistication of science gradually but decisively widened the gulf between men of science and the lay public. The Crystal Palace attempted, and failed, to bridge this gap. Charlotte Brontë's experience is emblematic. The exhibition came alive for her intellectually only when she had Brewster at her side, patiently providing explanation; otherwise, she found the cacophony of displays overwhelming and eventually wearisome, regardless of their "grandeur" (Gaskell 1900, 523–24, 548). In the end, organizers and supporters could not dictate what people experienced. Visitors could treat the Crystal Palace largely or exclusively as an entertaining spectacle rather than a site dedicated to rational education and investigation—and perhaps they had little choice, given the significant challenges its immensity and disjointed physical organization presented for serious lay study. The exhibition's philosophical mission required patient, long-term commitment, but, like the Egyptian Hall and other commercial entertainment enterprises whose marketplace survival depended upon quick-footed responses to consumer preference, that was a luxury it never enjoyed.

NOTES

1. See the *Oxford English Dictionary* for numerous examples from the early nineteenth century of *panorama* being used in its expanded sense.

2. William Hooker to George Bentham, July 1851, Sir W. Hooker's Letters to Mr. Bentham, f. 636; Maria Hooker to Dawson Turner, 30 May 185, Hooker Family Correspondence, Royal Botanic Gardens, Kew, Archives, London.

3. John Herschel to Alexander von Humboldt, 8 November 1851 (copy), Correspondence of Sir John Herschel, 23.110, Royal Society Archives, London.

REFERENCES

"Advertisement. Official Descriptive and Illustrated Catalogue." 1852. In *Reports by the Juries on the Subjects in the Thirty Classes into which the Exhibition was Divided.* London: William Clowes and Sons.

Aguirre, Robert D. 2002. "Annihilating the Distance: Panoramas and the Conquest of Mexico, 1822–1848." *Genre* 35:25–55.

Alborn, Timothy L. 1996. "The Business of Induction: Industry and Genius in the Language of British Scientific Reform, 1820–1840." *History of Science* 34:91–121.

Altick, Richard. 1978. *The Shows of London*. Cambridge, MA: Harvard University Press.

"American 'Notions.'" 1851. *Illustrated London News* 18:543–44.

"Anecdotes of London." 1850. *Chambers' Edinburgh Journal* 14:361–64.

[Askrill, Robert]. 1851. *The Yorkshire Visitors' Guide to the Great Exhibition, and Also to the Principal Sights of London*. Leeds: Joseph Buckton.

Auerbach, Jeffrey A. 1999. *The Great Exhibition of 1851: A Nation on Display*. New Haven, CT: Yale University Press.

Bailey, Isabel. 1996. *Herbert Ingram Esq. MP, of Boston: Founder of the "Illustrated London News" 1842*. Boston, UK: Richard Kay.

[Bayley, Frederick]. 1842a. "Preface." *Illustrated London News* 1:iii–iv.

[Bayley, Frederick]. 1842b. "Our Address." *Illustrated London News* 1:1.

Bellon, Richard. 2006. "Joseph Hooker Takes a Fixed Post: Transmutation and the 'Present Unsatisfactory State of Systematic Botany,' 1844–60." *Journal of the History of Biology* 39:1–39.

Brain, Robert. 1993. *Going to the Fair: Readings in the Culture of Nineteenth-Century Exhibitions*. Cambridge: Whipple Museum of the History of Science.

Brande, William Thomas, ed. 1842. *A Dictionary of Science, Literature, and Art: Comprising the History, Description, and Scientific Principles of Every Branch of Human Knowledge; with the Derivation and Definition of All the Terms in General Use*. London: Longman, Brown, Green, and Longmans.

[Brewster, David]. 1836. "Life and Works of Baron Cuvier." *Edinburgh Review* 62:265–96.

"The Canadian Timber Trophy." 1851. *Illustrated London News* 18:597–98.

Cannadine, David. 1999. *The Rise and Fall of Class in Britain*. New York: Columbia University Press.

Clarke, Henry Green. 1851a. *London as It is To-Day: Where to Go, and What to See, During the Great Exhibition*. London: H. G. Clarke.

———. 1851b. *London In All Its Glory; or, How to Enjoy London During the Great Exhibition*. London: H. G. Clarke.

———. 1851c. *London: What to See, and How to See It*. London: H. G. Clarke.

———. 1851d. *Londres et ce qu'on y voit*. London: H. G. Clarke.

"The Close of the Exhibition." 1851. *Illustrated London News* 19:441–42.

Cole, Henry. 1853. "On the International Results of the Exhibition of 1851." In *Lectures on the Results of the Great Exhibition of 1851*, 2nd ser., 419–51. London: David Bogue.

Comment, Bernard. 1999. *The Painted Panorama*, trans. Anne-Marie Glasheen. New York: Harry N. Abrams.

The Crystal Palace: A Little Book for Little Boys. 1851. London: James Nisbet.

Cunningham, Peter. 1851. *Modern London; or, London as It Is.* London: John Murray.

[Cunningham, Peter]. [1860] 2003. *Murray's Modern London: A Visitor's Guide.* Moretonhampstead, UK: Old House Books.

Davis, John R. 1999. *The Great Exhibition.* Thrupp, UK: Sutton.

De la Beche, Henry. 1852. "Mining, Quarrying, and Metallurgical Processes and Products." In *Lectures on the Results of the Great Exhibition of 1851,* 1st ser., 37–73. London: David Bogue.

Desmond, Ray. 1995. *Kew: The History of the Royal Botanic Gardens.* London: Harvill Press with the Royal Botanic Gardens, Kew.

"Diorama of Jerusalem and the Holy Land, Hyde-Park Corner / Taj Mehal." 1851. *Illustrated London News* 18:469–70.

Edwards, Steve. 2001. "The Accumulation of Knowledge, or, William Whewell's Eye." In *The Great Exhibition of 1851: New Interdisciplinary Essays,* ed. Louise Purbrick, 26–52. Manchester: Manchester University Press.

Ellis, Robert, ed. 1851. *Official Descriptive and Illustrated Catalogue of the Great Exhibition of the Work of Industry of All Nations.* 3 vols. London: Spicer Brothers.

"The Exhibition and Its Management." 1851. *Illustrated London News* 19:504.

"The Exhibition Jury Reports." 1852. *Fraser's Magazine* 46:491–502.

"The Exhibition of All Nations: Grand Banquet at the Mansion House." 1850. *Times* (London), March 22, 5.

Forgan, Sophie. 1999. "Bricks and Bones: Architecture and Science in Victorian Britain." In *The Architecture of Science,* ed. Peter Galison and Emily Thompson, 181–208. Cambridge, MA: MIT Press.

Gaskell, Elizabeth. 1900. *The Life of Charlotte Brontë,* with an introduction and notes by Clement K. Shorter. London: Smith, Elder.

Graham, Thomas. 1852. "Class I. Report on Mining, Quarrying, Metallurgical Operations, and Mineral Products." In *Reports by the Juries on the Subjects in the Thirty Classes into which the Exhibition was Divided,* 1–50. London: William Clowes.

"The Great Globe Itself." 1851. *Chambers' Edinburgh Journal* 16:118–19.

"The Great Industrial Exhibition." 1851. *Athenaeum,* no. 1232: 604–6.

"Guidebooks, etc." 1851. *Athenaeum,* no. 1234:654.

Hooker, Joseph. 1852. "Class III. Report on Substances Used for Food." In *Reports by the Juries on the Subjects in the Thirty Classes into which the Exhibition was Divided,* 51–67. London: William Clowes.

Howell, Thomas. 1850. *A Day's Business in the Port of London: A Lecture Delivered to the Clapham Athenaeum.* London: Simpkin, Marshall.

Humboldt, Alexander von. [1849] 1997. *Cosmos: A Sketch of the Physical Description of the Universe.* Vol. 2. Trans. E. C. Otté. Baltimore: Johns Hopkins University Press.

Hunt, Bruce J. 1997. "Doing Science in a Global Empire: Cable Telegraphy and Electrical Physics in Victorian Britain." In *Victorian Science in Context*, ed. Bernard Lightman, 312–33. Chicago: University of Chicago Press.

[Hunt, Robert, and Matthew Digby Wyatt]. 1851. "Review of *The Official Descriptive and Illustrated Catalogue.*" *Athenaeum*, nos. 1253–55: 1135–37, 1170–72, 1196–98.

Hyde, Ralph. 1988. *Panoramania! The Art and Entertainment of the "All-Embracing" View.* London: Trefoil Publications.

Inwood, Stephen. 1998. *A History of London.* New York: Carroll and Graf.

"London, Friday, May 2, 1851." 1851. *Times* (London), May 2, 4.

"London from the Crow's Nest." 1849. *Fraser's Magazine* 39:58–64.

"London, Thursday, May 1, 1851." 1851. *Times* (London) May 1, 4.

Mantell, Gideon. 1940. *The Journal of Gideon Mantell: Surgeon and Geologist.* ed. E. Cecil Curwen. Oxford: Oxford University Press.

Mayhew, Henry. 1851. *1851: or, the Adventures of Mr. and Mrs. Sandboys and Family who Came up to London to Enjoy Themselves and to See the Great Exhibition.* Illus. George Cruikshank. London: George Newbold.

"Memorabilia of the Exhibition Season." 1851. *Fraser's Magazine* 44:119–33.

Mogg, Edward. 1846. *Mogg's New Picture of London; or, Strangers' Guide to the British Metropolis.* 9th ed. London: E. Mogg.

Morrell, Jack, and Arnold Thackray. 1981. *Gentlemen of Science: Early Years of the British Association for the Advancement of Science.* Oxford: Clarendon Press.

"Mr. Wyld's Model of the Earth." 1851. *Illustrated London News* 18:511–12.

Murray, John Fisher. 1843. *The World of London.* 2 vols. Edinburgh: William Blackwood and Sons.

Noon, Patrick, ed. 2003. *Crossing the Channel: British and French Painting in the Age of Romanticism.* London: Minneapolis Institute of Arts in association with Tate Publishing.

Outram, Dorinda. 1996. "New Spaces in Natural History." In *Cultures of Natural History*, ed. N. Jardine, J. A. Secord, and E. C. Spary, 249–65. Cambridge: Cambridge University Press.

Owen, Richard. 1852. "On the Raw Materials from the Animal Kingdom." In *Lectures on the Results of the Great Exhibition of 1851*, 1st ser., 59–98. London: David Bogue.

Ruskin, John. [1839] 1903–12. "Remarks on the Present State of Meteorological Science." In *The Works of John Ruskin*, vol. 1, ed. E. T. Cook and Alexander Wedderburn, 206–10. London: G. Allen.

Schleiden, Matthias. 1848. *The Plant: A Biography.* Trans. Arthur Henfrey. London: Hippolyte Bailliere.

Secord, James A. 2004. "Monsters at the Crystal Palace." In *Models: The Third Dimension of Science*, ed. Soraya De Chadarevian and Nick Hopwood, 138–69. Palo Alto, CA: Stanford University Press.

Small, Helen. 1996. "A Pulse of 124: Charles Dickens and a Pathology of the

Mid-Victorian Reading Public." In *The Practice and Representation of Reading in England*, ed. James Raven, Helen Small, and Naomi Tadmor, 263–90. Cambridge: Cambridge University Press.

Solly, Edward. 1852. "The Vegetable Substances Used in the Arts and Manufacturers." In *Lectures on the Results of the Great Exhibition of 1851*, 1st ser., 245–89. London: David Bogue.

"Speaking to the Eye." 1851. *Economist* 9:533.

Tallis, John, ed. 1851. *Tallis's History and Description of the Crystal Palace.* 3 vols. London: John Tallis.

"Things Talked of In London." 1851. *Chambers' Edinburgh Journal* 15:13–15.

Timbs, John, ed. 1851a. *The Illustrated Year-Book: The Wonders, Events, and Discoveries of 1850.* London: Arthur Hall, Virtue.

———. 1851b. *The Year-Book of Facts in the Great Exhibition of 1851.* London: David Bouge.

———. 1867. *Curiosities of London, Exhibiting the Most Rare and Remarkable Objects in the Metropolis, with Nearly Sixty Years Personal Recollections.* New ed. London: J. S. Virtue.

"A Visit to the Museum of Practical Geology." 1851. *Fraser's Magazine* 43:618–30.

Whewell, William. 1840. *The Philosophy of the Inductive Sciences.* 2 vols. London: John W. Parker.

———. 1852. "General Bearings of the Great Exhibition." In *Lectures on the Results of the Great Exhibition of 1851*, 1st ser., 3–34. London: David Bogue.

Wyld, James. 1851. *Notes to Accompany Mr. Wyld's Model of the Earth, Leicester Square.* London: Model of the Earth, Leicester Square (privately printed).

Yanni, Carla. 1999. *Nature's Museums: Victorian Science and the Architecture of Display.* Baltimore: Johns Hopkins University Press.

Yeo, Richard. 1993. *Defining Science: William Whewell, Natural Knowledge, and Public Debate in Early Victorian Britain.* Cambridge: Cambridge University Press.

CHAPTER ELEVEN

"More the Aspect of Magic than Anything Natural": The Philosophy of Demonstration

IWAN RHYS MORUS

In April 1869, the Royal Polytechnic Institution's proprietors proudly invited London's press to a private showing. They wanted to unveil their latest acquisition to the world in suitably spectacular fashion before it was put on show to the public. The press duly assembled to see Professor Pepper's Great Induction Coil being put through its paces. Spectators gasped as Pepper manipulated the coil to produce what was reported as "a spark, or rather a flash of lightning, 29 inches in length and apparently three-fourths of an inch in width, striking the disk terminal with a stunning shock." The spark was so powerful that it could penetrate through five inches of plate glass. When the coil's terminals were brought close together, "the discharge appears to issue more slowly as a gush of waving flame, and this flame may be blown away in a broad sheet, leaving the actual line of discharge unaffected and visible by its different colour." The new instrument was set to "be a source of endless delight and wonder, and enable Professor Pepper to display effects, beautiful or terrible, such as have never been seen before." It was "a triumph of skill and knowledge of which English science may be justly proud" ("Great Induction Coil" [*Times*] 1869). There seems to have been no question in the breathless *Times* correspondent's mind that what they were witnessing was a demonstration of science in the making or that Pepper's Great Induction Coil was an instrument of discovery as much as demonstration. Looking at events such as this one and the sites where they took place should therefore provoke us to reexamine conventional distinctions between discovery and demonstration, or between backstage experimentation and public exhibition. It should also prompt us to examine more closely the ways these kinds of performances were managed and the resources that went into them.

This chapter investigates the role of performances like this in the process of constructing Victorian science. It takes seriously the notion that scientific performances such as Pepper's should be understood as being just that—*performances* in the broadest meaning of the term. Performances—making science and its products visible, pulling in the crowds and amazing them with nature's wonders—were part and parcel of the business of making science and its products real to their audiences. From the elegant lecture theaters of such august establishments as the Royal Institution to popular galleries of practical science and halls of science, exhibitionism was the order of the day. Visitors to London's exhibitions could see the microscopic magnified to gargantuan dimensions. They could see shocking displays of nature's powers. They could see apparitions walk around on stage. They were invited to marvel at such demonstrations as graphic examples of showmanship's capacities. Tracing the histories of these displays and the cultures that sustained them can tell us a great deal, not only about changes in Victorian science but also about the ways in which the Victorians went about making sense of the rapidly changing world about them. Exhibitions mattered because they made science's place in consumer culture materially explicit. To make sense of this culture of display, therefore, we need to think more about the material technology that underpinned these performances and the ways in which audiences responded to the experience. Doing so will help answer some deceptively simple questions: What did Victorians see at exhibitions? And how did they see them? Thinking about material culture and audience response draws attention to the heterogeneity of the elements—rhetorical, technical, and visual among others—that made for successful exhibition.

Hankins and Silverman have pointed not only to the survival of natural magic instrumental traditions well into the nineteenth century but also to the importance of those traditions in sustaining scientific culture (Hankins and Silverman 1995). The magic lantern shows and sophisticated optical illusions that drew audiences to London's exhibition halls of practical science and theaters were the direct descendants of natural magic repertoires. It was not uncommon for Victorian commentators to draw direct comparisons between the real, tangible achievements of their own scientific culture and the airy promises of medieval or Renaissance magi or contemporary oriental fakirs. Magic lantern shows might be, as the *Illustrated Polytechnic Review* expressed it, "the offspring of the foolish and unmeaning phantasms of the distant East, yet we now behold them no longer administering to the vulgar and depraved appetite, alternately exciting the laughter or terror of the beholders; but, assisted by the genius of philosophy and the pencil of art, they picture forth the truthful

representations of lovely and picturesque scenery, the holy temples of distant nations, and the heart-stirring scenes of our country's triumph" ("Dissolving Views" 1843, 97). The conceit was that science could deliver the
goods in ways that other traditions could not. The comparison worked to
provide an alluringly romantic gloss over the products of an increasingly
industrial economy as well (Morus 2000). Renaissance natural magicians
used optical tricks and sleight of hand to underline their virtuosity and
their access to hidden powers. Their nineteenth-century equivalents did
the same. The difference was that nineteenth-century audiences knew (or
were meant to know) it was a trick and were invited to decipher the performance at the same time as they applauded it. There was a crossover
between the natural magic tradition and a new industrial sensibility, but
Victorian audiences were invited to see illusions as demonstrations of the
possible rather than the impossible. At one level they were examples of
what Neil Harris in his biography of P. T. Barnum has described as an "operational aesthetic." Audiences were challenged, in Barnum's own words,
to find out "how does he do it?" (Harris 1973, 61–89). By displaying the
showman's ingenuity, they pandered to their audiences' sense of their own
superiority—their sense that they were the kind of people who could be
depended upon to see through the smoke screen of effects.

     Behind the glossy front-stage production, these displays required a
great deal in terms of effort, labor, and material resources. The technologies involved in exhibitions like these were by no means trivial. Neither
were they, as a rule, one-man productions. Even a straightforward magic
lantern show more often than not required careful backstage preparation
and collaboration. This was part of the point, of course. Part of what was
being displayed, as well as the ingenuity of the front-stage showman, was
the superior technological acumen and the managerial capacity needed to
get the show going (Gooding 1985; Sibum 1995; Morus forthcoming). Participants in such shows needed to prepare their roles carefully. A wrong
move, even quite literally a foot out of place, could fatally compromise the
integrity of a performance. It was this kind of backstage management that
the audience was in a sense being invited to deconstruct and simultaneously wonder at. Turning an experiment or an optical illusion into a stage
performance was a difficult and complex business. Scientific exhibitors
needed to walk a fine line between demonstrating the powers of nature
and demonstrating their own power over nature—and the audience. The
showman's skill and the ingenuity of his apparatus was certainly one of
the things on show, but making too much of the backstage artistry visible
could destroy the illusion on which a successful performance depended.

If the performance appeared too magical, then the moral of the display would be lost. Audiences needed to be able to know and see just enough to be impressed.

This chapter opens with a brief survey of the sites of metropolitan exhibition culture where these kinds of displays took place. It draws attention to the importance of locating those sites within Victorian London's cultural geography and of tracing the networks of resources—material and social—from which exhibitions were put together. By focusing on the geography of display in this way, we can circumvent discussion about the epistemological status of popular or elite science and attend instead to the ways in which audiences might have understood the social (and therefore epistemological) status of different scientific performances in different places. The bulk of the chapter then examines a number of particular examples of scientific displays in rather more detail. The magic lantern and its extension, the oxyhydrogen microscope, provide an opportunity to look at the careful management and preparation needed for successful exhibition. They furnish good examples as well of the ways in which technologies of display were integrated into other kinds of performances, such as lectures. Gargantuan electrical machines such as Armstrong's hydroelectric machine during the 1840s and Pepper's induction coil at the end of the 1860s draw attention to the spectacular scale of Victorian technologies of display. Instruments such as these were at the cutting edge of Victorian electrical technology and pushed at their audiences' sense of the boundaries of the possible. Finally, the optical illusion known as Pepper's Ghost, in which the Royal Polytechnic Institution's entire Lecture Theater was turned into a gigantic box of optical tricks, emphasizes how such technologies of display were used to play artfully with the boundary between the real and the unreal. Looking at how these pieces of showmanship worked—how they were put together, managed, and performed—tells us a great deal about the ways in which science and its artifacts were made part of Victorian culture.

Cultures of Display

Entertainment was big business in mid-Victorian London. There was a whole range of venues offering everything from the frivolous to the serious to the discerning public prepared to pay its money at the door. Natural philosophy and the mechanical arts were part and parcel of this culture of display. Scientific lectures jostled with gothic melodramas for the attention of the theatergoing public. Working models of the latest industrial machinery or experimental apparatus rubbed shoulders with collections of exotic curios and historical memorabilia of all sorts (Altick 1978; Morus, Schaffer, and Secord 1992). This was a culture within which a scientific lecturer could become a celebrity or an enterprising exhibitor of experiments could hope to make a fortune. Exhibition was part of the business of invention. Not everyone was happy with the light this culture cast on discovery and invention. The *Mechanic's Magazine* was scornful of the way in which such exhibitions competed for the attention of "curious idlers." The *Times* sardonically questioned the extent to which shows of the "studies and labours of the industrious fleas" followed by doses of laughing gas all round, might legitimately take place at a self-styled "Gallery of Practical Science" ("National Repository" 1829, 60; "Adelaide Gallery" 1843). Enthusiasts had nothing but praise for such places. The *Illustrated Polytechnic Review*, while strenuously denying any formal link with its near-namesake, lauded the Royal Polytechnic Institution for its high "moral influence" on the public mind ("Hall of the Polytechnic Institution" 1843, 209). It is undeniable, however, that this is where most Londoners who came across such things at all were likely to encounter natural philosophy and its products (Morus 1996; Morus 1998, 70–98; Beauchamp 1996).

The Adelaide Gallery—the original "National Gallery of Practical Science" and the object of the *Times*'s complaint—was one of the earliest on the scene. The gallery, with its "clever Professors" and "galvanic traps" laid for the unwary, was established in 1832 by the American entrepreneur Jacob Perkins, initially as a showcase for his own inventions (Bathe and Bathe 1943). It was soon expanded, however, to include exhibitions from other hopeful impresarios as well. It offered to "promote . . . the adoption of whatever may be found to be comparatively superior, or relatively perfect in the arts, sciences or manufactures" and to show off "specimens and models of inventions and other works &c. of interest" to the public (quoted in Altick 1978, 377). Its highlights included Perkins's own steam gun shooting high-velocity bullets at a distant target at regular intervals, a pair of electric eels, and a working Jacquard loom, the inspiration for Charles Babbage's analytical engine. The Adelaide Gallery developed a regular routine of public lectures and demonstrations, magic lantern shows and dioramas, feeding time for the eels, and musical soirées. With the arrival of daguerreotype in the early 1840s, a studio was set up at the gallery, prompting a visit from the Queen Dowager, the wife of the late King William IV, who professed herself "much pleased" with the outcome of her sitting ("Court Circular" 1842).

The Adelaide Gallery's proprietors offered their customers a carefully contrived blend of entertainment and edification. The main program of events for January 1844, for example, featured comic performances by the Infant Thalia, preceded by a display of Monck Mason's Aerial Machine, interspersed by musical interludes and followed on alternate days of the week by either the Adelaide Wizard performing his illusions; an oxyhydrogen microscope display; or a demonstration of laughing gas. Afternoon and evening sessions ended with "the magnificent transparent dissolving views" ("Programme of Morning Attractions" 1844)—in other words, a magic lantern show. The gallery's doors opened half an hour before the performance commenced so that visitors could take the opportunity to tour the exhibits. There is every indication that this, or a similarly heterogeneous blend, had been the gallery's routine of offerings from the outset. The proprietors were well aware that they needed to attract a broad clientele if their venture was to remain successful. As the *Magazine of Popular Science* (itself published by the gallery's proprietors) noted, "there will, of necessity, exist some points which shall be highly attractive to one class of visitors, whilst they will be regarded as comparatively unimportant by others. Thus the Persian rope-dancer, which, with its fairy-like music and elegant movements, is a never failing source of admiration to the young, may, by others, be held in light estimation; unless, indeed, a love of science shall lead them to examine and inquire into its ingenious and elaborate mechanism" ("Gallery of Practical Science" 1836, 10). Ideally, at least, there was to be no clear distinction between "entertaining" and "edifying" exhibits or performances.

The Adelaide Gallery's competitors seem to have followed its crowd-pulling format closely. The Polytechnic Institution, which opened its doors on Regent Street in 1838, offered a range of exhibits and entertainments that was almost identical to that available at its older rival (Wood 1965). Its highlight was a full-size diving bell in which visitors could descend to the bottom of a tank of water in the Great Hall (figure 11.1). The Polytechnic also soon boasted "the largest electrical machine in the world" ("Polytechnic Institution" 1842). Like the Adelaide, it offered a range of lectures and soirées as well. Two visitors from Bombay marveled that "we are quite sure that there is not in any country to be procured so much intellectual amusement for a shilling, as is to be had at the Polytechnic Institution; for you can be constantly amused and your mind improved, from half-past ten in the morning until half-past four, and only pay one shilling, or you can go from seven until nearly eleven at night, for a similar sum. A band of music plays daily from three to five o'clock, and in the evening from about eight until the exhibition closes" (Nowrojee and Hirjeebhoy 1841, 116). The

Figure 11.1. Illustration of the Great Hall of the Royal Polytechnic Institution. The famous diving bell can be seen suspended above a tank of water at the left. Reprinted from "The Hall of the Polytechnic Institution" 1843, 210.

Royal Panopticon of Science and Art, which opened its doors in Leicester Square in 1850, stuck also to what was clearly becoming a tried-and-tested formula for such exhibitions, though in this case with rather less commercial success ("Panopticon of Science and Art" 1855).

These exhibition halls shared more than just a range of similar offerings. They were part of the same networks as well. George Cayley, the aviation pioneer who helped finance the Polytechnic Institution and was chairman of its board of directors, had been involved with Jacob Perkins at the Adelaide Gallery (Pritchard 1961). Charles Payne, the Adelaide's former supervisor, was the new gallery's manager and had played a key role in finding financial backing for the venture. Edward Marmaduke Clarke, the philosophical instrument-maker-turned-entrepreneur who founded the Royal Panopticon, ran an instrument maker's shop on the Lowther Arcade just across the street from the Adelaide Gallery and had designed apparatuses for both the older institutions before setting up on his own account. He claimed credit for major magic lantern innovations at both the Adelaide and the Royal Polytechnic during the early 1840s ("Dissolv-ing Views" 1842). Lecturers and exhibitors moved to and fro between the competing galleries. Similarly, the rival proprietors borrowed or purloined ideas from one another and drew on the skills of much the same commu-nity of mechanics and instrument makers to provide the raw materials and the know-how that made the exhibitions work. They advertised them-selves in the same way to hopeful inventors eager for ways of bringing their productions to public attention ("Royal Polytechnic Institution" 1848). All of these places operated as competing public faces for a community that increasingly identified itself through shared scientific knowledge, inven-tive skill, and practical entrepreneurial acumen.

Galleries like these were unambiguously commercial enterprises. Their main aim—unlike elite institutions such as the Royal Institution, for example—was to make money for their financial backers. The stan-dard charge for entry to these and similar exhibition halls was a shilling, a price that seems to have remained stable for most of the century. The Poly-technic also offered an annual subscription of a guinea. The Polytechnic charged visitors an additional penny to descend into the tank in its famous diving bell. Actual numbers of visitors are difficult to assess. The *Maga-zine of Popular Science* maintained in 1836 that 80,375 paying visitors had passed through the Adelaide Gallery's doors the previous year ("Direction of the Society for the Illustration and Encouragement of Practical Science" 1836, 12). Some years later, the *Times* reported that on Monday, Decem-ber 27, 1841, as many as 5,400 people visited the Polytechnic Institution's attractions ("The Polytechnic Institution" 1842). Much later in the cen-tury, during the height of John Henry Pepper's reign at the Polytechnic, one commentator estimated that as many as "two thousand per diem" came through the door on a regular basis (Hepworth 1888, x). Some scribbled early estimates of financial receipts for the Polytechnic indicate that its promoters hoped for three hundred paying visitors a day, which, along with other sources of revenue, they anticipated would realize about ten thou-sand pounds annually before expenses ("Accounts for 1837" [1837]).

The Polytechnic's geographic location on Regent Street—as with the Adelaide on the Strand and the Panopticon on Leicester Square—was an in-dicator of cultural location as well. It was a prime site well placed to attract business in a part of London that was already largely devoted to catering for those in pursuit of entertainment, rational or otherwise. At the same time, it was within reasonable distance of districts such as Clerkenwell where the skilled instrument makers needed to sustain its exhibits might be found. It was not that far removed either from the more professorial philosophical circuit encompassing University and King's Colleges, the Royal Institution, and the Royal Society. Behind the Polytechnic's imposing exterior facade, the building's inner space was organized around the impressive Great Hall, housing indoor canals, the diving bell, and its tank. From there visitors could branch out to visit more-specialized areas such as the Hall of Manufactures, a Lecture Theater with seating for five hundred, and even a basement laboratory under the direction of the chemist J. T. Cooper. As with its competitors, its interior architecture was a physical instantiation of the place its proprietors imagined that natural philosophical display occupied in a broader culture of industrial consumption and entertainment (Morus 1998, 75–83).

As well as its diving bell and gargantuan electrical machine, the Polytechnic had a distinct repertoire of entertainments and exhibits. In addition to its regular and permanent showpieces, the gallery had a constant turnover of the latest inventions. Various telegraphs and electromagnetic clocks, for example, were on display throughout the 1840s. The *Times* in 1843 hailed the arrival at the Polytechnic of an "addition to the numerous specimens of machinery in the collection. . . . . which will well reward the inspection of the curious." The "machine for carding cotton" exhibited by its inventor, Mr. Bodmer, was "one of the most perfect and beautiful pieces of mechanism ever employed in manufacturing operations" ("The Polytechnic Institution" 1843). Lectures took place at regular intervals, drawing impressive audiences, interspersed with musical soirées and entertainments. Like the Adelaide Gallery, the Polytechnic quickly grasped the possibilities of the new art of photography, opening a studio where clients might attend and have their image captured on the spot. Increasingly however, from the 1840s onward, the Polytechnic's stellar attractions were its optical displays—its magic lantern shows, dissolving views, oxyhydrogen microscope demonstrations, and ultimately its ghostly stage apparitions.

The Grand Oxyhydrogen Microscope

Magic lanterns already had a long history by the middle of the nineteenth century. Their origins were to be found in Renaissance and early modern natural magic practices and technologies. Christiaan Huygens, diplomat and grandfather of his more philosophically eminent namesake, in the 1600s had encountered what sounds like a kind of magic lantern when he accompanied the English ambassador to The Hague to the court of James I and encountered his magician, Cornelius Drebbel, who could make ghosts appear in the clouds. Athanasius Kircher in his 1646 *Ars Magna Lucis et*

Umbrae also claimed authorship of a device that could project ghostly images onto a screen, and it was to him that most Victorian writers attributed
the lantern's invention. The younger Christiaan Huygens was apparently
the first to use the term *lanterna magica* to describe a version of the apparatus of his own invention, although mindful of its natural magical par-
entage, he professed to regard it as little more than a source of frivolous entertainment (Hankins and Silverman 1995, 43–47). Instruments like these
were designed to show off the virtuosity of their inventors by displaying
their ability to produce phenomena their audiences could not understand.
Increasingly throughout the eighteenth century they became a staple of
popular entertainment, with exhibitors vying with one another to produce
new and ever-more-spectacular and inexplicable effects as they tried to
stay one step ahead of their audiences' appetites for the macabre or the
incomprehensible (Mannoni 2000). By the mid-nineteenth century, such
showplaces as the Adelaide Gallery and the Royal Polytechnic could offer
their audiences a highly sophisticated range of visual extravaganzas. They
offered phantasmagoria in which ghostly images appeared to rush toward
the audience. Audiences could marvel at dissolving views in which different scenes appeared to fade into one another. Magic lantern shows were
often incorporated into pantomimes and other theatrical showpieces. The
Polytechnic, for example, was particularly celebrated for the magic lantern
displays in its Christmas pantomime (Ryan 1986).

These displays required a significant outlay in terms of time, resources,
and labor. The best magic lantern slides required a great deal of skill for
their execution and were accordingly jealously guarded. The former Polytechnic lanternist Thomas C. Hepworth recalled how, after each performance, one particularly ingenious mechanical effect slide, the astrometeoroscope, was promptly placed under lock and key to make sure that
no rival lanternists could examine it to discover its mechanism. When
the Royal Polytechnic Institution closed and its magic lantern equipment
was auctioned, the astrometeoroscope's inventor had to bid heavily for it
himself to prevent the secret from falling into other hands (Hepworth 1888,
180–81). A successful show needed careful and well-rehearsed coordination. Far more was involved than simply pointing a projector at the screen.
The magic lanterns themselves needed diligent management to keep the
light source in order and to adjust the lenses for different images. Behind-
the-scenes assistants toiled to provide the spectacular special effects that
accompanied each performance. Even a simple illustrated lecture needed
careful preparation and choice of the appropriate screen and lens combination (Clarke 1842; Wright 1895). Putting on a successful performance was a

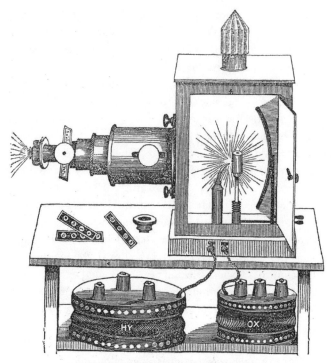

THE OXY-HYDROGEN MICROSCOPE.

Figure 11.2. Illustration of the oxyhydrogen microscope. The body of the lantern
is open, showing the oxyhydrogen light inside. Underneath are leather bags
full of hydrogen and oxygen gas to fuel the light source. Reprinted from
"Oxy-hydrogen Microscope" 1841, 313.

highly skilled and labor-intensive activity. At the same time, of course, it
had to appear from the audience's perspective to be a seamlessly effortless
piece of showmanship.

The oxyhydrogen microscope was among the most technically sophisticated of the Adelaide Gallery's and the Royal Polytechnic Institution's
magic lantern extravaganzas. These microscopes were specially adapted
magic lantern boxes in which the usual arrangement of lenses for projection was replaced by a combination of lenses that hugely magnified the
image (see figure 11.2). Later in the century, microscope attachments that
could be added on to standard magic lanterns were also commercially
available. The key to the oxyhydrogen microscope's power was the oxyhydrogen light, which had been devised during the 1820s by Henry Drum-
mond, a lieutenant in the Royal Engineers as a means of facilitating work

on the Irish ordnance survey. The intense white light was produced by heating a cylinder of lime with a flame of combined oxygen and hydrogen gas. Managing the *limelight,* as it was also called, could be quite tricky. The gases were often stored in leather bags and fed into the microscope at a constant pressure. If the pressure changed and the wrong mix emerged from the nozzle, the results could be explosive. The hydrogen gas could sometimes be replaced with the ordinary gas used for lighting purposes, though as E. M. Clarke pointed out, the ordinary gas was not always of the same quality. He expressed his preference for London gas over that available in Edinburgh or Glasgow, for example (Clarke 1842, 29). Oxyhydrogen microscopes were used in just the same way as ordinary magic lanterns to cast projected images onto a screen, except that in this case, what the audience was seeing was (usually at least) a hugely magnified natural object rather than an artificial painted image. According to one account, at least, the first oxyhydrogen microscope, properly speaking, was used by George Birkbeck to accompany a lecture on optical instruments at the London Mechanics' Institution in 1824 ("Oxy-hydrogen Microscope" 1841).

Visitors to oxyhydrogen microscope shows marveled—and presumably felt more than a little nauseated—at seeing the countless tiny creatures that inhabited a drop of Thames water magnified a thrilling "3,000,000" times. When the Polytechnic unveiled a new microscope in 1842, the *Times* could hardly contain its enthusiasm after its correspondent was "admitted yesterday evening to a private view of the most powerful apparatus of its kind in existence—a new microscope, constructed by Mr. Cary, the optician." The reporter dutifully relayed its vital statistics to the newspaper's readers: "The microscope in question consists of 6 powers, ranging from 130 times to 74,000,000 times. The second magnifying power magnifies the wings of the locust to 27 feet in length. The fourth power magnifies the sting of the bee 27 feet. By the sixth power, the eye of the fly, which is said to contain 750 lenses, is so magnified that each lens appears to be 14 inches in diameter; the human hair is magnified 18 inches in diameter, or 4 feet in circumference. Nothing can exceed the beauty with which insect architecture is developed under the influence of this enormous power" ("Royal Polytechnic Institution" 1842 [*Times*]). E. M. Clarke described the effect produced by showing iron filings falling into a pattern between the poles of a magnet, as seen through an oxyhydrogen microscope, as having "more the aspect of magic than anything natural," waxing lyrical about the way in which the effect seemed "to spread like the branches of a tree, bursting into its fulness of growth in seconds, instead of seasons" (Clarke 1842, 51). Another favorite was the decomposition of water by electricity,

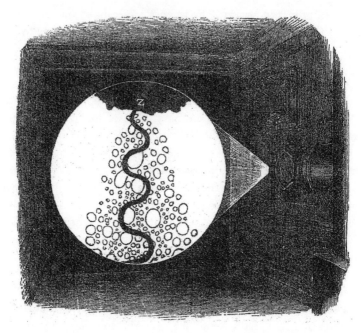

Figure 11.3. Illustration of the decomposition of water, as seen through the oxyhydrogen microscope. Reprinted from Clarke 1842, 54.

seen through the microscope (see figure 11.3), a demonstration devised by the electrician William Sturgeon and widely hailed as "one of the great attractions" at both the Polytechnic and the Adelaide Gallery ("Scientific Exhibitions" 1841). These displays were exhibitionist set pieces carefully designed to make the seemingly intangible appear spectacularly and tangibly visible.

The oxyhydrogen microscope is a particularly good example of the way in which such displays were intended to combine entertainment and edification. Audiences were meant to marvel at the wonders of nature and the ingenuity that went into making those wonders visible. As far as the Adelaide Gallery's or the Polytechnic's proprietors were concerned, "entertainment and edification" or "amusement and instruction" were not opposite ends of the spectrum. Magic lantern dissolving views were expected to edify as well as entertain. Oxyhydrogen microscope demonstrations were expected to amuse as well as instruct. The microscope was used both as demonstration aid in lectures on other topics—E. M. Clarke, for example, recommended his electromagnetic microscope displays as "a beautiful course of illustrations for lectures on magnetism and magnetic electricity"

(Clarke 1842, 53)—and as a subject for a lecture or demonstration in its own right. Such demonstrations were quite routinely incorporated as set pieces into the repertoire of performances offered at the Adelaide Gallery's or the Polytechnic's evening shows, listed as attractions just as the Infant Thalia's or the Adelaide Wizard's performances might be listed. In terms of the ways in which the oxyhydrogen microscope was presented to its audiences, at least, no discrimination was made between it and other exhibits with regard to its potential to entertain or edify. It was expected to do both.

The decomposition experiment is a good example of the careful preparation needed to make the oxyhydrogen microscope a convincing piece of showmanship. Not only did the microscope and its oxyhydrogen limelight itself have to be managed, but also the electrical decomposition apparatus had to be carefully set up and regulated. Without due care, the "putty often gave way under the effect of the violent action going on within, and the acid solution oozed out, to the great injury of all the parts of the microscope with which it came into contact" (Clarke 1842, 54). Even less-complex shows were the result of acquired technique and constant practice. The oxyhydrogen light had to be adjusted according to the type of slide being shown and the magnifying power being used; in particular, the position of the lime had to be adapted to suit the different powers. Experience was needed to know just how it needed to be positioned with respect to the optical condensers to produce the best result. These were tricks of the trade "without which no projection microscope can do its best with varying powers" (Wright 1895, 198). Practice could make the procedure appear seamless, however. As one demonstrator from the late Victorian era comforted his readers: "This all sounds rather formidable; but practically the demonstrator does it all without thinking about it, in a second or two. It is done so easily that, although my experience has been only that of rare and exceptional occasions, I have found no difficulty in exhibiting fifty slides representing organs of insects, with powers varying from 300 to 1,500 diameters, during a lecture of an hour and a half, performing all manipulations whilst explaining the slides, and with no assistance whatever beyond that of a friend to hand me each slide in order, wiped clean, in exchange for the one just used" (Wright 1895, 199). At the same time, there were plenty of cautionary tales of oxyhydrogen microscopes (as well as ordinary magic lanterns) quite literally blowing up in the hands of inexperienced practitioners who failed to treat the oxyhydrogen limelight with appropriate respect (Sanger 1908).

The business of successful performance was the same—as was the technology—whether the oxyhydrogen microscope was being used for "scientific" or "popular" display. E. M. Clarke certainly made no distinction in his exhortations to his readers concerning diligent preparation and management—it was all part of the same combination of entertainment and edification. Making the invisible visible was the result of careful choreography in either case. Producing exhibitions like these that made something tangible out of a phenomenon as fleeting as a gas discharge or as ephemeral as a fly's wings needed seamless showmanship. It mattered that the demonstrator's own performance and the technology's action appeared effortless. At the same time, however, there was an important sense in which just those invisible performances and actions were what was on show as well. The oxyhydrogen microscope celebrated what science (and the showman) could do—the modern capacity to put ancient magicians to shame and turn their fantasies into reality. In that sense, it mattered that the audience did have at least some sense that a great deal of work was going on, or had gone into the production of such evanescent phenomena, even if what made that work apparent was its seeming unobtrusiveness.

Extraordinary Electrical Machines

On August 26, 1843, the *Illustrated Polytechnic Review* announced to its readers that its namesake, the Royal Polytechnic Institution, would be shutting its doors for the coming fortnight so that "due preparation" could be made for the introduction of a new exhibit: the hydroelectric machine ("Electricity Generated through the Agency of Steam" 1843). The Royal Polytechnic Institution had already gathered a reputation for itself as a purveyor of gargantuan electrical machines and displays to the public. It prided itself on its exhibition of the world's largest electrical machine, powered by a steam engine. The old electrical machine would, however, be eclipsed by the new exhibit, while at the same time acting as a benchmark against which the prodigious electrical output of the hydroelectric machine could be made measurable to its visitors. Size clearly mattered in the world of London's scientific exhibitions. The Polytechnic was not alone in celebrating its electrical machine as the largest known. The same accolade was claimed for the machine on display in the "Department of Natural Magic" at the Colosseum in Regent's Park (Morus 1998, 82). The hydroelectric machine's arrival at the Polytechnic and the fanfare that accompanied it provide a telling example of the importance of such displays in the cultural economy of scientific showmanship during the 1840s. Closing the entire building for a fortnight was not, presumably, a decision taken lightly, and the fact that it happened at all is good evidence of the Polytechnic Institution managers' aspirations for their latest offering.

The Polytechnic's hydroelectric machine, invented by the Newcastle lawyer (and later industrialist, pioneering gun manufacturer, and first Baron Armstrong of Cragside) William George Armstrong, produced static electricity by the friction of steam escaping from a boiler through a series of small nozzles. Armstrong's first public account of the phenomenon had appeared in a letter to the *Philosophical Magazine* in 1840, in which he recounted how a laborer at a local colliery accidentally electrified himself while standing in a jet of steam discharged from a leaking boiler. In a series of further communications over the following few years, Armstrong described his efforts to isolate the phenomenon, understand the origins of the electricity of steam, and, finally, convert the transient occurrence into a reliably reproducible and powerful new source of electricity (Armstrong 1840–41, 1841, 1842). Michael Faraday also devoted his eighteenth series of experimental researches to investigating the phenomenon (Faraday 1843). Armstrong as well as other investigators constructed a number of pieces of experimental apparatus as they tried to refine the phenomenon and understand its origins. By 1843, however, Armstrong clearly felt confident enough of the phenomenon and his control over it to have something more ambitious in mind: "Each jet [of steam] affords quite as much electricity as a good electrical machine of ordinary dimensions; and when it is considered that a boiler of evaporating power equal to that of a locomotive engine would be adequate to sustain hundreds of such jets, an idea may be formed of the prodigious evolution of electricity which it is practicable to obtain by the agency of steam" (Armstrong 1843a).

By this stage, Armstrong was already in communication with the Royal Polytechnic Institution's managers, through the agency of one of the Polytechnic's directors and a personal friend of Armstrong's, Captain L. Boscawen Ibbetson (Hackmann 1993). What he offered them was a new exhibit that could massively outperform their existing "colossal plate machine" in those tried-and-tested experiments such as the charging and discharging of Leyden jars or the length of spark produced that early nineteenth-century electricians typically used to assess the respective merits of their apparatus (Armstrong 1843b; Morus 1988). Armstrong's letter to the *Philosophical Magazine* describing the new instrument (and calling it a "hydro-electric machine" for the first time) was dated August 12, 1843. By this stage, the instrument was probably already on its way to London. It was a substantial piece of equipment. The boiler for producing the steam, based on the usual design for a locomotive engine, was seven and a half feet

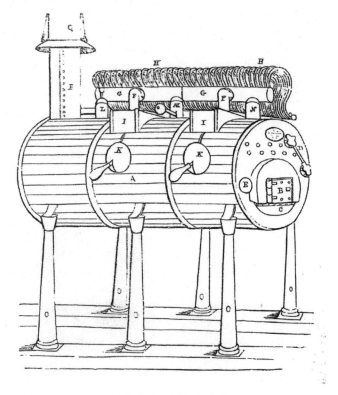

Figure 11.4. Diagram of the hydroelectric machine. Reprinted from
"The Hydro-electric Machine" 1843, 162.

long and three and a half feet in diameter and was made from iron plate
five-eighths of an inch in thickness (see figure 11.4). One report of its appearance drew particular attention to the aesthetics of its construction. It
"presented an appearance of considerable beauty, from the graceful curves
of a series of tubes through which the steam passed" ("Royal Polytechnic
Institution" 1843; "Royal Polytechnic Institution, The Hydro-electric Machine" 1843).

With the hydroelectric machine firmly ensconced, the Polytechnic re-
opened its doors with some fanfare on September 15. Distinguished scientific witnesses and members of the press were invited along to see the
new hydroelectric machine being put through its paces. George Bachhoffner, one of the Polytechnic's resident lecturers, operated the machinery,
having spent all that day practicing his routine. The *Morning Chronicle*'s
correspondent could hardly contain his enthusiasm over the machine's

capacities: "Instead of sixty spontaneous discharges in one minute [as was achieved by the old plate glass machine], the hydro-electric machine produced 140. . . . A constant stream to all parts of the boiler was kept up, and with this increased power it may well be supposed that all the former electrical experiments were greatly increased in magnificence. The passage of the electricity over the tinfoil on the tubes was far more brilliant, and the aurora borealis exceeded in intensity and beauty anything we had ever witnessed; the violet colour was brighter, and at the same time deeper, and the exhausted receiver showed more plainly the progress of the electric spark" ("Royal Polytechnic Institution" 1843). Similar reports appeared in the *Times* and elsewhere (Hydro-electric Machine" 1843; "Royal Polytechnic Institution, The Hydro-electric Machine" 1843). The reports suggest strongly that what mattered in terms of establishing the new apparatus's reputation was its ability to reproduce in suitably spectacular fashion a well-recognized and familiar repertoire of experimental effects. Aficionados presumably knew very well what newspaper reports were talking about as they referred to the "passage of electricity over tinfoil" or the "aurora borealis."

Experiments such as the "aurora borealis" had a history stretching back to the previous century (Hackmann 1995). The aurora consisted of a glass vessel or tube with a central conductor (usually a brass rod) running into it from one end and the outer surface opposing the conductor coated in foil. A valve allowed it to be evacuated using an air pump, and when attached to the prime conductor of an electrical machine, a colored glow could be seen inside the glass. It was a typical example of late eighteenth- and early nineteenth-century electricians' strategy of showing off their powers over nature by constructing apparatuses that seemingly reproduced natural phenomena (Morus 1998, 130–34). By bringing nature into the exhibition hall in this fashion, electrical showmen were demonstrating their own ingenuity as much as they were exhibiting nature or educating their audiences concerning the cause of natural phenomena. Like all experiments, productions like these needed to toe a fine line between making it look as if nature were speaking for itself and underlining the expertise of the performer (Gooding 1985; Morus 1992). The fact that these experiments were part of a standard repertoire rather than something novel or original suggests that what was being displayed was not nature but the machine itself. The standard experiments were being trotted out precisely because they were familiar and could therefore be trusted to validate the power of the instrument being used in their production.

The focus of attention in this kind of exhibit was therefore the apparatus itself and the work and ingenuity that went into building and operating it successfully, as much as the natural phenomena it ostensibly exhibited. The shocks and sparks that accompanied a successful performance were a means of making that operational success visible rather than being the ends of the show in themselves. Just as displays of such shocks and sparks were used by electricians to assess the relative performances of particular batteries or electromagnetic instruments, they could be used by exhibition audiences to gauge the relative superiority of particular shows (Morus 1988). This is one context at least in which the sheer size of a display was clearly an important criterion. A bigger machine and a bigger display translated into more technical skill and ingenuity to produce and control such power. Accounts of the hydroelectric machine's performance show that their authors were well aware of the time and effort that went into putting together a good show. Both the *Morning Chronicle* and the *Illustrated Polytechnic Review* drew their readers' attention to the amount of time taken to prepare the exhibit. The *Illustrated Polytechnic Review* commented that though "the machine was necessarily displayed under many disadvantages, owing to the novelty of the subject, few could have believed, from the masterly way Mr. Bachhoffner treated the subject, that only at eleven o'clock that day had he been able to commence the trials" ("Royal Polytechnic Institution, The Hydro-electric Machine" 1843). As with the oxyhydrogen microscope, visitors watching the hydroelectric machine in action were being invited to marvel at the ingenuity that made the display possible.

Just as Armstrong's hydroelectric machine had been an expensive proposition, the Great Induction Coil, which the Polytechnic acquired in 1869, was another significant investment—and one which continued the tradition of putting on gargantuan electrical extravaganzas. Induction coils had been a staple of the electrical technology of display since their invention in the 1830s by Nicholas Callan. They feature in the idealized electrician's laboratory pictured in the frontispiece to Henry Noad's *Lectures on Electricity*, published in 1844, along with Armstrong's newly invented hydroelectric machine as a centrepiece (see figure 11.5). Consisting of two concentric coils of wire, with the inner coil attached to a battery, they were a convenient source of relatively high-intensity electrical currents that were particularly useful for administering shocks and producing spectacular shows of electrical sparks. Induction coils were useful sources of electricity for medical purposes as well as important components of telegraph technology. With the development of new and more powerful induction coils by the German instrument maker Heinrich Ruhmkorff and others

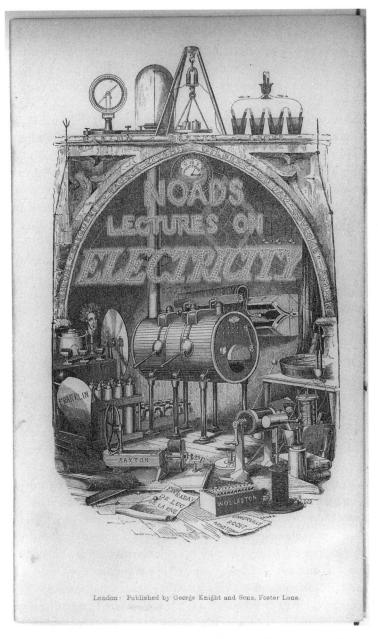

Figure 11.5. Illustration from the frontispiece to Henry Noad's *Lectures on Electricity*. The hydroelectric machine is in the foreground powering a spectacular electrical display. Reprinted from Noad 1844.

during the 1850s, electricians soon established a whole new repertoire of effects. By the end of the 1860s, when Pepper's Great Induction Coil went on show at the Polytechnic, induction coils were vital tools for investigating novel electrical phenomena. As we shall see, Pepper certainly represented his latest acquisition as a powerful tool for both demonstration and discovery.

The Polytechnic's huge induction coil was built by the London instrument maker Alfred Apps, based on the Strand, who was already establishing a name for himself as a maker of large induction coils (Noad 1879, 365). The machine that Apps built for the Polytechnic was nine feet ten inches in length and two feet in diameter. The primary coil weighed 145 pounds and was made of 3,770 yards of copper wire wound six thousand times around the central iron core. The secondary coil consisted of 150 miles of wire. The electrical current for the primary coil was delivered by a Bunsen battery of forty cells. Since induction coils were usually of dimensions that could be fitted quite comfortably onto an average-size table, these were impressive figures and were duly and faithfully reported in press accounts of the Polytechnic's new showpiece ("Great Induction Coil" 1869 [Times]). As with the hydroelectric machine a quarter of a century earlier, size clearly mattered. The "monster coil"—as Pepper called it—could be expected to produce commensurately gigantic effects, and scale was accordingly a major focus of attention (see figure 11.6). As the Times, presumably following Pepper's own showman's patter, noted, "every marked increase in the power of scientific apparatus has been followed by a corresponding increase in the growth of knowledge. The coil will not only amuse audiences, but will be diligently used at other times to promote the researches of electricians and physiologists" ("Great Induction Coil" 1869 [Times]).

During the 1850s and 1860s, electrical showmen had developed a particularly distinctive catalog of effects to demonstrate the capacities of their improved induction coils. With the high-tension currents produced by the coils, not only could electricians put on the traditional displays of shocks and sparks, but they could exhibit spectacular discharge effects as well. Effects such as these had been developed in Bonn and London by experimenters and instrument makers including John Peter Gassiot, William Robert Grove, Heinrich Geissler, and Julius Plücker. These effects derived from investigations of the properties of sparks and the observation that different metal electrodes and different gaseous media produced sparks and discharges of different colors. When electricity passed between electrodes inside partially evacuated and sealed glass tubes or flasks, spectacular glowing discharges became visible. Geissler in particular made a name for

Figure 11.6. Illustration of Professor Pepper putting the Polytechnic's Great Induction Coil through its paces. Reprinted from "Great Induction Coil" 1869 [*Illustrated London News*].

himself producing a range of glass tubes in a variety of complex shapes and filled with various gases that glowed in assorted colors under the effect of electricity from an induction coil. Displays such as these were the mid-Victorian equivalent of the aurora borealis experiments of the previous generation. The glowing discharges could be manipulated by holding a magnet, or even the demonstrator's finger, up against the glass tube (Noad 1879; Hackmann 1995).

One of the most spectacular of these discharge experiments, devised by

Figure 11.7. Illustration of Gassiot's cascade. Reprinted from Noad 1879, 371.

Gassiot, was the cascade (see figure 11.7). In this experiment, a glass cup lined with tinfoil was placed inside an air pump with an electrode placed at its mouth. When the terminals of the induction coil were attached and the pump evacuated, "at first a faint clear blue light appears to proceed from the lower part of the beaker to the plate; this gradually becomes brighter until by slow degrees it rises, increasing in brilliancy, until it arrives at that part which is opposite or in a line with the inner coating, the whole being intensely illuminated. A discharge then commences from the inside of the beaker to the plate of the pump in minute but diffused streams of blue light; continuing the exhaustion, at last a discharge takes place in the form of an undivided continuous stream, overlapping the vessel as if the electric fluid were itself a material body running over. . . . streams of lambent flame appear to pour down the sides of the plate, while a continuous discharge takes place from the inside coating" (Gassiot 1854). Gassiot's cascade experiment was often described as "one of the most beautiful that can be made with the Induction Coil" (Noad 1868). Shortly after the Great Induction Coil's arrival at the Polytechnic, the medical lecturer Benjamin Ward Richardson used it in conjunction with Gassiot's cascade to spectacular effect in displaying the capacity of various kinds of animal tissue to conduct electricity (Morus 2002, 100–2).

Induction coils and their attendant technologies of display were the latest manifestation of a tradition of electrical performance that stretched back into the middle of the previous century (Schaffer 1983). These were technologies designed to show off both the powers of nature and the power of their manipulators to dominate nature. Practical electricians of the kind that plied their trade at venues such as the Royal Polytechnic shared a tradition of producing and using apparatuses that made nature's powers visible on a grand scale (Morus 1992). But it also clearly seems to be the case with the Polytechnic's Great Induction Coil, just as with Armstrong's Hydro-electric Machine, that the technology itself was an important focus of attention. Part of the point of putting the Great Induction Coil through a recognized and familiar repertoire of effects was to emphasize the machine's superior capacity to produce them. The routine's very familiarity simply served to emphasize the unfamiliarity of the result, which may provide one possible explanation for the remarkable continuity in electrical showmanship from the middle of the eighteenth through to the middle of the nineteenth centuries and beyond. The phenomena being witnessed were means to an end rather than being an end in and of themselves. But a technology such as the induction coil was at the cutting edge of Victorian physics as well. Professor Pepper and other performers like him were adept at using technologies such as these to push at the boundaries of the familiar and the possible. Performances with the induction coil challenged its audiences' perceptions of the boundary between fact and artifact.

Professor Pepper's Ghost

If the oxyhydrogen microscope and similar magic lantern extravaganzas were the crowd-pullers of the 1840s, they were replaced in the limelight in the 1860s by another optical illusion. Like its magic lantern competitors, Pepper's Ghost, as it was popularly known, had a distinguished natural magic pedigree. The phenomenon itself was straightforward—the audience saw a ghost appear on stage, capable of moving around and interacting with living performers. It was the latest and most spectacular of a long line of ghostly stage apparitions, including the magic lantern–based phantasmagoria with which showmen had sought to bedazzle their audiences. Similar optical tricks were part of the stock-in-trade of magicians and illusionists and were extensively treated in David Brewster's magisterial early nineteenth-century *Letters on Natural Magic* (Brewster 1832, 160–63). Optical tricks that made the viewer see something that was not really there were popular throughout the nineteenth century. Brewster's own kaleidoscope is one example. Charles Wheatstone's stereoscope, in which a two-dimensional image was made to appear three-dimensional, is another (Hankins and Silverman 1995, 148–77). The various nineteenth-century optical arrangements, such as John Beale's choreutoscope, that gave the appearance of movement to a series of still images were part of the same genre of optical misdirection (*Art of Projection* 1893, 158). In that respect, Pepper's Ghost was already familiar to its audience—they knew about optical illusions like this. The challenge, of course, was to decipher the details.

The Ghost made its first stage appearance at the Royal Polytechnic Institution in 1862 to a select audience, playing the starring role in a dramatization of Edward Bulwer-Lytton's gothic novel, *A Strange Story* (Lamb 1976, 43–50). Its inventors were the Polytechnic's new proprietor, John Henry Pepper, whose activities at the Polytechnic are more comprehensively dealt with by Lightman elsewhere in this volume, and the patent agent and prolific scientific author and lecturer Henry Dircks. The Ghost could interact with living actors, move around the stage, appear or disappear at will, and walk through solid objects. It could do everything a ghost should be able to do. It was a fine handle from which to hang a morality tale. Dramatic scenarios played with the Ghost on stage featured misers haunted by the ghosts of their miserable debtors, villainous barons haunted by the specters of violated servant girls, and vile murderers exposed by apparitions from beyond the grave. In one playlet shown under license at the Britannia Theatre, titled "Faith, Hope and Charity," the audience saw the ghost of "a clergyman's widow who has been murdered by a baronet and accordingly haunts him at midnight. The baronet, to the great surprise of the audience, thrusts his sword through the apparition" ("Musical and Dramatic Gossip" 1863, 562). The audience's reaction was stunning: "The effect was to entrance them as if spell-bound, and the majority were evidently unable to explain the cause of so extraordinary an appearance" ("Musical and Dramatic Gossip" 1863, 562).

The cause of the extraordinary appearance was deceptively simple. A large sheet of plate glass at an angle of forty-five degrees was suspended between the audience and the stage. In a pit between the stage and the auditorium stood the actor who played the Ghost, out of sight of the audience. When a bright light (usually an oxyhydrogen or even an electric light) was played on the actor, his or her reflection on the plate glass screen made it appear to the audience as if he or she were actually onstage (see figure 11.8). The actor could then work his or her way through the usual ghostly repertoire of appearances and disappearances achieved by turning the illumination on and off as necessary. The appearance of simplicity is itself deceptive, however. The illusion needed careful choreography to be

Figure 11.8. Illustration of Pepper's Ghost performing onstage, showing the plate glass and hidden magic lantern light that made the effect possible. Reprinted from frontispiece of *Magic Lantern* 1876.

convincing. The sheet of plate glass had to be carefully placed to avoid detection by the audience. The pit and the light source illuminating the Ghost had to be similarly discreet. Even the plate glass was something of a novelty. Glass manufacturing processes were undergoing rapid innovation during the 1860s. The capacity to produce glass sheets of sufficient size and clarity to make the apparition convincing was very recent (Barker 1960, 138–41).

Recalling the first time he demonstrated the Ghost at the Polytechnic Institution, Pepper remembered that the "effect of the first appearance of the apparition on my illustrious audience was startling in the extreme, and far beyond anything I could have hoped for and expected, so much so that, although I had previously settled to explain the whole *modus operandi* on that evening, I deferred doing so, and went the next day to Messrs. Carpmael, the patent agents, and took out a provisional patent for the ghost illusion, in the names, at my request, of Dircks and Pepper" (Pepper 1890, 3). The illusion's value on this account was to be derived from its shattering. Dircks said something similar as well, suggesting to the *Athenaeum* that "optical illusions" like the Ghost "are of a character calculated to disabuse the public mind in regard to the vulgar errors respecting apparitions, a belief in which has deprived many of reason; but which, as in this instance, may be realized by art, and, when occurring naturally, are mostly traceable to generally known physical causes" (Dircks 1863). To its makers, by these accounts, the Ghost's virtue lay in the challenge it posed to the audience of deciphering the trick behind its production. As with Barnum's productions, the appeal lay in figuring out "how does he do it!" (Harris 1973).

An account of a visit to the Polytechnic by the Prince of Wales and his entourage gives us some sense of the way in which exhibiting the Ghost fitted in with the institution's repertoire of performances. The party's entertainment started with a tour of the galleries before they were conducted into the large lecture theater to hear Professor Pepper deliver his "ghost lecture." Following the lecture, they went behind the scenes and "examined with much interest the machinery and appliances required to produce the Polytechnic 'ghost.'" They then returned to the main gallery to see some experiments with the diver and the Polytechnic's famous diving bell. The evening reached its climax when the royal group had the opportunity of seeing the Ghost itself in action, with a performance of "the incantation and other chief scenes of Von Weber's opera *Der Freischütz*" ("Court Circular" 1863, 9). While it is unlikely that the Polytechnic's humbler visitors received the personal attention accorded the prince, it seems plausible, at least, to suggest that what he experienced was the same kind of range of performances that others would have encountered. Even a display as spectacular as the Ghost was intended to be seen as one of a range of entertaining and edifying performances rather than as a single unitary event.

The Ghost provided a way of reconstruing virtuosity. Whilst past virtuosos might have used optical tricks and illusions to cast a spell over their audiences by making them see things that were not there, their modern equivalents by this account were putting their skills to work in shattering the illusion. This was Brewster's rationale for his *Letters on Natural Magic* as well. The magical was recast as the ultimate in rationality by making its mechanism transparent. Shows like the Ghost, however, also provided a means of making the rational magical. Many Victorian commentators were keen to draw favorable analogies between the capacities of their own industrial culture and the dreams of past magicians and prophets. The telegraph was "a thousand times more than what all the preternatural powers which men have dreamt of and wished to obtain were ever imagined capable of doing," according to the bishop of Llandaff, for example (Copleston 1851, 169; Morus 2000). Playing with the boundary between the magical and the rational was a good way of establishing one's own modern virtuosity. The *Times* urged its readers to "go and see this curious performance for themselves, and if they don't—as few do—believe in them as optical illusions, they must at least admit that any other solution of the mystery but this involves such a series of tricks as would almost be as wonderful as these most real but unsubstantial phantoms" ("Royal Polytechnic Institution" 1862). Pepper and Dircks were casting themselves as expert showmen who could wow their audiences by satisfying their desire for the bizarre and inexplicable while simultaneously delivering to them the cold hard facts of Victorian industrial culture's ineluctable superiority.

There is a striking similarity between what Dircks or Pepper had to say about the way they could use their authority over the Ghost as a way of affirming their (and their audience's) superiority over the deluded and the antimesmeric and antispiritualist campaigns of such physiologists as William Benjamin Carpenter. They certainly invoked a very similar rhetoric (Winter 1998). Carpenter could put his superior knowledge of the hidden workings of the mind to use in seeing off the claims of spiritualist fantasists: they could see what was not there because their brains were badly wired (Morus 2000; Noakes 2002; Fichman 2004, 165–67). Pepper, on the other hand, could let his audience see what was not there and then carefully explain to them that what they were seeing after all was nothing. Even Madame Blavatsky acknowledged the power of Pepper's Ghost as an antispiritualist instrument, while stoutly maintaining that what "the Pepper ghosts pretended to do, genuine disembodied human spirits, when their reflection is materialized by the elementals, can actually perform" (Blavatsky 1877, 1:359). In the Ghost's case, by looking at it, they were simultaneously failing to see the mechanism for its production that really was there at center stage. The Ghost therefore held a moral tale for its audience beyond the rather hackneyed ones that were being played out on the Polytechnic Institution's stage. It reminded them that the true object of exhibition was not always what seemed to be on display. By doing so, it served to bolster its authors' claims to ultimate authority over what was and was not real.

Conclusion

Examples such as the ones discussed here serve as reminders that there was a continuum of technologies and tactics of display that united seeming hucksters the likes of P. T. Barnum with their apparently more respectable counterparts the likes of Professor Pepper. From the audience's perspective, there was not necessarily that much difference between marveling at the oxyhydrogen microscope, gasping at shows of electrical power, wondering at the apparition of Pepper's Ghost, and trying to work out where the join was on Barnum's Fejee Mermaid (Harris 1973, 62–67). In that sense, any dichotomy between quackery and scientific showmanship is itself an illusion. Both work in the same way and share in the same ideology of display. It may be ironic, then, that exhibits such as Pepper's Ghost—according to their authors' accounts, certainly—were seemingly directed at challenging the charlatan, the quack, or the plain unorthodox. While mounting that challenge, they shared in the appeal and served as well to highlight their exhibitors' ingenuity and skill in just the same way. Efforts such as these to debunk quackery by exposing the "real" nature of things—just like Edgar Allan Poe's exposure of the true mechanism of von Kempelen's notorious Chess-Playing Turk—ended up by highlighting the exposer's claim to possession of "real" expertise as well (Schaffer 1996).

This claim to authority over the "real" exposes another continuity (or symbiosis) as well. All such shows pushed at their audiences' expectations over what might or might not plausibly be regarded as a real component of the natural world. The Victorian age was a period when the boundaries of the real were highly contested. Audiences were faced with machines that could think, pictures that drew themselves, instruments that seemingly transmitted messages instantaneously over vast distances, and devices that made the insignificant appear impossibly huge. Exhibitions that appeared to push at the limits of the possible served to make the Victorians' world simultaneously more and less explicable. They made it more explicable by debunking mysteries and exposing the true nature of things. At the same time, they expanded the horizon of possibilities as to what might plausibly exist in the natural world and by doing so expanded the sphere of the mysterious and unknown as well. This was where their appeal lay. Such exhibitions operated by reconfiguring their audiences' sense of the boundaries between the possible and the impossible, the natural and the artificial, the genuine article and the fake. They worked by pushing at the boundaries of the real.

All the examples of technologies of display discussed in this chapter were expensive in terms of labor, material resources, and performing skill. Acquiring Armstrong's hydroelectric machine or the Great Induction Coil required significant outlay, and the resources required to turn the Polytechnic's Large Lecture Theater into a haunted house were by no means trivial either. These were shows that required careful preparation and diligent rehearsal if the illusion of seamlessness was to be maintained. One relatively banal observation, therefore, is that these shows could not take place just anywhere. They could be done well only in places where the performers had relatively easy access to the resources they needed. In this respect, exhibitions such as these were, like other experiments, local affairs.

In Britain, certainly, there were relatively few places outside the metrop-olis that possessed the sort of networks of material and social resources that were required. Even something like the oxyhydrogen microscope that might at first glance appear relatively portable, in practice required very particular resources for its best performances—if only a relatively clean source of hydrogen gas for the limelight (Clarke 1842, 29–30). Careful col-laboration among performers was essential if the effortlessness of display was to be properly sustained. Magic lantern operators in complex produc-tions developed codes to make sure that all participants knew what was go-ing on at any particular time during a performance. With a large electrical apparatus such as the induction coil, elaborate precautions were needed to ensure the performers' safety ("Great Induction Coil" 1869 [*Times*]).

Looking at scientific performances and the places in which they took place also provides one way of circumventing debates about the identity of popular science (Cooter and Pumfrey 1994). Rather than seeing different individuals, institutions, or practitioners as being emblematic of "popular" as opposed to "elite" (or "professional" or "specialist") science, we might more productively view them in terms of their position in particular kinds of networks. By viewing them in this way, we might then understand the varying epistemological or social status of a range of practices as being the outcome of their participants' location in these networks. What mattered as much as anything for audiences' understanding of the epistemological loca-tion of particular practices was quite literally the geographic (and therefore cultural) locations of the places where they took place (Livingstone 2003). Practices and performances that might pass muster in one place could very easily fall flat in another. It is as difficult to imagine Professor Pepper at the Royal Institution as it is to imagine Michael Faraday performing at the Polytechnic Institution. The outcome of successful performance—one that suited its location and its audience's understanding of what such per-formances might be expected to achieve—was authority.

If we can understand a performer's authority as being an outcome of the site where the performances took place, then we can start investigating the ways in which particular audiences expected different kinds of experiences in different places. Audiences visited the Royal Polytechnic Institution or the Adelaide Gallery because those places offered a particular repertoire of experiences. Their experiences and expectations of what constituted scientific knowledge were formed by what they encountered in these set-tings. For audiences such as these, scientific knowledge was experienced in the context of shows ranging from magic lantern extravaganzas to conjur-ers' illusions. For these audiences, simply speaking, that is where science belonged. Its practices were part of the culture of wonder. For other kinds of audiences in other settings (the Royal Institution or the Royal Society, say), the context of scientific performances, and therefore their experiences of those performances, might be very different. As a result, those audiences' notions of what counted as science might be very different too. For the performers, sites and their places in particular networks mattered as well. Successful performance meant tailoring their shows to the expectations of their audiences and the contexts in which they took place. It also follows that what audiences took to be real might be understood as being contingent on the places where they witnessed scientific performances and their sense of what might or might not be plausible within those settings.

By pushing at their audiences' sense of what might or might not plausibly be, scientific performers were establishing their own claims to expertise over just that issue. Pepper's Ghost was a superlative piece of stagecraft demonstrating optical principles, rather than a rather tasteless exercise in necromancy, because Pepper (and his audience) knew there were no such things as ghosts. That at least was one of the lessons the audience was expected to infer. In a similar way, as audiences thrilled at the oxyhydrogen microscope's capacity to make monsters out of midgets, or at the shocks and sparks produced with Armstrong's hydroelectric machine or Pepper's Great Induction Coil, they were also ingesting the message that the demonstrator knew something about the laws of nature that they did not. This message leads to a final moral about what was or was not visible about performances like these. It is by now a truism that successful public experimentation demands the effacement of the experimenter (Gooding 1985). A good performance has to be so seamless that the audience fails to see that it is there. In shows like these, however, making the skills and technologies that went into a successful performance invisible was paradoxically to enhance their visibility to the audience. The less the audience could see of what was really going on, the more they were convinced of what a clever chap the performer really was.

REFERENCES

"Accounts for 1837." [1837]. University of Westminster Archive, RPI R22a.
"The Adelaide Gallery of Practical Science." 1843. *Times* (London), December 15, 5.
Altick, Richard. 1978. *The Shows of London.* Cambridge, MA: Harvard University Press.

Armstrong, W. G. 1840–41. "On the Electricity of Effluent Steam." *Philosophical Magazine* 17:452–57; 18:50–57; 19:25–27.

———. 1841. "On the Electrical Phenomena attending the Efflux of Condensed Air and Steam." *Philosophical Magazine* 18:328–37.

———. 1842. "On the Cause of the Electricity of Effluent Steam." *Philosophical Magazine* 20:5–8.

———. 1843a. "On the Efficiency of Steam as a Means of Producing Electricity, and on a Curious Action of a Jet of Steam upon a Ball." *Philosophical Magazine* 22:1–5.

———. 1843b. "Account of a Hydro-electric Machine constructed for the Polytechnic Institution, and of some Experiments Performed by its Means." *Philosophical Magazine* 23:194–202.

The Art of Projection and Complete Magic Lantern Manual. 1893. London: Beckett.

Barker, T. C. 1960. *Pilkington Brothers and the Glass Industry.* London: George Allen and Unwin.

Bathe, Dorothy, and Greville Bathe. 1943. *Jacob Perkins: His Inventions, His Times and His Contemporaries.* Philadelphia: Historical Society.

Beauchamp, Ken. 1996. *Exhibiting Electricity.* London: Institute of Electrical Engineers.

Blavatsky, H. P. 1877. *Isis Unveiled.* 2 vols. New York: J. W. Bouton.

Brewster, David. 1832. *Letters on Natural Magic, Addressed to Sir Walter Scott.* Edinburgh: John Murray.

Clarke, Edward M. 1842. *Directions for Using Philosophical Apparatus in Private Research and Public Exhibition.* London: E. M. Clarke.

Cooter, Roger, and Stephen Pumfrey. 1994. "Separate Spheres and Public Places: Reflections on the History of Science Popularization and Science in Popular Culture." *History of Science* 22:237–67.

Copleston, W. J. 1851. *Memoir of Edward Copleston, D.D., Bishop of Llandaff.* London: John W. Parker and Son.

"Court Circular." 1842. *Times* (London), July 12, 6.

———. 1863. *Times* (London), May 20, 9.

Dircks, Henry. 1863. "The Spectre Drama," *Athenaeum* 1:585.

"Direction of the Society for the Illustration and Encouragement of Practical Science." 1836. *Magazine of Popular Science* 1:1–12.

"Dissolving Views." 1842. *Mirror of Literature, Amusement and Instruction* 1:97–101.

"Dissolving Views." 1843. *Illustrated Polytechnic Review* 1:97–98.

"Electricity Generated through the Agency of Steam." 1843. *Illustrated Polytechnic Review* 2:111.

Faraday, Michael. 1843. "Experimental Researches in Electricity, Eighteenth Series." *Philosophical Transactions* 134:17–32.

Fichman, Martin. 2004. *An Elusive Victorian: The Evolution of Alfred Russel Wallace.* Chicago: University of Chicago Press.

"Gallery of Practical Science." 1836. *Magazine of Popular Science,* 1:9–12.

Gassiot, John P. 1854. "On Some Experiments made with Ruhmkorff's Induction Coil." *Philosophical Magazine* 7:97–99.

Gooding, David. 1985. "In Nature's School: Faraday as an Experimentalist." In *Faraday Rediscovered: Essays on the Life and Work of Michael Faraday,* ed. David Gooding and Frank James, 105–35. London: Macmillan.

"The Great Induction Coil at the Polytechnic Institution." 1869. *Illustrated London News* 54 (April 17):401–2.

"The Great Induction Coil at the Polytechnic Institution." 1869. *Times* (London), April 7, 4.

Hackmann, Willem. 1993. "Electricity from Steam: Armstrong's Hydro-electric Machine in the 1840s." In *Making Instruments Count: Essays on Historical Scientific Instruments Presented to Gerard l'Estrange Turner,* ed. R. G. W. Anderson, J. A. Bennett, and W. F. Ryan, 146–73. Aldershot: Variorum Press.

———. 1995. "Instrument and Reality: The Case of Terrestrial Magnetism and the Northern Lights (Aurora Borealis)." *Philosophy and Technology.* Supplement to *Philosophy* 38:29–51.

"The Hall of the Polytechnic Institution." 1843. *Illustrated Polytechnic Review* 1:209–10.

Hankins, Thomas L., and Robert J. Silverman. 1995. *Instruments and the Imagination.* Princeton, NJ: Princeton University Press.

Harris, Neil. 1973. *Humbug: The Art of P. T. Barnum.* Boston: Little, Brown.

Hepworth, Thomas C. 1888. *The Book of the Lantern.* London: Wyman and Sons.

"Hydro-electric Machine." 1843. *Times* (London), September 15, 7.

"The Hydro-electric Machine." 1843. *Illustrated Polytechnic Review* 2:162–63.

Lamb, Geoffrey. 1976. *Victorian Magic.* London: Routledge.

Livingstone, David. 2003. *Putting Science in its Place: Geographies of Scientific Knowledge.* Chicago: University of Chicago Press.

The Magic Lantern: How to Buy and How to Use It; Also How to Raise a Ghost. 1876. London: Houlston and Sons.

Mannoni, Laurent. 2000. *The Great Art of Light and Shadow: Archaeology of the Cinema.* Exeter: University of Exeter Press.

Morus, Iwan Rhys. 1988. "The Sociology of Sparks: An Episode in the History and Meaning of Electricity." *Social Studies of Science* 18:387–417.

———. 1992. "Different Experimental Lives: Michael Faraday and William Sturgeon." *History of Science* 30:1–28.

———. 1996. "Manufacturing Nature: Science, Technology and Victorian Consumer Culture." *British Journal for the History of Science* 29:403–34.

———. 1998. *Frankenstein's Children: Electricity, Exhibition and Experiment in Early Nineteenth-Century London.* Princeton, NJ: Princeton University Press.

———. 2000 "The Nervous System of Britain: Space, Time and the Electric

Telegraph in the Victorian Age." *British Journal for the History of Science* 33:455–75.

———. 2002. "A Grand and Universal Panacea: Death, Resurrection and the Electric Chair." In *Bodies/Machines*, ed. Iwan Rhys Morus, 93–123. Oxford: Berg.

———. 2007. "The Two Cultures of Electricity: Between Entertainment and Edification in Victorian Science," *Science and Education* 16:593–602.

Morus, Iwan Rhys, Simon Schaffer, and James A. Secord. 1992. "Scientific London." In *London—World City, 1840*, ed. Celina Fox. New Haven, CT: Yale University Press.

"Musical and Dramatic Gossip." 1863. *Athenaeum* 1:561–62.

"National Repository." 1829. *Mechanics' Magazine* 11:58–60.

Noad, Henry M. 1844. *Lectures on Electricity*. London: J. Churchill.

———. 1868. *The Inductorium, or Induction Coil*. London: J. Churchill.

———. 1879. *Student's Text-book of Electricity*. London: Lockwood.

Noakes, Richard. 2002. "Instruments to Lay Hold of Spirits: Technologizing the Bodies of Victorian Spiritualism." In *Bodies/Machines*, ed. Iwan Rhys Morus, 125–63. Oxford: Berg.

Nowrojee, Jehangeer, and Hirjeebhoy Merwanjee. 1841. *Journal of a Residence of Two Years and a Half in Great Britain*. London: W. H. Allen.

"The Oxy-hydrogen Microscope." 1841. *Magazine of Science and School of Arts* 2:313–15.

"The Panopticon of Science and Art." 1855. *Year-Book of Facts in Science and Art* 16:9–11.

Pepper, John Henry. 1890. *The True History of the Ghost; and all about Metempsychosis*. London: Cassell.

"The Polytechnic Institution." 1842. *Times* (London), January 3, 7.

"Polytechnic Institution." 1842. *Times* (London), August 15, 5.

"The Polytechnic Institution." 1843. *Times* (London), February 11, 5.

Pritchard, J. L. 1961. *Sir George Cayley, the Inventor of the Aeroplane*. London: Max Parrish.

"Programme of Morning Attractions. Royal Adelaide Gallery." 1844. January 8. Pamphlet held at the Theatre Museum, London.

"Royal Polytechnic Institution." 1842. *Times* (London), December 1, 5.

"Royal Polytechnic Institution." 1842. *Illustrated Polytechnic Review* 2:148.

"Royal Polytechnic Institution." 1843. *Morning Chronicle*, September 16, unpaginated.

"The Royal Polytechnic Institution." 1848. *Patent Journal and Inventor's Magazine* 4:411.

"Royal Polytechnic Institution." 1862. *Times* (London), December 27, 4.

"Royal Polytechnic Institution, The Dissolving Views." 1843. *Illustrated Polytechnic Review* 1:97–98.

"Royal Polytechnic Institution, The Hydro-electric Machine." 1843. *Illustrated Polytechnic Review* 2:162–63.

Ryan, W. F. 1986. "Limelight on Eastern Europe: The Great Dissolving Views at the Royal Polytechnic." *New Magic Lantern Journal* 4:48–55.

Sanger, George. 1908. *Seventy Years a Showman.* London: Arthur Pearson.

Schaffer, Simon. 1983. "Natural Philosophy and Public Spectacle in the Eighteenth Century." *History of Science* 21:1–43.

———. 1996. "Babbage's Dancer and the Impresarios of Mechanism." In *Cultural Babbage: Technology, Time and Invention,* ed. Francis Spufford and Jenny Uglow, 53–80. London: Faber and Faber.

"Scientific Exhibitions." 1841. *Year-Book of Facts in Science & Art* 3:1.

Sibum, Heinz Otto. 1995. "Reworking the Mechanical Value of Heat: Instruments of Precision and Gestures of Accuracy in Early Victorian England." *Studies in History and Philosophy of Science* 26:73–106.

Winter, Alison. 1998. *Mesmerized: Powers of Mind in Victorian Britain.* Chicago: University of Chicago Press.

Wood, E. M. 1965. *A History of the Polytechnic.* London: MacDonald.

Wright, Lewis. 1895. *Optical Projection: A Treatise on the Use of the Lantern in Exhibition and Scientific Demonstration.* 3rd ed. London: Longmans Green.

CHAPTER TWELVE

The Museum Affect: Visiting Collections of Anatomy and Natural History

Samuel J. M. M. Alberti

There were many opportunities to encounter the natural world on display in late nineteenth-century Britain, from fossils, plants, and taxidermic mounts to body parts in jars. Victorian towns abounded with collections of things that were once alive, including those within the grand architecture of city-center museums, commercial museums displaying oddities for sixpence, and teaching cabinets in the back rooms of schools. A large public museum could accommodate hundreds of thousands of visitors in a year, and a medical school collection might be used to train generations of surgeons. And yet while historians of science and medicine have paid careful attention to the readers of scientific texts, visitors to these museums remain largely silent. This chapter will engage with audiences by asking: Who visited these collections? What was their experience like, and how did they react? And what was the role of curators in this process?

To address these questions, this chapter draws from the established bodies of work in cultural theory, mass communication studies, and book history that view the communication process from both sides (Cooter and Pumfrey 1994; Darnton 1990; Freedberg 1989; Hall 1986; Hay, Grossberg, and Wartella 1996; Machor and Goldstein 2000; Topham 2000). Postmodern reception studies and theories of response in history of art are moving away from canonical texts to examine the circulation and meaning of marginalized texts, images, television, and other mass media, but they have yet to reach the (nonart) exhibition. In museum studies, meanwhile, visitor theory and contemporary surveys are replacing the passive audience with active participants in the construction of meaning, but seldom has the historical visitor been awarded the same courtesy (Carroll 2004; Forgan 2005; G. Fyfe and Ross 1996; Haynes 2001; Hooper-Greenhill 1994; Hudson 1975; Longhurst, Bagnall, and Savage 2004; Macdonald 2005). Visitors were not vessels waiting to be filled but autonomous agents with their own agendas. Just as reader-response theorists are seeking to recover not only the meaning of texts but the practices of reading, so this chapter sets out to examine not only the intentions of curators but also, as far as possible, the sensations of visiting. It will come as no surprise that the two did not always tally, and my aim is to explore the tension between visitors' appropriation of meaning and the explicit aims of the museums themselves (that is, their keepers, directors, honorary curators, trustees, and committees). Like the other chapters in this volume, I aim to supplement notions of the "public understanding of science"—with its connotations of a transmission of arcane knowledge from active expert to an inert public—with an account of their *experience.* And because I assert that the Victorians' experience was different from our own, such a study must account for the historical specificity of the visit.

There can be no doubt that experiences and responses concerning any form of display are as varied as the individuals viewing them. As Macdonald found in her recent study of late twentieth-century visitors, however, "amidst the variety were also certain patterns which . . . could be seen as part of a repertoire of prevalent interpretations" (Macdonald 2002, 220). Similarly, among the variety of recorded responses to Victorian anatomy and natural history displays, some groups of common experiences and responses are discernible. Certain aspects of the experience were privileged, and certain ways of expressing the visiting experience were afforded more validity than others.

The mid-nineteenth century was a critical period in the construction of the public for museums (Bennett 1995). Museum campaigners presented newly opened collections as serious places for enlightenment and for quiet, hands-off contemplation. Municipal museums were attracting visitors from across the social spectrum on an unprecedented scale, while medical schools and the new university colleges were closing the doors of their museums, seeking to cleanse them of their associations with the circus and fair, and attempting to construct a more elite audience from the ranks of students and professionals. In both closed and open sites, museum authorities sought to narrow the range of acceptable affective responses to their collections, to render the museum a purely scopic site, to condition the visit.

The first task of this chapter will therefore be to chart audience constituencies over the course of the Victorian era, linking changes in class, age, and gender to curatorial aims. In the second section I examine the range of sensations that had been involved in the museum visit—olfactory, tactile, and aural as well as visual—which will set into context the late-century restrictions imposed upon visitors. There follows discussion of two of the possible emotional reactions to displays of dead things, wonder and disgust. I argue throughout that despite the attempts to condition the visit, museums continued to be multisensory sites in which visitors were confused and delighted in equal measure (A. Fyfe this volume).

The "museum effect" is that phenomenon observed by museologists whereby an object is radically dislocated from its point of origin, wrenched from its context and rendered a frozen work of art in the surrounds of the museum (Alpers 1991; Benjamin [1936] 1999; Vogel 1991). The poor pun of my title is an expression of my wish to present a complete account of this process, including not only the institution's role in the museum effect but also the emotional and sensory experiences of the visitors—that is to say, my topic is the "museum affect."

Access

In general, museum audiences were inclusive at the end of the Victorian era. This was a stark contrast to a century earlier, when most museums had barred general access—even the purportedly public British Museum had been run by a ticket-of-entry scheme. Other sites for natural display steadily emerged but without necessarily an expansion in audience constituency. The number and size of hospital museums (which comprised comparative, healthy, and morbid anatomy) grew quickly at the beginning of the century, but access was limited to students and staff of the hospitals and esteemed visitors with suitable introductions. From the 1820s, voluntary associations such as literary and philosophical societies established provincial natural history museums (Alberti 2002), and they granted their members and families free entry; access to nonmembers was at the keepers' discretion. Those collections that remained in private hands were accessible generally only to fellow virtuosos and appropriately introduced travelers.

Nevertheless, there were opportunities for the nonelite to experience natural objects. Anne Secord has demonstrated the thriving working-class cultures of botanical collecting and display (A. Secord 1994; Mosley 1928; Percy 1991). Proprietors of commercial anatomy shows were eager to admit any visitors who would pay their small fee (see figure 12.1): showmen-physicians such as Joseph Kahn, Signor Sarti, and J. W. Reimers proclaimed "know thyself" to the workingman (Altick 1978; Burmeister 2000; G. S. 1840; Kahn 1851; Knox 1834). Whether or not this constituted a genuine

THIS MUSEUM
CONTAINS
1000 Models and Diagrams
of the Human Body.

ILLUSTRATIVE OF HEALTH
AND DISEASE.

LIVERPOOL
MUSEUM OF ANATOMY
29, PARADISE STREET,

ADMITTED BY ALL
TO BE
AN INTERESTING STUDY
AND A
PUBLIC ADVANTAGE.

OPEN DAILY. For GENTLEMEN from 10–0 a.m. until 7 p.m.
ADMISSION SIXPENCE.
For LADIES—On Tuesdays and Fridays 2 until 5 p.m.

Figure 12.1. Illustration of the Liverpool Museum of Anatomy. Reprinted
from Liverpool Museum of Anatomy 1877, back cover. Courtesy of the
Wellcome Library, London.

democratization of medical knowledge, such shows were certainly popular, expanding significantly in the 1840s, advertising widely, opening in
the evenings, and admitting both men and women (albeit often at different times). "It cannot be but salutary to step aside for an hour," claimed
Reimers, "and quietly view the mechanism of the human frame" (Goulder
1853, 4). While doing so, one could conveniently purchase remedies for various popular ailments (especially spermatorea, the dreadful consequence
of onanism).

Middle-class museum reformers were alarmed by these and other such
unsavory sites for working-class leisure, and so they sought to present
their wholesome collections as alternatives. While mechanics' institutes
arranged exhibitions of industrial arts and the physical sciences (Inkster
1985; Moore 1991), natural history and anatomy displays were supplied by
other learned societies and educational establishments. Already in 1832
the Royal College of Surgeons Museum in Edinburgh was opened to the
wider public upon application to the conservator. In the first five years,
fifty thousand people visited, "of which three-fourths have been non-
professional, of both sexes and of all classes. None have [sic] been refused
admittance, excepting a very few (not more than five) persons in a state

of intoxication" (MacGillivray 1837). Although the college later had to restrict access as a cost-cutting measure, by midcentury other museums were seeking to open their doors wider as part of the middle-class program of "rational recreation" for the laboring classes. They permitted occasional access to workingmen on specific days or allowed general access subject to a small entrance fee.

Meanwhile, another administrative locale was emerging, eventually to become the most prevalent form of collection control in provincial England: the *public* museum. Parliament gave the increasingly powerful local authorities the power to levy a penny rate to maintain museums and free libraries in the Museum Acts of 1845 and 1850: over the following century, many society museums in Britain were absorbed by corporation-run museums, of which there was an exponential increase between 1850 and 1900 (Bennett 1995; Hill 2000, 2005; Lewis 1989; Pearce 1992). With the genesis of the municipal museum, the projected audiences for natural history museums were restructured. At the opening of an extension to the Leicester Museum, John Burns, president of the Board of Trade, proclaimed that "museums were absolutely essential if they were to provide for the great mass of people a nobler method of spending their leisure time than the public house" (Howarth 1913, 34). These noble intentions were asserted not only by those who ran town museums but also by those voluntary associations that continued to maintain collections. In Newcastle, the Hancock Museum remained the property of the Natural History Society, but its aims were similar to those of municipal institutions. In the words of Albany Hancock, "one of the primary objects of the Natural History Society has always been to disseminate a taste for natural history through-out the masses, and thus, so far, to educate and improve them" (Hancock 1864, 13).

In order to benefit from the museum as an educational tool, however, visitors needed to learn how to use it. As the Hull Literary and Philosophical Society (1860) expounded in the guide to its collections:

> Visitors to Museums often fail in deriving the instruction and pleasure which such collections are calculated to afford them, from a sense of confusion, which is the result of suddenly finding themselves in the midst of so vast a number of unknown and miscellaneous articles. Not knowing what a collection contains, and perhaps without a definite purpose in looking through its treasures, ignorant of the distinct char-acter and arrangement of its different parts, they overlook not a few of the things, which, if understood, would most interest them. For this

reason, as this volume aims to be a *Guide,* it will proceed to direct the
visitor to the different departments of interest in the Museum.

Just as nineteenth-century art exhibitions began to offer catalogs to assist in the viewing of their contents, natural history museum guidebooks
suggested routes around the collection (A. Fyfe this volume; Koven 1994;
Matheson 2001). Both serve to shape the visit, to construct what a visitor is
and does. The impact of such literature was compounded by the architectural regulation to which new buildings subjected visitors (Bennett 1995;
Forgan 1994; MacLeod 2005). Text, arrangement, and building worked together to condition the visit: the York Museum included a trail through
geologic time (Keeping 1881); in Hull, the collections were arranged from
simple to complex animal life (Hull Literary and Philosophical Society
1860); in Leeds, a route map was provided for schoolchildren (figure 12.2).
By parity in the National Gallery, the trustees ordered a new chronological
display of the collection that was designed to facilitate the "correct" reading of the collection.

Members of the public were to be admitted to museums, but they were
to be carefully managed. Self-regulation was to be encouraged among the
working classes, so long as the appropriate conduct was learned. Rigorous
behavior codes in museums served to transform the many-headed mob

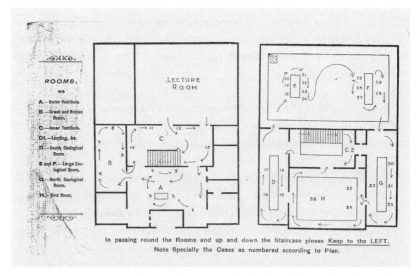

Figure 12.2. Route map for schoolchildren to follow around the Leeds Philosophical
Museum. Reprinted from Leeds Philosophical and Literary Society 1902–3, end matter.

into an orderly crowd, to promote appropriate tastes and generate a genteel, mixed-sex environment (Bennett 1995; Bourdieu 1984). To effect this atmosphere, museum staff had to wrench their institutions from the cultural locale of the festival and fair, and visitors had to be trained to appreciate the exhibits. As virtuosos ceased to personally guide visitors to collections, and before the emergence of the formal tour in the later century (discussed later in the chapter), police and later attendants or warders were stationed around the museum (Forgan 1999). In the first few weeks of general access at the Sheffield Public Museum, visitors' conduct was "most exemplary," and the curator observed "very few instances of disorderly or improper conduct" (Borough of Sheffield 1875, 19; Borough of Sheffield 1876, 7). In the later nineteenth century, opening hours were extended as far as possible in order to encourage working-class attendance, and electric lights were installed, with great fanfare, to this end (Hoyle 1898). Many museums dared Sunday opening, a contested practice that was hugely successful.

Municipal museums were not short of visitors on any day of the week. A small provincial museum might expect anywhere between two thousand and fifty thousand visitors per year, of which the majority were probably local. Those collections that remained within the remit of a philosophical society rarely exceeded these figures. Larger, public institutions, by contrast, attracted hundreds of thousands per annum, depending on opening hours and transport links (Greenwood 1888). An astounding 350,000 visited the Sheffield Public Museum in its first year, which rivaled even the mighty British Museum's half million (Borough of Sheffield 1875; Wilson 2002). Even taking into account multiple visits, it is clear that museums were a very common way for the public at large to experience natural history—alongside books, periodicals, and firsthand experience.

From the 1880s onward, curators of natural history museums were expanding their constituencies even further by seeking ever-younger audiences. Henry H. Higgins, first president of the Museums Association, was the earliest (or if not, the loudest) advocate of schoolchildren's use of museums, implementing a scheme of loans to local schools from the Liverpool Free Museum in 1884 (Allen 1994; Jenkins 1981). "Apart from the use of collections by specialists," the Northumberland Naturalists reported, "there is certainly no better purpose to which they could be put than that of interesting school children in natural history" (Natural History Society of Northumberland 1911–12, 256). By the turn of the century, curators sought not only to take the specimens to schools but also to bring schools to the museum. Previously, children had largely been denied admittance, especially if unattended, except on specified holidays; and fairly stringent

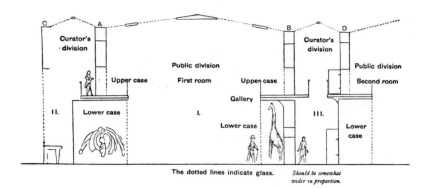

Figure 12.3. Illustration of the "new museum" idea. Reprinted from Huxley [1868] 1896, 127. Courtesy of the Manchester Museum.

conditions continued. (In Sheffield, children had been permitted only on certain days for one hour.) Gradually, however, natural history museums began to thrive in this role; lectures for schools were a great success, and by the First World War, the museum-school link was so established that museums were used as schoolhouses when the school buildings were taken up by hospitals (Kavanagh 1994).

These shifts to inclusivity notwithstanding, certain visitors continued to be more privileged than others. In the late nineteenth century, museum reformers such as Thomas Huxley sought to distinguish between those collections for public display and those for "serious study" by advocating the notion of the "new museum" (Huxley [1868] 1896). Research collections were to be kept in separate rooms from the public displays, in row after row of densely packed cabinets (figure 12.3). As well as saving space, this meant that access to the collections was limited to those deemed suitable by the curator, thus regimenting museum research and elevating the status of keepers. Through the application of the new museum idea, curators sought once more to limit access—this time to the research collections, which should "be used only for consultation and reference *by those who are able to read and appreciate their contents*" (Flower 1889, 14, emphasis added). The Manchester Museum had been transferred to Owens College by the beleaguered Manchester Natural History Society in 1868. When it finally opened two decades later, for the first year it was generally open to the public three days per week and to the college only for a further two. Even the Sheffield Public Museum was closed on Fridays to all except for suitable "students" (meaning those engaged in study in general, rather than college students). Such visitors were allowed to use museums' reserve collections and libraries and to remove items from the cases (Howarth 1889; Sheppard 1904).

Other natural collections were reserved entirely for exclusive access. William Hunter's collection, bequeathed to his alma mater, the University of Glasgow, was generally accessible only to medical students. The public was charged admission, and even arts students were allowed only a ticket or two per annum (Brock 1980; Coutts 1909, 515; Murray 1925). Not until 1907 were all students and the public admitted freely, by which time the anatomy collections had been removed to the departmental museum. The Hunterian and university museums such as those in Oxford and Cambridge were joined in the late century by collections in the new civic colleges (Alberti 2005; Forgan 1994, 1998a, 1998b). Museums comprised important architectural features of college buildings in many departments, from metallurgical and industrial to pathological and physiological. At Firth College's Weston Bank site, adjacent to the Sheffield Public Museum, for example, Alfred Denny, professor of biology, kept an extensive collection, allegedly one of the finest in the country (Chapman 1955; Denny 1898).

While some departmental collections were open to the public, generally their use was restricted to faculty and students—but the demographics of such visits are difficult to judge. Rare glimpses into historical visitor profiles are provided by surviving visitor books (Macdonald 2005), which documented the presence of esteemed persons gaining access to collections throughout the nineteenth century. Originating in the private cabinets of early modern collectors, in Victorian museum and Renaissance cabinet alike, such tomes served to proclaim the value of the collection and its national and international significance. Whereas staff and students of hospitals and colleges used the museums without leaving traces, autograph registers capture the presence of esteemed external visitors; in the late nineteenth as in the seventeenth century, education, nobility and distance traveled qualified them to sign the books. The museum thereby collected notables to add kudos to the collection. Visitors came from all reaches of the globe. The Guy's Hospital Museum visitor book reads like a gazetteer and includes the Duke of Devonshire (twice), Lord Bowingdon, Mungo Park the African explorer, the Duke of Wellington, Lord Brougham, and on one notable occasion a Chinese "minister" and his entourage, likely to have been the general and statesman Li Hung-Chang (Guy's Hospital Museum 1862; Guy's Hospital Museum 1900). At the turn of the twentieth century, the Manchester Museum was graced by the presence of Mary Kingsley, the Duke and Duchess of Devonshire, and a host of donors, politicians, "men of science," and curators (Manchester Museum Owens College 1891;

Manchester Museum 1938). The demographic—professionals, nobles, scholars, and so forth—is surprisingly similar to that of the recorded visitors to Ulisse Aldrovandi's museum in Bologna three centuries earlier (Findlen 1994). And like their early modern counterparts, it is likely that such prestigious visitors benefited from personal tours from curators.

The late nineteenth century, then, was marked by twin developments in access criteria for museums of natural objects. On the one hand, a broader public was constructed for municipal and national museums, as part of the program of rational recreation and science education. On the other, the educated and elite were accorded privileged access to these same public collections; and medical and other specialized museums remained behind closed doors. Those that did not were condemned for their permeability: the *Lancet* considered Kahn's anatomy collection to be of "certain professional interest, but totally unfit for general exhibition" ("The Action against Kahn" 1857, 175). It is no accident that these developments occurred during the construction of professional communities in the sciences and in museums, as demonstrated by the formation of the Museums Association in 1888 (Desmond 2001; Lewis 1989). Both the expansion and the restriction of access can be understood as attempts by these new professionals and others to regulate the museum visit, to train the visitor. United in their ostensible dedication to pedagogy, the hospital museum was intended to be an encyclopedia of anatomy and disease for diagnostic purposes and the natural history collection a monograph of the district for serious study. But access was only the beginning of the visiting experience. In the following sections, I turn attention to the responses of the visitors themselves, to the sensory and emotive impact of natural collections. The regulation of access, I will argue, was accompanied by attempts to condition the very experience of the visit.

Senses

The first sense we now associate with the museum visit is sight. Theorists of visual culture have posited historically contingent "scopic regimes" and have characterized the high Victorian era as an age of panoramic spectacle (Brennan and Jay 1996; Crary 1990; Hyde 1988). Performative and graphic aspects, as Morus has argued, were integral features of technological and scientific developments in the Victorian era; a growing number of historians have also emphasized the importance of the visual to Victorian science and to its popular manifestations in particular (Lightman 2000; Morus 1998; Pang 1997; A. Secord 2002). Popularizers of science were at the forefront of the emerging visual culture, and they bombarded their audiences with a range of ocular technologies in print, in the lecture hall, and in the museum.

Not that visual bombardment was new to the museum world—cabinets had dazzled visitors since the Renaissance. But whereas miscellany had been a cause for delight in earlier periods, Victorian observers commended order, and confusion was cause for opprobrium. The Middlesex Hospital museum was "a tiny room" with "a motley collection of specimens . . . a conglomeration of jars crowded with specimens with which no one could keep pace" (Thomson 1935, 83–84). "The man who for the first time in his life makes an attempt at going over the museum," wrote a Guy's student with a sharp pen, "is apt to be somewhat disconcerted by the number of specimens he will have to look at . . . [such that] he feels inclined to exclaim, 'oo-er,' as the lady did on an historic occasion, and to seek a more congenial if equally 'thick' atmosphere in the Club smoking room" ("Guide to the Pathological Museum" 1898, 69).

In *Our Mutual Friend* (Dickens [1864–65] 1997, 86–88), the ghoulish taxidermist Mr. Venus greets Mr. Wegg, who has come into Venus's macabre place of business seeking the remains of his own leg. Mr. Venus says, "You're casting your eye round the shop, Mr. Wegg. Let me show you a light. . . . Bones, warious. Skulls, warious. Preserved Indian baby. African ditto. Bottled preparations, warious. Everything within reach of your hand, in good preservation. The mouldy ones a-top. What's in those hampers over them again, I don't quite remember. Say human warious. Cats. Articulated English baby. . . . Glass eyes, warious. Mummied bird. Dried cuticle, warious. Oh, dear me! That's the general panoramic view." Dickens's vision parodies the Victorian panorama by revealing its underbelly, a heterodox site for display that was soon to be eclipsed. In the expanding public museum sector in the late Victorian period, the emphasis shifted to transparent, ordered display and clear labeling, which were constituents of particular ways of displaying and viewing. Like General Pitt Rivers's archaeological exhibits, natural history displays were intended to "speak to the eyes" (Bennett 2004). Looking was the fastest and most effective way to learn—simply to place an object in view would be sufficient to embed it within the memory and understanding of the museum visitor. Such absorption, however, depended on a sustained gaze. The glance, the passing look, the wandering attention of the passing viewer—all likewise historically contingent—produced very different, often undesired, effects. The "transient observer" had a very different experience from that of someone with a trained eye (Bicknell 1846–47, 1:307). As the naturalist Eliza Brightwen advised the student of nature, "He must pass alone, from chamber to chamber, down corridor after corridor, until he discovers that sleeping princess, Knowledge, who is never found unless we industriously seek for her. All I can do is point out the difference between languidly strolling with vacant face between the glass walls of our great museums, and passing eagerly with intelligent interest from one cabinet of recognised treasures to another" (1892, 222). The museum visitor was intended to gaze intently upon the objects, in calm contemplation. "Went to Owens College Museum in afternoon," jotted Beatrix Potter in her journal in 1895, "very cool and quiet among the fossils" (Potter 1966, 396).

Not all visit(or)s were so quiet. Whether from personal tours, from the ever-popular museum lectures, or from the noisy bustle of the audience, the visit was an aural as well as a visual experience. The museum was still a site for sociability, as it had been up to the Enlightenment (Alberti 2003; Bennett 2004; Findlen 1994). Rather than social equals conversing with the collector, however, visitors were talked *at* in the museum. No doubt, however, they continued to share their responses with one another: as James Secord demonstrates in chapter 2 of this volume, conversation was a crucial component in the experience of science. In commercial anatomy shows in particular, the exhibits must have resounded with the ballyhoo of the proprietor as he touted his business and the exchanges of the visitors as they encountered the curious, the titillating, and the diseased. Admission to such establishments, in common with many other museums, involved a museum lecture; Joseph Kahn employed Dr. Sexton to lecture to men and Mrs. Sexton to women (Mason 1994). In the genteel surroundings of the British Museum in 1913, according to Cecil Hallett, an official guide, it was still the case that "the public, as a rule, are not given to the study of guide books, nor to the reading of labels . . . nothing can bring the general public and a museum into a right relation with each other so well as the living voice of a human expositor" (Hallett 1913, 200). Such museum guide programs were formalized in the early twentieth century to provide the public with this desired exposition.

Although museum lectures and talking guides continued throughout the century, gradually those who ran the collections sought to eradicate other noise—a move that can be seen as part of a broader movement toward collective noiselessness in the public sphere. Like the art gallery, theater, and the concert hall, the museum was no longer intended to be a space for conviviality but for silent contemplation (Gunn 1997; Sennett 1977; Stafford 1994). Silence as an ideal in museums and elsewhere, however, was not always realized. Curators' efforts—polite notices, for example (see the

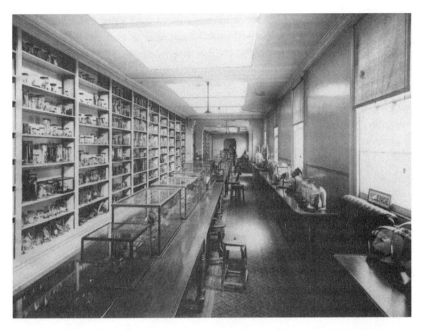

Figure 12.4. Photograph of the Guy's Hospital Museum in the late nineteenth century. Reprinted by kind permission of the Trustees of Guy's and St. Thomas' Charity.

sign at the right in figure 12.4)—tell as much of their failures as their successes in this endeavor. In hospital and college museums, for instance, the presence of medical students, a notoriously rowdy body, rendered silence unlikely. "On entering [the museum] it is well to get together a few stools," advised the cynical Guy's student, "(not necessarily to sit on, but as a guarantee of hard work) which should be dragged along the galleries, not carried. This will impress any junior man, which may be reading anatomy in the lower regions, with your importance as a pathologist" ("Guide to the Pathological Museum" 1898, 69). "The greatest and worst difficulty is that of noise," complained Hallett, because museum spaces "are particularly resonant, and seem to magnify the slightest sound; and persons who do not attach themselves to the Guide's party are as a rule utterly without regard for the convenience of anyone but themselves. It rarely occurs to them to lower their voices, or to attempt to walk quietly" (Hallett 1913, 194). It is not unreasonable to assume that the same was true in South Kensington as it was in Bloomsbury (Leonard 1914).

If the aural experience has been too often ignored in the history of museums, so too has the tactile (Hetherington 2003). Late in the eighteenth century, William Hunter ([1767] 1784, 112) instructed that "specimens must be sent round the company; that every student may examine them in his own hand." One of the principal roles of nineteenth-century institutional medical and biological museums alike was to supply the lecture courses in anatomy, physiology, and pathology with specimens and diagrams. They were to provide handling specimens and to act as anatomical store cupboards—which was how many collections originated. The collections supplied objects for the dreaded medical viva voce throughout the nineteenth century and beyond. Students fretted of the specimens that "any one of them may suddenly be hurled at him (like a bolt out of the pathological blue) by a merciless examiner in a place where catalogues are not and labels are unknown" ("Guide to the Pathological Museum" 1898, 69).

While students and other visitors with privileged access continued to be allowed to handle collections, specimens had to be protected from the masses of untrained visitors. At La Specola in Florence, where the anatomical collection was opened to the Tuscan public as early as 1789, visitors soon set about molesting the specimens—especially the wax genitalia (Düring, Didi-Huberman, and Poggesi 1999; Maerker 2005). At the Royal College of Surgeons of Edinburgh, by contrast, working-class visitors behaved impeccably:

> There has never been a disturbance in the Museum. Although upwards of 2,000 preparations have been exposed on open shelves, none of them has received injury from visitors. . . . Visitors of the lower classes, mechanics, sailors and soldiers, have uniformly been quiet, careful and most orderly. Indeed the only visitors who ever touch the preparations are the medical students, who, in their desire to inspect an object, sometimes forget that it is prohibited to handle it. Visitors of the lower classes seem to take more interest in the specimens than those of the higher, many of whom, especially ladies, merely walk through the room without looking at the objects particularly. (MacGillivray 1837)

Not all visitors were so obliging. Carroll details in her account of visiting Charles Waterton's collection at Walton Hall that Thomas Dibdin sought to gain entrance by assuring the housekeeper that he and his daughter "were peaceable and honest, and would touch nothing." Dibdin explained, "These were not idle words. On the contrary, they had too significant a meaning. Many visitors of this spot of enchantment had exhibited the too predominant, and too scandalous, propensity of their country, to defile and destroy much of what was placed before them" (Carroll 2004; Dibdin 1838, 1:148–49). He had a point. When the poet and civil servant Arthur Munby visited an anatomy exhibition in 1859, he gazed, fascinated, at the female of the two human skins on display. Finally, he "lifted the stiff hand, & touched the dusty hair" (Reay 2002, 36).

But just as the exhibits at La Specola had to be locked in display cases, Victorian collections were gradually removed from tactile range. The use of formal vitrines increased, standardizing museum displays and distancing the observer from the specimen. The objects were reified, rendered sacrosanct. They also became more mysterious—for to sense temperature is to distinguish living from dead. Munby wanted to confirm that the skin was dead not only with sight but also with touch. Sight is extremely suggestive but notoriously deceptive. Touching objects confirms suspicions or dispels fears or both; denied this tactile resolution, visitors were disconnected from the objects.

Students in Rudolf Virchow's pathology museum in Berlin were encouraged to learn from specimens not only by handling but also by smelling them (Matyssek 2002). To deny visitors touch and to remove sources of noise from the museum had some success. But still the museum experience could not be restricted to the visual, for objects that had once been alive tended to smell. The rich olfactory history of natural history and especially medical museums remains to be written, but some observations can be made. Certainly it seems that the aroma of such places remained in the visitor's mind. As David Freedberg notes in his history of response, viewers are even "arrested by size or color or the suggestion of smell, or nauseous agglomeration of surface texture" (Freedberg 1989, 435). Collections of wet anatomical specimens tended to be the most pungent. Ironically, it was the fluids involved in the very process of preservation that generated the distinctive smell of such museums, but nevertheless the aroma was associated with death and decay. The close association of smell and disease in the presence of so much visual evidence of the morbid was overwhelming. The miasmatic notion of aroma as pathogenic—Chadwick's insistence that "all smell is disease"—endured long after the advent of germ theory (Stallybrass and White 1986, 139). And death was never far way: pathological collections were commonly stored adjacent to the postmortem facilities. At Guy's Hospital, Thomas Hodgkin was curator and inspector of the dead, and even he acknowledged "the emphysematous state of the subject" and "the odour exhaled by the bodies of the dead" (Hodgkin 1828, 424). His successor Samuel Wilks described a specimen as a "horny excrescence, from the head of a woman. . . . When recent these were soft, waxy, and had a cheesy odour" (Wilks 1861, 53). It was common practice to smoke in the dissecting room to cover such noxious odors.

Preservation should have dispensed with these organic scents but was not always successful. The zoological department of the British Museum came under censure for its olfactory expansion: Antonio Panizzi, keeper of printed books and later chief librarian, complained that "the smell of putrifying [*sic*] animal matter extended to the north end of the Royal Library" (Gunther 1975, 104), and a student at Firth College in Sheffield similarly reported "a decidedly fish-markety aroma" in the biology collection ("Delta" 1897). The medical museum was, after all, as one student tastefully put it, a storehouse for "dead meat" ("Our Special Pathologist" 1898, 42). The association of dead specimens and food had been a common theme early in the century. At the Hunterian Museum in Glasgow, the preparations in the Hall of Anatomy were displayed on "presses"—that is, "A large (usually shelved) cupboard . . . in Scotland, also for provisions, victuals, plates, dishes and other table requisites" (*OED*, 2nd ed., s.v. "Press"; Laskey 1813). The bookselling firm Wheatley and Adlard, during the depression in the book trade in the 1820s, diversified into wine and medical specimens at the same time (De Chantilly 2001). As the firm advertised at one sale, "On the table is a superb assortment of various organs corroded: and almost unexampled injections of the hand and foot, inimitably dissected, with numerous other anatomical *chef-d'œuvres*"—a common culinary allusion (Kahn 1853; Wheatley and Adlard 1830, viii). The sale in question was that of the collection of the renowned anatomist Joshua Brookes, whose classroom was said to have smelled like a ham shop (Desmond 1989, 160) and whose collection was known to have made visitors recoil in horror at the stench (Bliss 1999).

That such complaints endured, despite flues and ventilation in late nineteenth-century custom-built premises, may be illustrative of the changing role of smell in modern society. During the Enlightenment, Alain Corbin has argued, the "olfactory revolution" relegated smell from the sacred to the profane, from the realm of the religious to that of the sensual, sentimental (Corbin 1986; see also Classen, Howes, and Synnott 1994). As cities became more fragrant thanks to sanitary reforms, and the revolution in civic cleanliness was accompanied by similar changes in the personal realm, so the middle classes became less tolerant of stench. That any smell remained in the museum was unsatisfactory as the ideal public space became free from odor.

By the close of the Victorian era, odor-free museums were intended to be about sight: the sense of reason, rationality, education, and the chosen sensory mode of late nineteenth-century scientific enterprise. Smell—emotive,

savage, intuitive—was to be avoided, as were other sensory pollutants: de-
structive touch and unruly, chaotic sound.

Responses

Over the course of the nineteenth century, then, accepted museum practice in the arts and the sciences was geared toward encouraging particular
codes of appropriately visual, silent, and hands-off behavior. As conditioned
as the sensory experience of museum visiting was intended to be, however,
a significant range of reactions endured, and seldom did visitors respond in
the way curators hoped or anticipated. And yet they responded, powerfully,
engagingly, and viscerally. Traces of these responses have already emerged
in the preceding discussion—from happiness to horror—and this section
focuses in more detail on particular kinds of reactions to natural history
and anatomy displays.

Useful here is the definition of response utilized by some historians
of art: that responses and their traces are the symptoms of the relationship between the object (or image) and the observer (Freedberg 1989, xxii).
This relationship is historically and culturally contingent, but it is never
one-way. However didactic and interpreted the exhibition, responses were
a combination of that which was elicited by the display and that which
came from within the visitor—things remembered and felt. Both memory and emotion have powerful roles to play in the visiting experience—
contributing to the construction of the museum effect and affect. Kavanagh
(2000) has explored their roles in the contemporary visit, but to extend her
approach to the Victorian visitor requires careful historical sensitivity.
Recent cultural history has demonstrated the contingency of affect—so
too emotional response is specific to era and culture (Bound Alberti 2006;
Bruno 2002). And so in setting out responses to natural collections, I do so
in light of the historical development of emotions.

Reactions are and were unpredictable and unruly, and to explore the
history of varied responses from a different era is necessarily to present a
rather messy picture that defies an overarching narrative. One could profit-
ably analyze the shifting emphasis on curiosity (Benedict 2001), fear (Ogden
1774), or sympathy (Brightwen 1892). Here, however, I shall concentrate on
two responses from either end of the moral spectrum—wonder and disgust.
As such I present only a sample of the evident reactions and leave, for now,
the typology of visitor response unwritten.

How, then, did Victorian visitors respond to natural collections upon

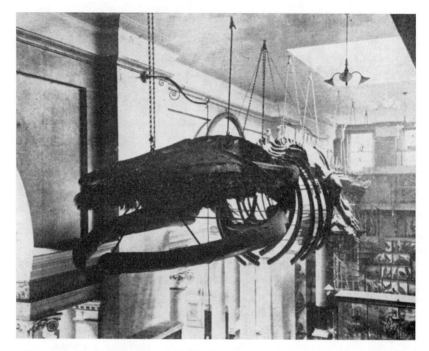

Figure 12.5. Photograph of the vast Sibbald's Rorqual at the Hull Municipal Museum, 1904. From Sheppard 1904, facing page 29. Reprinted by permission of the Brotherton Collection, Leeds University Library.

entrance to a museum? From the few personal traces left of museum attendance, it seems that visitors gravitated first to the most spectacular exhibit—usually the largest (Stewart 1993). In his printed guide to the Hull Museum, the curator Thomas Sheppard encouraged visitors to turn left into ethnography and antiquities as they entered. The visitor's eye, however, was nearly always captured by the whale on the right, suspended above case 14, and similar specimens at Edinburgh and South Kensington were just as captivating (Holdsworth 1904; Sheppard 1904; Thompson 1904). "Upon entering," wrote eleven-year-old Eva Nightscales, "the first thing that drew my attention was the large whale suspended from the ceiling" (Nightscales 1904, 30; see figure 12.5).

Massive specimens impressed young and old, expert and lay alike. The physician C. G. Carus, accompanying the king of Saxony on his tour of Britain, was shown around the London Hunterian. It was not the recent human remains that first attracted his attention but rather the ancient animals. He recorded his response upon entering the museum: "On the very entrance the attention is immediately arrested by the rarest fossil animals; on the right, the great armadillo from Buenos Aires (*clyptodon clavipes*), with its massive bony scales, almost like an immense egg, of the size of the largest drum. Opposite to it, on the left, is a gigantic creature of the sloth tribe (*mylodon robustus*), with its bird like pelvis, and rudely power- ful bones, set up as if to ascend the stem of a tree. At the end of the room, the skeleton of a magnificent elephant rises far above everything around" (Carus 1846, 60). Thirty years later, the display elicited a similar response from the physician Andrew Wynter: "The first things that strike the eye of the visitor are the enormous skeletons of different animals. . . . The mastadon [*sic*] and the hippopotamus can here also be seen in undress." He concluded, "The skeletons of the larger animals strike the visitor most forcibly" (Wynter 1874, 1:264–65).

Size invoked a powerful sense of wonder in the visitor, a response that was prompted more by skeletal fossil remains than their fleshed recent counterparts. Their difference, their remoteness from the visitors' here, now, and self shifted the displays from the real toward the fantastic. At the British Museum, "The student surveys these *leaves* of the book of nature with wonder and surprise, and having gazed upon these *giant* remains of by-gone times, he retires from them, scarcely believing the *reality* of that which he is *certain*" (Bicknell 1846–47, 1:309, emphasis in the original). Throughout the century, visitors of all qualifications relay their sense of awe and wonder at the gigantic (as they still do; see Stewart 1993). Vastness was integral to the romantic notion of the sublime, and even its subse- quent Victorian middle-class "taming" did not remove this aspect of the museum experience. Curators took advantage of the popularity of the im- mense: Richard Owen, for example, constructed the new order Dinosauria as iconic museum pieces (Rupke 1994; J. Secord 2004).

The whale that hangs from the ceiling or the elephant that loomed at the end of the hall were particular, removed from the rest of the collection by their size or position. Wonder, argues Greenblatt (1991), is prompted in museums by the arresting uniqueness of such objects. "Since the Enlight- enment, however," Daston and Park note, "wonder has become a disrepu- table passion in workaday science, redolent of the popular, the amateurish, and the childish" (1998, 14–15). Nevertheless, the sense of wonder that was the raison d'être of Renaissance cabinets of curiosity endured in distinct forms through the Victorian era. Wonderful objects stimulated an intense, enchanted gaze in the Victorian observer—the brighter side of the disci- plinary gaze that historians find so rampant in nineteenth-century cities (Brennan and Jay 1996; Crary 1990). As Sappol notes, museums of anatomy were known to "precipitate a joyous epiphany" (2002, 269) and natural history museums likewise. The language of response continued to be full of "wonderful," "marvellous" and (especially) "remarkable," emphasizing the unique properties of specimens even as display techniques were designed to set them within taxonomic schemes, removing them from the realm of the curious.

But visitors refused to be policed. Responses came not only from the objects on display but from the visitors themselves, their experiences, and their imaginations. Wonder led many visitors into flights of fancy. Thomas Hardy was particularly poetic: visiting the Albert Memorial Museum in Exeter in 1915, he stopped by a skeletal cast of the fossil bird *Archaeopteryx macrura*—which inspired the composition of a short poem, "In a Museum":

Here's a mould of a musical bird long passed from light,
Which over the earth before man came was winging;
There's a contralto voice I heard last night,
That lodges in me still with its sweet singing.

Such a dream is Time that the coo of this ancient bird
Has perished not, but is blent, or will be blending
Mid visionless wilds of space with the voice that I heard,
In the full-fugued song of the universe unending. (1976, 430)

The visitor's imagination rendered the present haunted by the past and filled the now-silent museum with noise. The experience was not always so pleasant, however. The writer Thomas de Quincey visited Charles White's medical museum in Manchester, and upon finding that the skeleton on display there had once been a highwayman, was prompted into a long imaginative contemplation upon the criminal's former activities (de Quincey 1862, 432–40).

The viewer's imagination, then, reanimated dead exhibits. Audiences brought their own experiences to the museum—which could engender discomfort as easily as it could pleasant feelings. Visitors described their visits in terms of fear and nightmares: and as such, they appropriated the language of a widely available cultural discourse, the Gothic, to articulate their emotional experience (Mighall 1999). The roots of Gothic literature were in the fascination with an imagined and terrible past, and visitors' engagement (with fossils in particular) may reflect this interest. By the late nineteenth century, medicoscientific discourses contributed to the somatic emphasis of Gothic fiction, which privileged racial degeneration, atavism, deviant sexualities, and monstrosity—examples of which could all be found in anatomical museums (Hurley 1996). The awe-some aspect of the wondrous was closely tied to the awe-ful responses such as disgust.

Disgust, indeed, that "most embodied and visceral of emotions," was a common reaction to ex-living displays in the nineteenth century (Menninghaus 2003; Miller 1997, xii). Arthur Munby is particularly detailed in his account of the lifelike specimen at an anatomy show: "The woman's skin was quite perfect from head to foot: it was slit down the back, and hung loosely on a wooden cross, in hideous mockery of the living figure. . . . It was horrible to look at in the face—it was like a leather mask, every feature perfect, yet hanging helpless & collapsed—the nose awry, the lips drooping, the eyes wide and empty" (Reay 2002, 35). Even the physician Andrew Wynter acknowledged that the Hunterian's display of morbid stomachs was "not very pleasant to contemplate" (1874, 1:268). Kahn's museum was variously considered "horrible" (Hibbert 1916, 41) and "revolting, filthy and disgusting" ("An Obscene Exhibition" 1856, 376–77). And it was not only human remains that turned the visitor's stomach. Students at Firth College in Sheffield dubbed the biological collection the "diabolical museum": as one student complained in 1897, "Miserable relics called specimens confront you everywhere" ("Delta" 1897).

Over the last three centuries the sight of the corpse and the internal workings of the human body have increasingly been deemed disgusting, and in the Victorian period the anatomy museum was one of the few remaining legitimate sites for its display. Increasing squeamishness in this respect was testament to the widespread expansion of disgust since the eighteenth century. "Disgust," writes Miller, "has a key role to play in the civilizing process, working as it does to internalise norms of cleanliness, reserve and restraint" (Miller 1997, 20; see also Elias 1994). While disgusting things were removed from the public sphere, that which constituted "disgusting" expanded. Disgust was in part a reaction against the ugly, the abnormal (themselves categories in the process of redefinition)—an emotion working to police the norms of modern society, elevating the moral, the middle-class, the clean, from the squalors of poverty, filth, and diseased deformity. Venereal diseases in wax and in jars were clearly moral offenses, especially when they disrupted the skin, where beauty should lie. Further, revulsion also situated other forms of ugliness on display within the moral realm—monsters and even exotic ethnographic specimens.

Disgust was rarely the explicit intention of those who constructed the displays. In the first half of the nineteenth century, the wax models discussed previously were widely employed as a means to *eliminate* the disgust associated with dissection. One anatomical gallery claimed that its model Venus could illuminate anatomy "without in the slightest degree wounding delicacy," and another "can be looked upon without any of those repulsive feelings that would strike many on inspecting a human subject dissected" (Burmeister 2000, 46); the *Athenaeum* advised its readers "to avail themselves of a few general ideas on the subject of anatomy" from Signor Sarti's wax museum, "which they may do so without labour or disgust" ("Our Weekly Gossip" 1839, 279; Mawhinney 1854). As such, they were even suitable for a female audience.

Common in London and the provinces, such commercial anatomical shows were applauded in the early nineteenth century: the *Lancet* had been "much gratified with the collection of anatomical and surgical curiosities" at Joseph Kahn's museum, for example ("Dr. Kahn's Anatomical Museum" 1851, 474). In the second half of the century, however, they were the objects of virulent criticism (Burmeister 2000; Sappol 2002). If the *Times* and the *Lancet* are to be believed, these were no longer sites for a young man to "know thyself," as they proclaimed, but for him to know female anatomy, and rather too intimately at that (Mason 1994; "The Museum Nuisance" 1864). One medical student recorded in his diary in 1860:

> DECEMBER 7ᵀᴴ: Went to Dr. Kahn's Museum in the afternoon. A decidedly indecent pseudo-scientific affair, founded by quack doctor. . . . Lecture on Deleterious Influences was of course disgusting.
>
> DECEMBER 8ᵀᴴ: Had a rather serious attack of diarrhœa, a malady to which I am seldom subject. Can it have been bestowed on me by an allwise Providence as a just punishment for my having visited that sink of iniquity, Dr. Kahn's Museum yesterday? Who knows? (Taylor 1927, 16)

Proprietors of such shows defended them as morally and scientifically instructive, and they duly reserved the most sensitive material for gentlemen or even "medical men" only ("Dr. Kahn's Anatomical Museum and Gallery of All Nations" 1851; "Dr. Kahn's Anatomical Museum" 1854). Their critics nevertheless condemned them as sensationalist and pornographic. "Anatomical museums," raged the same *Lancet* that had once applauded them, "offer to the sensual cravings of the more degraded members of the community genial recreation" ("Anatomical Museums" 1865, 600).

Rather, only within properly delimited spaces could appropriate audiences view anatomy soberly, without arousal or revulsion. The *Times* approved of Joshua Brookes's museum: "What on a first view is not only displeasing to the sight, but even disgusting, becomes a matter of interest and examination, when considered as the means of improving science and assisting the cause of humanity" ("Brookes's Museum" 1830, 3f). Sites for the display of anatomy and pathology were segregated, accessible only to those who knew how to react properly. By the end of the nineteenth century, there were far fewer places where the uninitiated could experience dead bodies and their representations. Gruesome specimens that were likely to provoke disgust were removed from the public, and especially female, view. Those with access knew that the appropriate external reaction was sober contemplation rather than disgust or arousal.

Conclusion

From this preliminary exploration of a variety of experiences, we can see a pattern not so much of changing reactions to collections but attempts by those displaying them to limit the range of acceptable responses. Just as extreme reactions to fine art were suppressed over the modern era and gallery audiences were trained in the appropriate tastes and behaviors, so too responses to displays of nature and anatomy were conditioned, both sensually and emotionally (Freedberg 1989; Hemingway 1995; Matheson 2001). The performative codes woven into the nineteenth-century public sphere applied in the museum.

By the late Victorian era, educators, curators, and men of science seeking to construct professional communities were working to arrange collections—whether open or closed—as ordered or taxonomic or both. Haphazard private collections were eclipsed by regulated public museums and dedicated educational collections. Those that did not conform to these standards—the taxidermist's shop, the lurid anatomy show—were deprived of credibility. Distinctions were reinforced between street shows and worthy museums, between the circus and the zoological garden, between low and high medicine, between unruly crowd and well-behaved visitors. They sought to replace the perceptual promiscuity of the cabinet of curiosities with a regulated gaze, presenting the museum as a site for remote, reasoned observation rather than gawking spectacle (Bennett 1995, 2004; Stafford 1991, 1994). In this period when audiences for cultural activities were being herded, civilized, and quieted, the museum visitor was subject to a new, strictly scopic regime, and the opportunity to experience the unruly sensations of the oral, tactile, and olfactory realms was restricted. In doing so, the range of emotional responses available to them was ostensibly narrowed. Mere curiosities were discouraged; wonder deemed inappropriate; that which might be construed as disgusting was removed from sight.

But visitors were not passive dupes, and approaching this study with an active audience model in mind has been profitable. Throughout this account there has been evident the tension between museum authorities controlling the visit and museum audiences constructing their experience and affective responses. Visitors chatted, touched things, and complained about the smell. That the objects had once been alive made reactions very difficult to standardize, to regulate, or even to gauge. As curators well knew when they displayed whales and dinosaurs, visitors walked straight past carefully worded labels and gaped at the largest specimen. Carefully classified fossils prompted flights of fancy to lands far away and times long ago. Visitors trembled with fear or winced in empathy; they were aroused by models and repulsed by corpses.

These reactions, and visitors' memories and imagination, were historically specific. Response theorists posit that reception is locally embedded within particular interpretive formations and social conditions (Bennett 1996; Machor and Goldstein 2000). The wonder and disgust felt by visitors in 1800 were different from those experienced by their counterparts a century later, working for different reasons, in a different climate. Affective responses to natural collections were transient, ephemeral, and if it is possible at all, much work remains to be done not only to recapture them but also to link them to gender and class and to culture, place, and era.

By studying curatorial intention alongside audience response, it has become clear that as didactic as displays may have become, the museum-visitor encounter was dialogic. The standardization of the museum space, the route maps, educational schemes, guidebooks, and attendant-guide programs were constructed to police this enduring diversity of behaviors and the emotive reactions of the gallery space. The increased conditioning of the museum visit that gave rise to the sanctity of the modern museum object—the museum *effect*—developed in response to the museum *affect*.

NOTES

Especially I would like to thank Fay Bound Alberti; I am also grateful to Aileen Fyfe, Barbara Gates, Dawn Kemp, Lawrence Keppie, Bernard Lightman, Bernadette

Lynch, Sadiah Qureshi, and Helen Rees Leahy for their help and suggestions. Writing this chapter was made possible by the Wellcome Trust, the Manchester Museum, and the University of Manchester Centre for Museology.

REFERENCES

"The Action against Kahn, of Coventry Street." 1857. *Lancet*, August 15, 175.

Alberti, Samuel J. M. M. 2002. "Placing Nature: Natural History Collections and Their Owners in Nineteenth-Century Provincial England." *British Journal for the History of Science* 35:291–311.

———. 2003. "Conversaziones and the Experience of Science in Victorian England." *Journal of Victorian Culture* 8:208–30.

———. 2005. "Civic Cultures and Civic Colleges in Victorian England." In *The Organisation of Knowledge in Victorian Britain*, ed. Martin J. Daunton. Oxford: Oxford University Press, 337–56.

Allen, David Elliston. 1994. *The Naturalist in Britain: A Social History*. 2nd ed. Princeton, NJ: Princeton University Press.

Alpers, Svetlana. 1991. "The Museum as a Way of Seeing." In *Exhibiting Cultures: The Poetics and Politics of Museum Display*, ed. Ivan Karp and Steven D. Lavine, 25–32. Washington, DC: Smithsonian Institution Press.

Altick, Richard D. 1978. *The Shows of London: A Panoramic History of Exhibitions, 1600–1862*. Cambridge, MA: Harvard University Press, Belknap Press.

"Anatomical Museums." 1865. *Lancet*, June 3, 600–601.

Benedict, Barbara M. 2001. *Curiosity: A Cultural History of Early Modern Inquiry*. Chicago: University of Chicago Press.

Benjamin, Walter. [1936] 1999. "The Work of Art in the Age of Mechanical Reproduction." In *Illuminations*, 2nd ed., trans. Harry Zorn, ed. Hannah Arendt, 211–44. London: Pimlico.

Bennett, Tony. 1995. *The Birth of the Museum: History, Theory, Politics*. London: Routledge.

———. 1996. "Figuring Audiences and Readers." In *The Audience and Its Landscape*, ed. James Hay, Lawrence Grossberg, and Ellen Wartella, 145–59. Boulder, CO: Westview.

———. 2004. *Pasts beyond Memory: Evolution, Museums, Colonialism*. London: Routledge.

Bicknell, W. I. 1846–47. *Illustrated London, or a Series of Views in the British Metropolis and Its Vicinity*. 2 vols. London: Brain.

Bliss, Michael. 1999. *William Osler: A Life in Medicine*. Oxford: Oxford University Press.

Borough of Sheffield. 1875. *Nineteenth Annual Report of the Committee of the Free Public Libraries and Museum*. Sheffield: Leader.

———. 1876. *Twentieth Annual Report of the Committee of the Free Public Libraries and Museum*. Sheffield: Leader.

Bound Alberti, Fay, ed. 2006. *Medicine, Emotion and Disease, 1700–1950*. London: Palgrave Macmillan.

Bourdieu, Pierre. 1984. *Distinction: A Social Critique of the Judgement of Taste*. London: Routledge.

Brennan, Teresa, and Martin Jay, eds. 1996. *Vision in Context: Historical and Contemporary Perspectives on Sight*. London: Routledge.

Brightwen, Eliza. 1892. *More about Wild Nature*. London: Unwin.

Brock, C. H. 1980. "Dr William Hunter's Museum, Glasgow University." *Journal of the Society for the Bibliography of Natural History* 9:403–12.

"Brookes's Museum." 1830. *Times* (London), February 26, 3f.

Bruno, Giulana. 2002. *Atlas of Emotion: Journeys in Art, Architecture and Film*. New York: Verso.

Burmeister, Maritha Rene. 2000. "Popular Anatomical Museums in Nineteenth-Century England." PhD diss., Rutgers University.

Butler, Samuel. [1881] 1986. *Alps and Sanctuaries of Piedmont & the Canton Ticino*. Gloucester: Sutton.

Carroll, Vicky. 2004. "The Natural History of Visiting: Responses to Charles Waterton and Walton Hall." *Studies in History and Philosophy of Biological and Biomedical Sciences* 35:31–64.

Carus, Carl Gustav. 1846. *The King of Saxony's Journey through England and Scotland in the Year 1844*. Trans. S. C. Davison. London: Chapman and Hall.

Chapman, Arthur W. 1955. *The Story of a Modern University: A History of the University of Sheffield*. London: Oxford University Press.

Classen, Constance, David Howes, and Anthony Synnott. 1994. *Aroma: The Cultural History of Smell*. London: Routledge.

Cooter, Roger, and Steven Pumfrey. 1994. "Separate Spheres and Public Places: Reflections on the History of Science Popularization and Science in Popular Culture." *History of Science* 32:237–67.

Corbin, Alain. 1986. *The Foul and the Fragrant: Odor and the French Social Imagination*. Leamington Spa, UK: Berg.

Coutts, James. 1909. *A History of the University of Glasgow*. Glasgow: Maclehose.

Crary, Jonathan. 1990. *Techniques of the Observer: On Vision and Modernity in the Nineteenth Century*. Cambridge, MA: MIT Press.

Darnton, Robert. 1990. *The Kiss of Lamourette*. London: Faber.

Daston, Lorraine J., and Katharine Park. 1998. *Wonders and the Order of Nature, 1150–1750*. New York: Zone.

De Chantilly, Marc Vaulbert. 2001. "Property of a Distinguished Poisoner: Thomas Griffiths Wainewright and the Griffiths Family Library." In *Under the Hammer: Book Auctions since the Seventeenth Century*, ed. Robin Myers, Michael Harris, and Giles Mandelbrote, 111–42. London: Oak Knoll Press and the British Library.

"Delta." 1897. "Our Occupations." *It*. Firth College students' MS periodical. Sheffield University Archives.

Denny, Alfred. 1898. "The Relation of Museums to Elementary Teaching." *Museums Association Report of Proceedings* 9:39–44.

de Quincey, Thomas. 1862. *Autobiographic Sketches 1790–1803*. Edinburgh: Black.

Desmond, Adrian. 1989. *The Politics of Evolution: Morphology, Medicine, and Reform in Radical London*. Chicago: University of Chicago Press.

———. 2001. "Redefining the X Axis: 'Professionals,' 'Amateurs' and the Making of Mid-Victorian Biology." *Journal of the History of Biology* 34:3–50.

Dibdin, Thomas Frognall. 1838. *A Bibliographical, Antiquarian and Picturesque Tour in the Northern Counties of England and in Scotland*. 2 vols. London: Richards.

Dickens, Charles. [1864–65] 1997. *Our Mutual Friend*. New ed. Ed. Adrian Poole. London: Penguin.

"Dr. Kahn's Anatomical Museum." 1851. *Lancet*, April 26, 474.

———. 1854. *Lancet*, June 24, 654.

"Dr. Kahn's Anatomical Museum and Gallery of All Nations." 1851. *Manchester Guardian*, October 29, 5.

Düring, Monika von, Georges Didi-Huberman, and Marta Poggesi. 1999. *Encyclopaedia Anatomica: A Complete Collection of Anatomical Waxes*. Cologne: Taschen.

Elias, Norbert. 1994. *The Civilizing Process: The History of Manners and State Formation and Civilization*. New ed. Trans. Edmund Jephcott. Oxford: Blackwell.

Findlen, Paula. 1994. *Possessing Nature: Museums, Collecting, and Scientific Culture in Early Modern Italy*. Berkeley and Los Angeles: University of California Press.

Flower, William Henry. 1889. "Address." *Report of the Meeting of the British Association for the Advancement of Science* 59:3–24.

Forgan, Sophie. 1994. "The Architecture of Display: Museums, Universities and Objects in Nineteenth-Century Britain." *History of Science* 32:139–62.

———. 1998a. "'But Indifferently Lodged . . . ': Perception and Place in Building for Science in Victorian London." In *Making Space for Science: Territorial Themes in the Shaping of Knowledge*, ed. Crosbie Smith and Jon Agar, 195–215. London: Macmillan.

———. 1998b. "Museum and University: Spaces for Learning and the Shape of Disciplines." In *Scholarship in Victorian Britain*, ed. Martin Hewitt, 66–77. Leeds: Trinity and All Saints.

———. 1999. "Bricks and Bones: Architecture and Science in Victorian Britain." In *The Architecture of Science*, ed. Peter Galison and Emily Thompson, 181–208. Cambridge, MA: MIT Press.

———. 2005. "Building the Museum: Knowledge, Conflict, and the Power of Place." *Isis* 96:572–85.

Freedberg, David. 1989. *The Power of Images: Studies in the History and Theory of Response*. Chicago: University of Chicago Press.

Fyfe, Gordon, and Max Ross. 1996. "Decoding the Visitor's Gaze: Rethinking Museum Visiting." In *Theorizing Museums: Representing Identity and Diversity in a Changing World*, ed. Sharon MacDonald and Gordon Fyfe, 127–50. Oxford: Blackwell.

Goulder, Charles V., ed. 1853. *Catalogue of J. W. Reimers's Gallery of All Nations and Anatomical Museum*. Leeds: Jackson and Asquith.

Greenblatt, Stephen. 1991. "Resonance and Wonder." In *Exhibiting Cultures: The Poetics and Politics of Museum Display*, ed. Ivan Karp and Steven D. Lavine, 42–56. Washington, DC: Smithsonian Institution Press.

Greenwood, Thomas. 1888. *Museums and Art Galleries*. London: Simpkin, Marshall.

G. S., ed. 1840. *Exhibitions of Mechanical and Other Works of Ingenuity*. London: Privately compiled. British Library.

"Guide to the Pathological Museum." 1898. *The Guyoscope* 2 (18):69–70, 98–100.

Gunn, Simon. 1997. "The Sublime and the Vulgar: The Hallé Concerts and the Constitution of 'High Culture' in Manchester c.1850–1880." *Journal of Victorian Culture* 2:208–28.

Gunther, Albert Everard. 1975. *A Century of Zoology at the British Museum through the Lives of Two Keepers, 1815–1914*. London: Dawsons.

Guy's Hospital Museum. 1862. *Visitor Book 1829–1862*. MS ledger. Gordon Museum Archive, Guy's and St. Thomas' Hospital, London.

———. 1900. *Visitor Book 1862–1900*. MS ledger. Gordon Museum Archive.

Hall, David. 1986. "The History of the Book: New Questions? New Answers?" *Journal of Library History* 21:27–36.

Hallett, Cecil. 1913. "The Work of a Guide Demonstrator." *Museums Journal* 13:192–202.

Hancock, Albany. 1864. *Statement of the Conditions of Agreement between the Natural History Society of Newcastle-Upon-Tyne, and Other Societies with Kindred Objects in the Town and Neighbourhood*. Newcastle-upon-Tyne: Bell.

Hardy, Thomas. 1976. *The Complete Poems*. Ed. James Gibson. London: Macmillan.

Hay, James, Lawrence Grossberg, and Ellen Wartella, eds. 1996. *The Audience and Its Landscape*. Boulder, CO: Westview.

Haynes, Clare. 2001. "A 'Natural' Exhibitioner: Sir Ashton Lever and his *Holosphusikon*." *British Journal for Eighteenth-Century Studies* 24:1–14.

Hemingway, Andrew. 1995. "Art Exhibitions as Leisure-Class Rituals in Early Nineteenth-Century London." In *Towards a Modern Art World*, ed. Brian Allen, 95–108. New Haven, CT: Yale University Press.

Hetherington, Kevin. 2003. "Spatial Textures: Place, Touch, and Praesentia." *Environment and Planning A* 35:1933–44.

Hibbert, H. G. 1916. *Fifty Years of a Londoner's Life*. London: Grant Richards.

Hill, Kate. 2000. "'Civic Pride' or 'Far-Reaching Utility'?: Liverpool Museum c. 1860–1914." *Journal of Regional and Local Studies* 20:3–28.

———. 2005. *Culture and Class in English Public Museums, 1850–1914.* Aldershot: Ashgate.

Hodgkin, Thomas. 1828. "On the Object of Post Mortem Examinations." *London Medical Gazette* 2:423–31.

Holdsworth, Edgar B. 1904. "A Visit to the Hull Museum during the Xmas Holidays: Essay I." *Hull Museum Publications* 19:28–30.

Hooper-Greenhill, Eilean. 1994. *Museums and Their Visitors.* London: Routledge.

Howarth, Elijah. 1889. *List of Plants Collected Chiefly in the Neighbourhood of Sheffield by Jonathan Salt and Now in the Sheffield Public Museum.* Sheffield: Sheffield Literary and Philosophical Society.

———. 1913. "Presidential Address." *Museums Journal* 13:33–52.

Hoyle, William Evans. 1898. "The Electric Light Installation in the Manchester Museum." *Report of the Proceedings of the Museums Association* 9:95–105.

Hudson, Kenneth. 1975. *A Social History of Museums: What the Visitors Thought.* London: Macmillan.

Hull Literary and Philosophical Society. 1860. *A Guide to the Museum of the Literary and Philosophical Society, Hull.* Hull: Plaxton.

Hunter, William. [1767] 1784. *Two Introductory Lectures, Delivered by Dr. William Hunter, to his Last Course of Anatomical Lectures, at His Theatre in Windmill-Street.* London: Johnson.

Hurley, Kelly. 1996. *The Gothic Body: Sexuality, Materialism, and Degeneration at the Fin de Siècle.* Cambridge: Cambridge University Press.

Huxley, Thomas Henry. [1868] 1896. "Suggestions for a Proposed Natural History Museum in Manchester." *Report of the Proceedings of the Museums Association* 7:126–31.

Hyde, Ralph. 1988. *Panoramania! The Art and Entertainment of the "All-Embracing" View.* London: Trefoil.

Inkster, Ian, ed. 1985. *The Steam Intellect Societies: Essays on Culture, Education and Industry Circa 1820–1914.* Nottingham: Department of Adult Education, University of Nottingham.

Jenkins, Edgar W. 1981. "Science, Sentimentalism or Social Control? The Nature Study Movement in England and Wales, 1899–1914." *History of Education* 10:33–43.

Kahn, Joseph. 1851. *Catalogue of Dr. Kahn's Anatomical Museum, Now Exhibiting at 315, Oxford Street, near Regent Circus.* London: Golbourn.

———. 1853. *Catalogue of Dr. Kahn's Celebrated Anatomical Museum.* London: Golbourn.

Kavanagh, Gaynor. 1994. *Museums and the First World War: A Social History.* Leicester: Leicester University Press.

———. 2000. *Dream Spaces: Memory and the Museum.* Leicester: Leicester University Press.

Keeping, Walter. 1881. *A Popular Hand-Book to the Natural History Collection in the Museum of the Yorkshire Philosophical Society, York.* York: Sampson.

Knox, George. 1834. *Description of an Artificial Anatomical Figure, Constructed by the Chevalier Auzoux, M.D. Exhibited in 1832 before the King, in London*. Madras: Mission Press.

Koven, Seth. 1994. "The Whitechapel Picture Exhibitions and the Politics of Seeing." In *Museum Culture: Histories, Discourse, Spectacle*, ed. Daniel J. Sherman and Irit Rogoff, 22–48. London: Routledge.

Laskey, J. C. 1813. *A General Account of the Hunterian Museum, Glasgow*. Glasgow: Smith.

Leeds Philosophical and Literary Society. 1902–3. *The 83rd Annual Report*. Leeds: Jowett and Sowry.

Leonard, J. H. 1914. "A Museum Guide and His Work." *Museums Journal* 13:234–46.

Lewis, Geoffrey. 1989. *For Instruction and Recreation: A Centenary History of the Museums Association*. London: Quiller.

Lightman, Bernard V. 2000. "The Visual Theology of Victorian Popularizers of Science: From Reverent Eye to Chemical Retina." *Isis* 91:651–80.

Liverpool Museum of Anatomy. 1877. *Descriptive Catalogue of the Liverpool Museum of Anatomy, 29, Paradise Street*. Liverpool: Matthews.

Longhurst, Brian, Gaynor Bagnall, and Mike Savage. 2004. "Audiences, Museums and the English Middle Class." *Museum and Society* 2:104–24.

Macdonald, Sharon. 2002. *Behind the Scenes at the Science Museum*. Oxford: Berg.

———. 2005. "Accessing Audiences: Visiting Visitor Books." *Museum and Society* 3:119–36.

MacGillivray, William. 1837. Unidentified journal clipping, 21 November. Royal College of Surgeons of Edinburgh Archive.

Machor, James L., and Philip Goldstein, eds. 2000. *Reception Study: From Literary Theory to Cultural Studies*. New York: Routledge.

MacLeod, Suzanne, ed. 2005. *Reshaping Museum Space: Architecture, Design, Exhibitions*. London: Routledge.

Maerker, Anna Katharina. 2005. "Model Experts: The Production and Uses of Anatomical Models at La Specola, Florence, and the Josephinum, Vienna, 1775–1814." PhD diss., Cornell University.

Manchester Museum. 1938. *Autographs 24 November 1897—9 September 1938*. MS ledger. Manchester Museum Archive.

Manchester Museum Owens College. 1891. *Visitors Book 3 December 1889—24 August 1891*. MS ledger. Manchester Museum Archive.

Mason, Michael. 1994. *The Making of Victorian Sexuality*. Oxford: Oxford University Press.

Matheson, C.S. 2001. "'A Shilling Well Laid Out': The Royal Academy's Early Public." In *Art on the Line: The Royal Academy Exhibitions at Somerset House 1780–1836*, ed. David H. Solkin, 38–54. New Haven, CT: Yale University Press.

Matyssek, Angela. 2002. *Rudolf Virchow: Das Pathologische Museum; Geschichte einer wissenschaftlichen Sammlung um 1900.* Darmstadt: Steinkopff.

Mawhinney, W. 1854. *Anatomical and Physical Description of the Late Signor Sarti's New Florentine Venus, Together with the Causes, Symptoms, and Treatment of the Diseases of the Principal Organs.* 7th ed. London: Mallett.

Menninghaus, Winfried. 2003. *Disgust: The Theory and History of a Strong Sensation.* Trans. Howard Eiland and Joel Golb. Albany: State University of New York Press.

Mighall, Robert. 1999. *A Geography of Victorian Gothic Fiction: Mapping History's Nightmares.* Oxford: Oxford University Press.

Miller, William Ian. 1997. *The Anatomy of Disgust.* Cambridge, MA: Harvard University Press.

Moore, Kevin. 1991. "'Feasts of Reason?' Exhibits at the Liverpool Mechanics' Institution in the 1840s." In *Museum Languages: Objects and Texts,* ed. Gaynor Kavanagh, 157–77. London: Leicester University Press.

Morus, Iwan Rhys. 1998. *Frankenstein's Children: Electricity, Exhibition, and Experiment in Early-Nineteenth-Century London.* Princeton, NJ: Princeton University Press.

Mosley, Charles. 1928. "Inn-Parlour 'Museums.'" *Museums Journal* 27:280–81.

Murray, David. 1925. *The Hunterian Museum in the Old College of Glasgow.* Glasgow: Jackson, Wylie.

"The Museum Nuisance." 1864. *Lancet,* August 27, 243–44.

Natural History Society of Northumberland, Durham, and Newcastle-upon-Tyne. 1911–12. *Transactions.* Newcastle: Privately printed.

Nightscales, Eva. 1904. "A Visit to the Hull Museum during the Xmas Holidays: Essay II." *Hull Museum Publications* 19:30–31.

"An Obscene Exhibition." 1856. *Lancet,* May 5, 376–77.

Ogden, James. 1774. *A Poem, on the Museum at Alkrington, Belonging to Ashton Lever, Esq.* Manchester: Ashton.

"Our Special Pathologist." 1898. *Guyoscope* 2 (18):42.

"Our Weekly Gossip." 1839. *Athenaeum,* April 13, 278–79.

Pang, Alex Soojung-Kim. 1997. "Visual Representation and Post-Constructivist History of Science." *Historical Studies in the Physical and Biological Sciences* 28:139–71.

Pearce, Susan M. 1992. *Museums, Objects and Collections: A Cultural Study.* Leicester: Leicester University Press.

Percy, John. 1991. "Scientists in Humble Life: The Artisan Naturalists of South Lancashire." *Manchester Region History Review* 5 (1): 3–10.

Potter, Beatrix. 1966. *The Journal of Beatrix Potter from 1881 to 1897.* Ed. Leslie Linder. London: Warne.

Reay, Barry. 2002. *Watching Hannah: Sexuality, Horror and Bodily De-Formation in Victorian England*. London: Reaktion.

Rupke, Nicolaas A. 1994. *Richard Owen: Victorian Naturalist*. New Haven, CT: Yale University Press.

Sappol, Michael. 2002. *A Traffic of Dead Bodies: Anatomy and Embodied Social Identity in Nineteenth-Century America*. Princeton, NJ: Princeton University Press.

Secord, Anne. 1994. "Science in the Pub: Artisan Botanists in Early Nineteenth-Century Lancashire." *History of Science* 32:269–315.

———. 2002. "Botany on a Plate: Pleasure and the Power of Pictures in Promoting Early Nineteenth-Century Scientific Knowledge." *Isis* 93:28–57.

Secord, James A. 2004. "Monsters at the Crystal Palace." In *Models: The Third Dimension of Science*, ed. Nick Hopwood and Soraya De Chadarevian, 138–69. Stanford, CA: Stanford University Press.

Sennett, Richard. 1977. *The Fall of Public Man*. Cambridge: Cambridge University Press.

Sheppard, Thomas. 1904. *Guide to the Municipal Museum, Royal Institution, Albion Street, Hull*. Hull: Hull Museum.

Stafford, Barbara Maria. 1991. *Body Criticism: Imaging the Unseen in Enlightenment Art and Medicine*. Cambridge, MA: MIT Press.

———. 1994. *Artful Science: Enlightenment, Entertainment, and the Eclipse of Visual Education*. Cambridge, MA: MIT Press.

Stallybrass, Peter, and Allon White. 1986. *The Politics and Poetics of Transgression*. London: Methuen.

Stewart, Susan. 1993. *On Longing: Narratives of the Miniature, the Gigantic, the Souvenir, the Collection*. Durham, NC: Duke University Press.

Taylor, Shephard T. 1927. *The Diary of a Medical Student during the Mid-Victorian Period, 1860–1864*. Norwich: Jarrold.

Thompson, Charlotte. 1904. "A Visit to the Hull Museum during the Xmas Holidays: Essay IV." *Hull Museum Publications* 19:33–36.

Thomson, H. Campbell. 1935. *The Story of the Middlesex Hospital Medical School*. London: Murray.

Topham, Jonathan R. 2000. "Scientific Publishing and the Reading of Science in Nineteenth-Century Britain: A Historiographical Survey and Guide to Sources." *Studies in History and Philosophy of Science* 31:559–612.

Vogel, Susan. 1991. "Always True to the Object, in Our Fashion." In *Exhibiting Cultures: The Poetics and Politics of Museum Display*, ed. Ivan Karp and Steven D. Lavine, 191–204. Washington, DC: Smithsonian Institution Press.

Wheatley, Benjamin, and George Adlard. 1830. *Museum Brookesianum. A Descriptive and Historical Catalogue of the Remainder of the Anatomical and Zootomical Museum, of Joshua Brookes, Esq. F.R.S. F.L.S. F.Z.S. &c.* London: Taylor.

Wilks, Samuel. 1861. *Pathological Catalogue of the Museum of Guy's Hospital. Diseases of the Nervous System, Integument, and Organs of the Senses.* London: MacKenzie.

Wilson, David M. 2002. *The British Museum: A History.* London: British Museum Press.

Wynter, Andrew. 1874. *Peeps into the Human Hive.* 2 vols. London: Chapman and Hall.